Nucleic Acid
Geometry and Dynamics

Pergamon Titles of Related Interest

Nucleic Acid Geometry and Dynamics

Edited by

Ramaswamy H. Sarma

Director
Institute of Biomolecular Stereodynamics
State University of New York at Albany

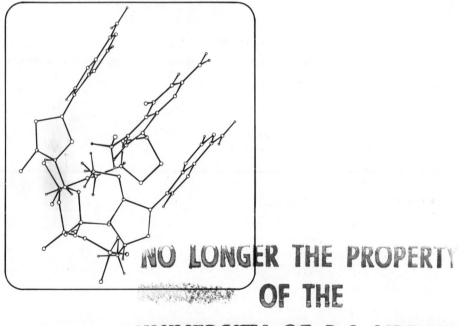

Pergamon Press
New York ☐ Oxford ☐ Toronto ☐ Sydney ☐ Frankfurt ☐ Paris

Pergamon Press Offices:

U.S.A.   Pergamon Press Inc., Maxwell House, Fairview Park,
                     Elmsford, New York 10523, U.S.A.

U.K.   Pergamon Press Ltd., Headington Hill Hall,
                     Oxford OX3 0BW, England

CANADA   Pergamon of Canada, Ltd. Suite 104, 150 Consumers Road,
                     Willowdale, Ontario M2J 1P9, Canada

AUSTRALIA   Pergamon Press (Aust.) Pty. Ltd., P.O. Box 544,
                     Potts Point, NSW 2011, Australia

FRANCE   Pergamon Press SARL, 24 rue des Ecoles,
                     75240 Paris, Cedex 05, France

FEDERAL REPUBLIC Pergamon Press GmbH, Hammerweg 6, Postfach 1305,
OF GERMANY   6242 Kronberg/Schönberg, Federal Republic of Germany

Cover Illustration: One complete turn of A-RNA which
is similar in spatial configuration to A-DNA. Illustration
copyright © by Irving Geis

Library of Congress Cataloging in Publication Data
Main entry under title:

Nucleic acid geometry and dynamics

Bibliography: p.
Includes index.
1. Nucleic acids. 2. Biochemorphology.
3. Structure-activity relationship (Pharmacology)
I. Sarma, Ramaswamy H., 1939-
QP620.N77 1980   547.7'9   80-10620
ISBN 0-08-024631-1
ISBN 0-08-024630-3 pbk.

Printed in the United States of America

Preface

An attempt is made in this volume to present the frontier developments in nucleic acid geometry and dynamics that have recently taken place. The book is aimed at advanced undergraduates, graduate students and scientists in the discipline. In the beginning of the text the nomenclature and detailed discussions of the methodologies of NMR spectroscopy and single crystal crystallography are presented with emphasis on nucleic acid structural studies. This is followed by a section on the details of the various forms of DNA double helices such as A, B, C, D, forms as well as the new ones—the propeller twist, alternating B, Z-DNA and the vertically stabilized double helix. An understanding of the material in this section is necessary to appreciate the rest of the text because many authors in these sections have taken this information for granted.

Intimate details of the spatial configuration of oligonucleotides so far reported by single crystal crystallographic studies have then been presented by Nadrian C. Seeman. This is followed by a discussion of NMR studies of the structural dynamics of olignucleotides and synthetic DNA and RNA duplexes by Ramaswamy H. Sarma and Dinshaw J. Patel respectively. These studies reveal that nucleic acids are flexible structures and are *conformationally pluralistic* in solution. While flexibility is allowed and alternate conformations are accessible, these molecules nevertheless show preferences for certain conformations. The composition of the conformational blend and the preferred intramolecular order are largely determined by the constitution and sequence which control the conformationally interconnected torsional movements about the various bonds via *stereochemical domino effects*. The concept of flexibility is further taken by Neville R. Kallenbach who examines the equilibrium and dynamics of transient base-pair breakage by following proton exchange rates in tRNA and polynucleotide duplexes. The rotational dynamics and torsional flexibility of DNA, as revealed by EPR spectroscopy is then discussed by Leonard S. Lerman. From an examination of the crstallographic data on short olignucleotides, their intercalated complexes and from a detailed anatomy of tRNA structure, Alexander Rich elaborates on the intrinsic flexibility of nucleic acids: the sugar ring is the elastic element and it enables the polynucleotide to adopt unusual conformations associated with chain extension and changes in chain direction.

The stereodynamical aspects of drug-nucleic acid complexes are then discussed by Henry M. Sobell, Helen M. Berman and Stephen Neidle, Donald M. Crothers, and Thomas R. Krugh respectively. The corner stone of Sobell's interpretation is a dynamic DNA with mixed sugar pucker as the elastic element. He advocates a unified stereodynamical structural theory for drug intercalation, a theory, he contends is also capable of providing a molecular rationalization for such phenomena as DNA breathing motions, conformational transitions and protein-DNA recognition. However, following Sobell's opus, Berman and Neidle provide computer modelling studies of drug-intercalation and single crystal data in which they indicate that the cardinal tenet in Sobell's model, i.e., the mixed sugar pucker, is unimportant. In the next article, Crothers, based on electric dichroism data, questions the idealized B-DNA model and concludes that DNA bound drug chromphores are not perpendicular to the helix axis and are tilted by an average of about 20°. Krugh presents hard kinetic and inhibition data which unequivocally show that the sequence specificity for a given drug can be considerably modifed by the presence of an appropriate second drug. In an earlier chapter on synthetic RNA and DNA duplexes, Patel discusses the NMR studies of drug-nucleic acid complexes, and there he attempts to arrive at the geometic relationships between the base-pairs and the drugs from NMR parameters.

To this point the volume is concerned with experimental studies. The last three contributions deal with theoreticl approaches. It is gratifying to note that Suse Broyde and Brian Hingerty show that their theoretical model for the coil form of poly(rU) agrees well with both hydrodynamic and NMR data. In the next chapter Wilma K. Olson uses polymer chain statistics to arrive at the flexible DNA double helix. The volume ends with the enormously important, but little explored subject of accessibility and surface structure of nucleic acids. Here Sung-Hou Kim attempts to arrive at the dynamically accessible surface areas by considering the van der Walls radii of the atoms in the nucleic acid.

The volume is illustrated by a large number of diagrams, many of which are stereoscopic. Particularly note-worthy are the twenty four-color processed plates from both the computer graphics of Richard Feldman of NIH and the artistic renditions of nucleic acid structures by Irving Geis.

In editing this book, while I have attempted to establish connections among the various contributions, I have also attempted to retain the individuality of each chapter as written by the authors. I have not removed some of the repetitious matter such as structural formulas for drugs or definitions of nomenclature etc., because a reader may be interested in the subject matter in a particular article and he/she should be able to read it without interruptions to refer elsewhere for formulas, etc.

Even though the volume covers recent developments in nucleic acid geometry and dynamics, it does not have a detailed section devoted to the structure of transfer RNA. Because of the availability of a large number of articles on this subject and particularly because of the recently published monograph *Transfer RNA, Structure, Properties and Recognition* (Eds: Schimmel, Soll and Abelson, Cold Spring Harbor Laboratory), it was decided to cover just certain aspects of tRNA in the overall context of the volume. Thus we cover such subjects as transient base-pair opening and closing in tRNA (Kallenbach), conformational fluctuations of the CCA terminus (Sarma), chain extension, changes in polynucleotide direction and self-intercalation in tRNA (Rich) and surface accessibility in tRNA (Kim); in addition the volume contains an artistic rendition of the structure of yeast tRNA^{Phe} (Plate 1) by Irving Geis.

This volume grew out of a Conversation in the Discipline "Stereodynamics of Molecular Systems" held at the State University of New York at Albany in April, 1979. The subject of nucleic acid structure was so thoroughly covered at the conference that these deliberations received extensive reporting in *Nature* (S. Neidle, Nucleic Acid Statics and Dynamics, *Nature 279* 474-476, 1979). In fact about thirty percent of the pages for the present volume are taken verbatim from the proceedings of the conference (*Stereodynamics of Molecular Systems,* Ed., Sarma, R.H., Pergamon Press, 1979).

This book was set in type by a high speed, third generation CRT typesetter—the University's new Compugraphic Videosetter Universal, driven by the UNIVAC 1100/82. I thank Mr. Steve Rogowski and Ms. Katie Huxford of the Computing Center for their efficient services in typesetting. I greatly appreciate the services of Ms. Barbara Hale who patiently fed the manuscript through the terminal to the UNIVAC.

I thank the authors for contributing to this volume. I am indebted to Alexander Rich for providing a preprint of his paper on Z-DNA and this enabled us to present a fairly detailed story about the left-handed helix. I take this opportunity to thank Richard Feldmann and Irving Geis for the sophisticated color plates in this volume. I wish to acknowledge the assistance and help from my colleagues, particularly Nadrian C. Seeman, Mukti H. Sarma, M.M. Dhingra and C.K. Mitra.

Ramaswamy H. Sarma
Albany, New York
December 12, 1979

CONTENTS

7. Structure, Fluctuations and Interactions of Base Pairs in Nucleic Acids Monitored by NMR, Tritium Exchange and Stopped-Flow Hydrogen Deuterium Exchange Measurements

Neville R. Kallenbach, C. Mandel, S.W. Englander

8. Torsional Flexibility of DNA As Determined by Electron Paramagnetic Resonance

I. Hurley, B.H. Robinson, C.P. Scholes, Leonard S. Lerman

The Use of Nuclear Magnetic Resonance Spectroscopy to Determine Nucleic Acid Geometry and Dynamics

Ramaswamy H. Sarma
Institute of Biomolecular Stereodynamics
State University of New York at Albany
Albany, New York 12222

Introduction

Albert Szent-Gyorgye[1] in his opus on the electronic theory of life states "Life is a miracle, but even miracles must have their underlying mechanism. The archives in which the basic blueprints of this mechanism are preserved and xeroxed are the nucleic acids, while the business of life is carried on by the proteins." Even though Szent-Gyorgye believes that these large biopolymers are merely the stage in which the drama of life is acted out and that the real actors are the electrons, to comprehend the chemical basis of life at the molecular level and then at the electronic level, it is imperative to know the dynamic structures of these molecules.

In this volume, an attempt is made to present the diverse points of view about the geometry and dynamics of nucleic acids with a discussion of the methodologies that are being applied to study them.

Constitutional Features of Nucleic Acids

Nucleic acids have three main structural features. (i) a five-membered sugar ring which is ribose for RNA and deoxyribose for DNA, (ii) the heterocyclic bases, adenine (A) guanine (G), cytosine (C), uracil (U) and thymine (T) attached to the C1' of the sugar ring in the β-configuration, (iii) the 3'5' phosphodiester linkage joining the individual nucleoside units. The combination of the base and sugar is called a nucleoside to which one may attach a phosphate group at the 5' or 3' position to generate a 5' or a 3' nucleotide. In Figure 1 is shown a single-stranded nucleic acid segment CpApUpG. Note the direction of the chain, i.e., 5' → 3'. Double-stranded nucleic acids are formed by complementary hydrogen bonding between G and C, and A and T between chains running in opposite directions. The hydrogen bonding usually encountered is the Watson-Crick type shown in Figure 2. In addition, several types of non-WC hydrogen bondings have been detected in nucleic acid systems. These are discussed elsewhere.[2/3] Complementary hydrogen bonding between long segments of single-stranded nucleic acids usually generates a double helix. Space filling and line drawings of some of these structures in stereo are illustrated in color in Plates 2-16. In Figure 3 is shown an exploded view of the "miniature double helix" made by a simple self-complementary dimer, pCpGp.

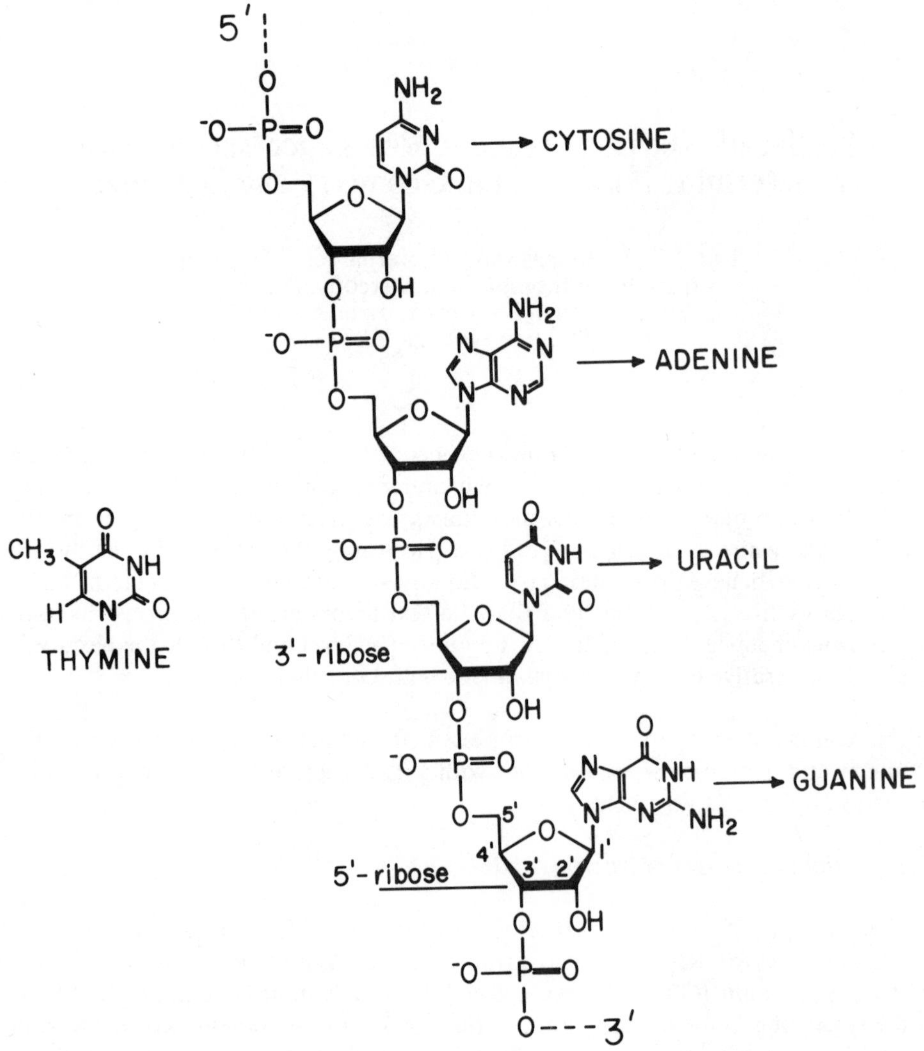

Figure 1. Structure of the nucleic acid segment pCpApUpGp which is an RNA oligonucleotide. In the corresponding DNA oligomer, the ribose will be replaced by 2′ deoxyribose, i.e., the 2′OH in the above structure will become H and uracil will be replaced by thymine. The numbering for the individual units is shown only for the bottom residue.

Determination of these molecular geometries is an exciting area of chemical research and the principal tool in solution work is NMR spectroscopy and that in solid state is X-ray diffraction, details of the latter are presented later in the volume. In the discussion below an elementary knowledge of NMR spectroscopy and stereochemistry is assumed.

In Figure 4 is shown the structure of the nucleic acid segment Adenylyl (3'5')
adenosine i.e. 3'5' ApA. The numbering scheme for various atoms and the
nomenclature for various torsion angles are also shown. The different torsion angles
are defined as:

O4'-C1'-N9-C8	χ (purines)
O4'-C1'-N1-C6	χ (pyrimidines)
C4'-C3'-O3'-P	α
C3'-O3'-P-O5'	β
O3'-P-O5'-C5'	γ
P-O5'-C5'-C4'	δ

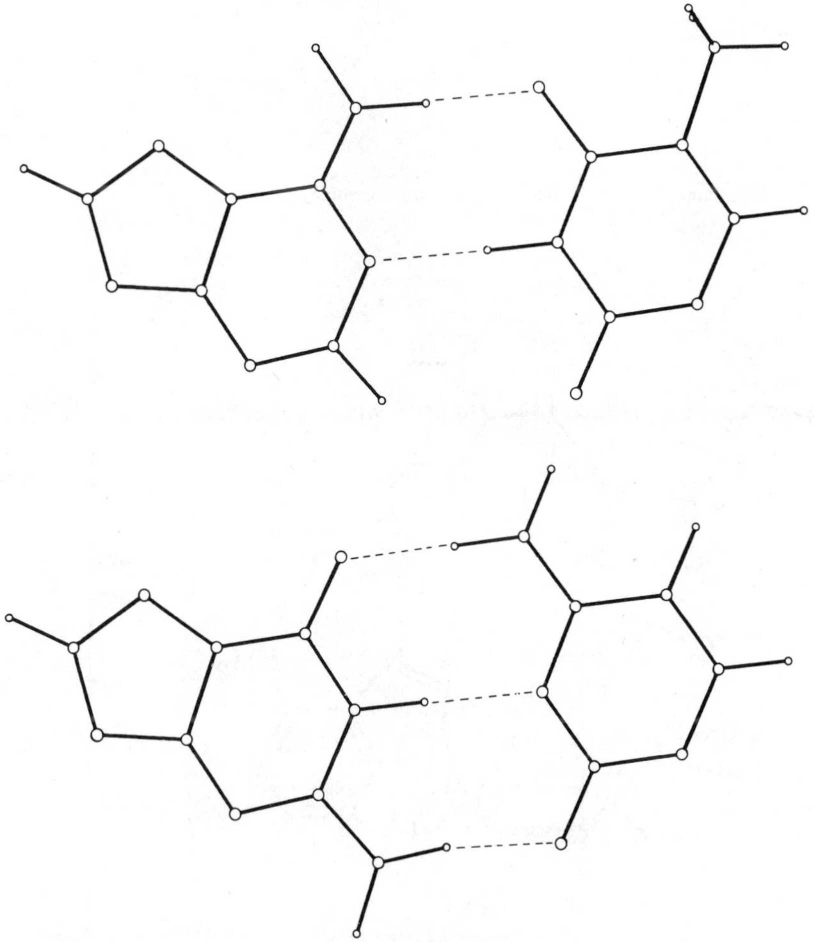

Figure 2. Watson-Crick hydrogen bonding between AT (top) and CG (bottom) base pairs.

$$O5'-C5'-C4'-C3' \qquad \varepsilon$$
$$C5'-C4'-C3'-O3' \qquad \zeta$$

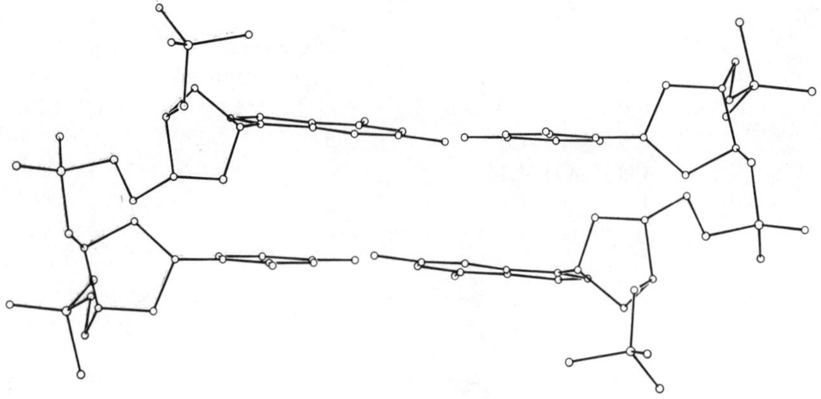

Figure 3. The "miniature" double helix of a self-complementary pCpGp.

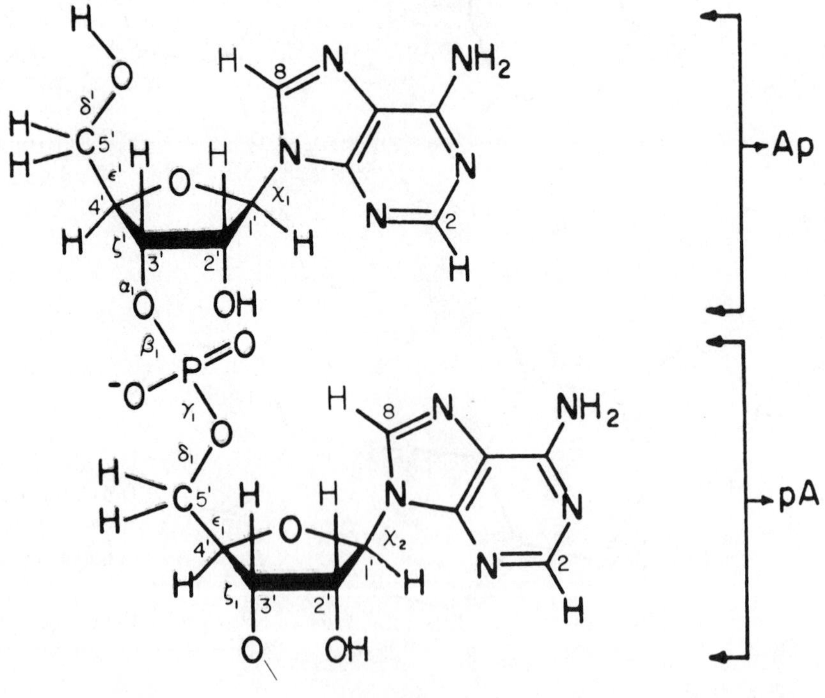

Figure 4. Structure of 3'5' ApA, the numbering scheme and the proposed IUPAC-IUB nomenclature for the various torsion angles. See Figure 6 for the old nomenclature.

The value of the torsion angle is zero for a cis planar arrangement and clockwise rotation results in a positive value (see later Figures 12-14, 18-21). A complete conformational analysis involves the determination of the preferred orientation or the minimum energy conformation about these bonds. For a dinucleoside monophosphate (Figure 4) this includes the determination of (i) the two sugar base torsions, i.e., glycosidic torsions χ_1 and χ_2, (ii) the mode of pucker of the two sugar rings; (iii) torsion angles α_1, β_1, γ_1, δ_1, ε_1 and (iv) torsion angles of the free exocyclic CH_2OH group, i.e., ε' and δ'. In the case of a trimer, the situation is more complex, i.e., three glycosidic torsions, mode of pucker of three sugar rings and thirteen other torsion angles. The determination of these conformational features essentially involves assignment and analysis of NMR spectra, accurate extraction of chemical shifts and coupling constants and translation of these NMR parameters to conformational angles. The methodology is extremely time consuming, complex and difficult, sometimes bordering on the torturous. However, a knowledge of the approaches to solve the NMR spectra of nucleic acid systems should make such studies of organic and inorganic molecules very easy.

Assignments and Analysis of the NMR Spectra of Nucleic Acids

The application of NMR spectroscopy to delineate the spatial configuration of nucleic acid structures involves a step-by-step analysis of the spectra. The foremost step is the unambiguous assignment of the various resonance lines which becomes complex as the number of nucleotidyl units increase. One may succeed to assign the spectral lines which originate due to couplings by appropriate homo and heteronuclear decoupling experiments (*vide infra*). However, the unambiguous assignment of the various single resonances (chemical shifted) observed particularly in oligonucleotides is still a formidable task. This is because one has to determine from which protons the lines originate and to which residue the protons belong. For example in ApA (Figure 4), there are four well separated downfield resonances which arise from a pair of H8 and a pair of H2. The distinction between H8 and H2 is achieved on the basis that H8's are relatively more acidic and hence exchange with deuterium in D_2O at high temperatures.[4/5] They can also be distinguished from spin lattice relaxation time measurements,[6] H2 having a much longer relaxation time than H8.

There have been several approaches to assign the H8 to individual segments. Ts'o, *et al.*[7] achieved the assignment of the H8 to the 5' or 3' nucleotidyl unit by comparing the chemical shifts in ApA and pApA. They showed that the phosphate group at the 5' position will deshield the H8 of -pA in ApA and hence assigned[8] the lower field resonance to H8 of -pA and the higher field one to the H8 of Ap- in ApA. Chan and Nelson[9] assigned the H8 resonances based on Mn^{++} ion binding to the phosphate groups. The paramagnetic ions are known to broaden proton resonances because of the electronic spin nuclear spin dipole dipole interactions. The effectiveness of a paramagnetic center in broadening a nuclear resonance varies as the square of the paramagnetic moment and as the inverse sixth power of the separation between the nucleus and the paramagnetic center. Hence the methodology enables to locate the

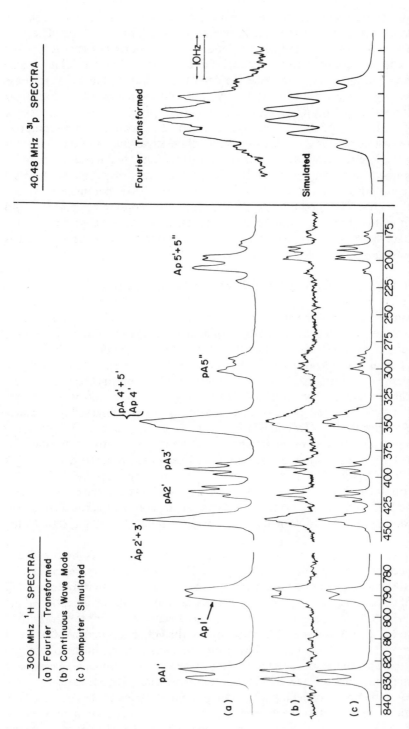

Figure 5. *Left* The diagrams a and b respectively represent the 300 MHz ¹HNMR spectra of ApA (0.05 M, pD 7.5, temp. 27°C) obtained by Fourier transform (FT) methods, (a) and in continuous wave (CW) modes (b). The diagram c is the computer simulated one. Only the ribose region is shown. The FT spectra were superior in S/N ratio but suffered from a poor resolution of 0.7 Hz because only 8K of the total 16K memory was available to transform a band width of 2500 Hz. The CW spectra was superior in resolution (0.2 Hz) but had only fair S/N ratio. By carefully studying both FT and CW spectra peaks were identified and finally the spectra were simulated. The chemical shifts are expressed in Hz, 300 MHz system, upfield from internal tetramethyl ammonium chloride. *Right* The Fourier transformed (top) and computer simulated (bottom) ³¹P NMR spectra (40.48 MHz) of dApdA 0.05 M, pD 7.5 and temp. 27°C. From references 12 and 13.

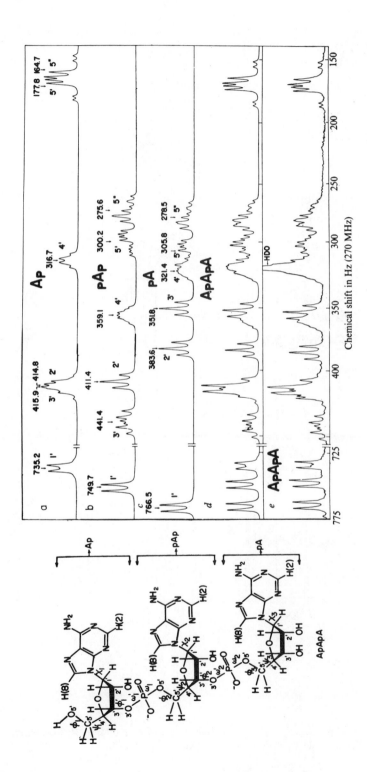

Figure 6. *Left* Structure of ApApA. Note that the old IUPAC-IUB nomenclature is used. Compare this nomenclature with the proposed IUPAC-IUB nomenclature in Figure 4 for easy conversions. *Right* Computer simulation of the Ap- (a), - pAp- (b) and -pA (c) parts of ApApA; (d) Combination of the above three parts into one to produce ApApA simulation; (e) The 270 MHz ¹H experimental NMR spectrum of ApApA at 72°C, pD 7.0, 0.02 M, shifts are up-field from tetramethylammonium chloride. From reference 15 with slight modification.

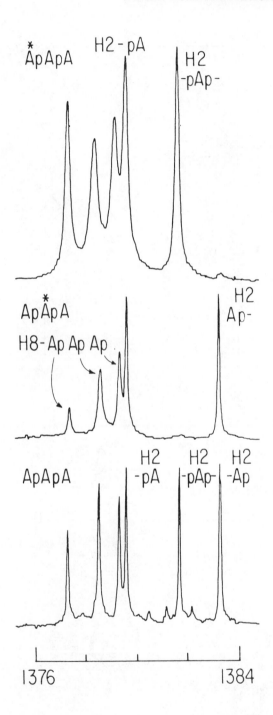

Figure 7. The base proton region of A*pApA, ApA*pA and ApApA, 270 MHz ¹HNMR system, the chemical shifts are upfield from internal tetramethylammonium chloride, 19°C.

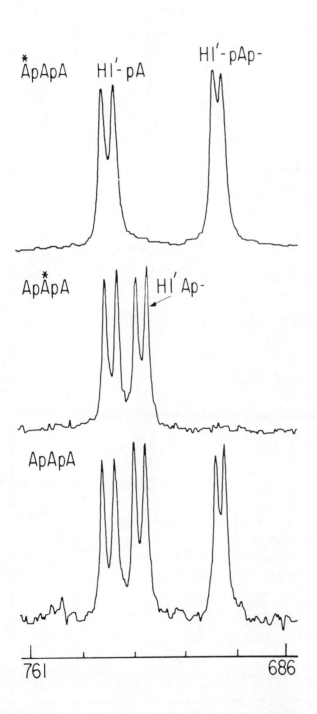

Figure 8. The H1′ region of A*pApA, ApA*pA and ApApA. The rest of the details as in Figure 7.

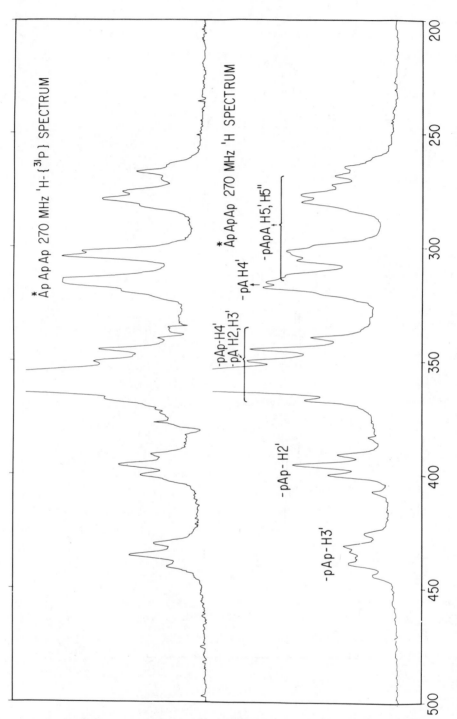

Figure 9. 270 MHz ¹HNMR spectra of A*pApA taken under conditions in which phosphorus-31 was decoupled (top) and coupled (bottom). The shifts are up-field from tetramethylammonium chloride and the employed temperature is 50°C. Comparison of spectra *top* and *bottom* immediately allows recognition of signals from protons coupled to the backbone phosphorus atoms. Unpublished data (Sarma, Danyluk et al)

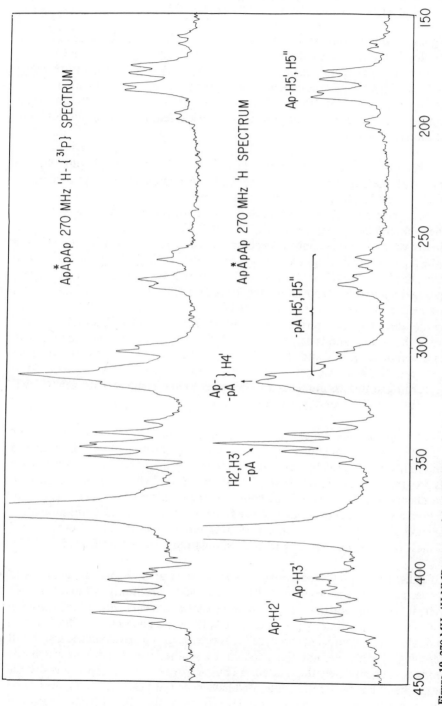

Figure 10. 270 MHz ¹H NMR spectra of ApA*pA taken under conditions in which phosphorus-31 was decoupled (top) and coupled (bottom). Temperature, 40°C and compensation for heating due to decoupling was not made and hence HDO peak position is affected. Comparison of spectra in Figures 9 and 10 makes possible the assignment of protons coupled to the backbone phosphorus -31 as well as protons from the common residue - pA, the resonances of which (for example, 2' and 3') are highly temperature sensitive. Unpublished data (Sarma, Danyluk et al)

proton relative to the site of Mn^{++} binding, i.e., the phosphate backbone. Their assignments[6] in ApA were identical to that proposed by Ts'o, *et al.*[7]

The assignment of H2 from individual segments is a difficult task because of its lack of sensitivity to the nature of the ribose phosphate backbone. Chan and Nelson[9] have proposed their assignments based on their sensitivity to the temperature induced destacking of ApA, and consequent reduction in the ring current shielding. These methods of assignments of H8 and H2 to individual segments are by no means unambiguous because they assume special geometrical features for the molecules to arrive at the assignments.

At the level of a dinucleoside monophosphate such as ApA (Figure 4), the assignments of the sugar protons can be achieved unambiguously by a series of homo and heteronuclear decouplings. For example, Ap- H5' and H5" of ApA occurs at the highest field (Figure 5), decoupling of which will enable to locate Ap-H4'. Then this can be decoupled to locate Ap- H3'. This location can be further independently verified by phosphorus decoupling. Once H3' is assigned, the H2' and H1' of the Ap- residue is assigned by appropriate decouplings. Once the H1' of Ap-is assigned, the H1' of -pA is automatically assigned because the H1's usually appear in a region well separated from other sugar protons. The decoupling of H1' of -pA will assign H2' and which in turn can be used to assign H3', H4', etc. of the -pA unit. The series of decouplings described here have been employed by Cheng and Sarma[10] to arrive at the unambiguous assignments in 2'-0-methyl CpC. No individual unambiguous assignments of H5' and H5" have been made so far. Generally, the low field hydrogen from the ABX system has been assigned[11] to H5' and the higher field to H5". According to this assignment, the H5' refers to that geminal hydrogen which is *gauche* to H4' and the ether oxygen of the ribose.

Once the assignments are done, the next step involves the extraction of the accurate NMR parameters, i.e., chemical shifts and coupling constants from the observed spectrum. This is achieved by computer synthesizing the observed spectrum using various programs such as LAOCN III, LAME and BIGNMR. By simulation techniques the chemical shifts and coupling constants can be obtained with an accuracy of ±0.005 ppm and ±0.1 Hz respectively. The first complete assignment and simulation of a regular dinucleoside monophosphate was reported from this laboratory.[12/13] One of these original spectra along with the simulation is shown in Figure 5.

From the discussion presented above, it appears that analysis and assignments of NMR spectra of dimer segments of nucleic acids are fairly straightforward. However, it becomes a difficult problem in the case of a trimer. From the inspection of the structure of the trimer ApApA (Figure 6, left), it is apparent that one is facing a formidable task to completely analyze the spectrum of ApApA. Nucleic acid NMR spectroscopists have overcome this problem by synthesizing ApApA in which certain residues are completely deuterated.[14] Thus A*pApA and ApA*pA in which the residue marked * has been fully deuterated have been synthesized.[14] In Figures 7 and 8, we illustrate the base proton and the H1' region of the 270 ¹H NMR spectra of

A*pApA, ApA*pA and ApApA and they immediately provide unambiguous information about the assignments of the base protons and H1' region in ApApA. In Figures 9 and 10 are illustrated the high field 270 MHz spectra of A*pApA and ApA*pA taken under conditions in which phosphorus-31 was coupled and decoupled. The assignments indicated in the figures are derived from phosphorus decouplings as well as from a series of step-by-step closely interconnected homonuclear decoupling experiments. Once the assignments of the protons of the individual residues at a convenient temperature is achieved, the complete spectra can be computer synthesized by separately synthesizing spectra for the Ap-, -pAp- and -pA residues and thus adding them together. Complex simulation of such magnitude was achieved for the first time in this laboratory for ApApA and was reported[15] in 1976 and this is reproduced in Figure 6 with a small change. This small change involves the exchange of the Ap- and pAp- parts of H1', H2' and H3'. This is because in the original report[15] there was a computer mix-up of the H1', H2' and H3' region of Ap- and -pAp- regions; however, this does not affect any interpretations.[15] It may be noted that phosphorus-31 and homonuclear decouplings were carried out at 40°C for ApA*pA and 50°C for A*pApA. This was the lowest temperature at which the resonances began to separate so that decoupling experiments could be carried out meaningfully. Once the assignments are made, by following the shift trends with temperature carefully, complete computer simulation at most temperatures can be achieved. Comparison of the spectra of the H1' at 19°C (Figure 8) with that of 72°C (Figure 6) indicates that H1' of Ap- and -pAp- cross over with temperature. Also a comparison of spectra of A*pApA (50°C), ApA*pA (40°C) and ApApA (72°C) clearly shows that elevation of temperature causes significant changes in chemical shifts for several protons; particularly noteworthy are the changes on -pA H3', H2' as well as -pApA H5', H5'' region. The computer simulation provides accurate values for coupling constants and chemical shifts which can then be translated into conformational parameters. In the case of deoxy nucleic acid systems assignments and analysis of spectra even at the level of a trimer can be done without selective deuteration. This is because the complex set of resonances from the 2'2'' region shift upfield from the envelope containing the remaining sugar protons. Here again the assignments are based on careful step-wise decoupling experiments and computer simulation. Details on this has been presented by Cheng, *et al.*[16] In Figure 11 are shown the experimentally recorded and the completely analyzed spectra of the deoxy trimers d-TpTpT and d-TpTpC.

The NMR data accurately derived from computer simulation can be translated into conformational details. For example, the three bond vicinal coupling constants between various nuclei (^{1}H-^{1}H, ^{1}H-^{31}P, ^{1}H-^{13}C, ^{13}C-^{31}P) can provide the conformational preferences about the various single bonds in nucleic acid structures. Karplus[17] has related the observed vicinal coupling constants to the dihedral angle ϕ between vicinal nuclei. The analytical expression for ^{1}H-^{1}H vicinal coupling is given by Equation 1:

$$J_{HH} = A \cos^2\phi + B \tag{1}$$

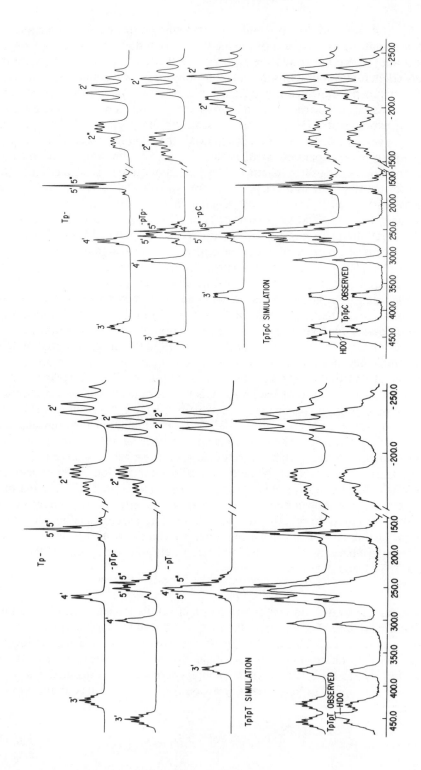

Figure 11. The observed and computer simulated 270 MHz ^{1}H NMR spectra of d-TpTpT (left) and d-TpTpC (right) at 20°C. As is clear the regions from the three nucleotidyl units were separately simulated and then combined to produce the total spectra of the trimers. The shifts are in Hz from tetramethyl ammonium chloride.

where A and B are constants and A has different magnitude for the ranges $0° < \phi < 90°$ and $90° < \phi < 180°$. Similar equations for other vicinal couplings J_{HCOP}, J_{CCOP}, J_{HCCC} has been derived (*vide infra*). One uses all these expressions to derive geometry information about nucleic acid structures.

With the availability of high frequency FT NMR spectrometers which provide well-resolved and dispersed proton spectra, a number of long range coupling constants have been measured and they can provide information about the spatial arrangement in many a nucleic acid systems. Long range coupling constants in both saturated and unsaturated systems involving hetero atoms have been well-reviewed.[18-20]

Four bond couplings like three bond vicinal couplings are sensitive to stereochemistry. When the coupling path is a planar zig-zag "W" observably large couplings results.[18/19] Four bond J_{HCCCH} couplings have been observed for the sugar ring protons of a few nucleic acid derivatives,[21/23] but they are of such small magnitude (0.3-0.7Hz) that they contain little geometry information. Under favorable conditions one is able to measure long range five bond couplings. Maximum coupling usually results when the five bond coupling path is an inplane extended zig-zag path.[24-25]

Conformation of the Sugar Rings

It is well known that the ribose and deoxyribose rings exhibit a non-planar puckered conformation which can be described by ring torsion angles, by least square planes[26] and by pseudo rotation parameters.[27] Single crystal data[28/29] on several nucleic acid components show that the furanose ring exists in a majority of cases in a conformation in which either C2' or C3' atom is displaced toward C5'from the plane of the other ring atoms of the sugar moiety. The conformation in which C2' is displaced is designated as ^2E (C2'-*endo*) and the one in which C3' is displaced as ^3E (C3'-*endo*). These conformations are shown in Figure 12. In addition a few other twist and envelope conformations[28/29] have been reported for pentose rings in nucleic acids.

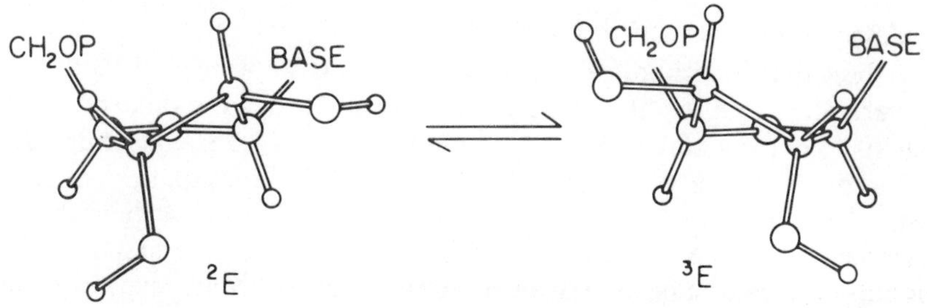

Figure 12. The two major conformations of the sugar ring.

These solid state findings are helpful to determine the solution stereodynamics of pentose rings in nucleic acids. The dihedral angles between the C-H bonds on adjacent carbons in a furanose ring are determined by the mode and the extent of the ring puckering and thus a knowledge of the vicinal H-H coupling constants will permit determination of its stereochemistry,[30/31] i.e., the observed coupling constants can be related via Karplus equation to apparent values of $\phi_{1'2'}$ and $\phi_{3'4'}$ in the relevant H-C-C-H fragments. However, there are difficulties in using Karplus equation to distinguish between the small differences in dihedral angles that exist among the various possible furanose ring conformations. This has been discussed elsewhere[32] in detail.

It has been proposed[33] that the best way to handle the pentose coupling constants is to treat them as arising from an equilibrium blend of ^2E and ^3E conformations. The concentration dependence of the coupling constants $J_{1'2'}$, $J_{2'3'}$, and $J_{3'4'}$ in 5'AMP indicates that $J_{1'2'} + J_{3'4'}$ essentially remains constant.[33] This pattern of variation in the endocyclic proton-proton coupling constants is consistent with the existence of ^2E and ^3E conformations in equilibrium, the relative population of these states being determined by the magnitude of $J_{1'2'}$ and $J_{3'4'}$. To illustrate this point in Table I are listed the projected coupling constants for various envelope conformations. These

Table I

Calculated Coupling Constants in Hz
in a Pentose Ring for Various Envelope Conformations

	$J_{1'2'}$	$J_{1'2''}$	$J_{2'3'}$	$J_{2''3'}$	$J_{3'4'}$
C2'-exo	0.2	5.0	4.9	11.0	8.2
C3'-endo	0.1	7.4	4.9	11.0	10.3
C4'-exo	2.4	9.1	7.0	9.0	10.3
C1'-exo	10.1	5.5	7.6	0.1	2.9
C2'-endo	10.1	5.5	5.5	0.1	0.2
C3'-exo	7.9	7.6	5.5	0.1	0.0

are calculated using the modified Karplus equation[27] $J_{HH} = 10.5 \cos^2\phi_{HH} - 1.2 \cos\phi_{HH}$. The data show that for ^2E and ^3E the sum $J_{1'2'} + J_{3'4'}$ has a value of 10.3 to 10.4 Hz and $J_{2'3'}$ varies from 5.5 to 4.9. For a pure ^2E state in solution the value of $J_{1'2'}$ would be about 10.1 Hz for a pure ^3E state $J_{3'4'}$ would be about 10.3 Hz. Note that $J_{3'4'}$ for ^2E and $J_{1'2'}$ for ^3E are practically zero. Any deviation from these pattern of coupling constants suggest deviation from the ideal $^2E \rightleftharpoons {}^3E$ equilibrium such as changes in the phase angle of pseudorotation as well as in the amplitude of pucker.[27] The Karplus approach cannot be used to determine precisely the conformational features of the pentose ring,[33] and we believe that the treatment of the endocyclic coupling constants on the basis of a $^2E \rightleftharpoons {}^3E$ or a $^2T_3 \rightleftharpoons {}^3T_2$ equilibrium (where T = twist chair) gives reasonably satisfactory results. Theoretical calculations[34/35] do also support such an approach. Operationally ratio of $J_{1'2'}$ to the sum $J_{1'2'} + J_{3'4'}$ gives the fractional population of ^2E conformers; $100 -$ percent $^2E =$ percent ^3E.

Conformation About the C4'-C5' (ε)
and C5'-O5' (δ) Bonds

The symbol ε defines the torsion about C4'-C5' and ε is zero when C4'-C3' is cis planar to the C5'-O5' bond, i.e., the fully eclipsed conformation. The minimum energy rotamers about C4'-C5' are shown in Figure 13. They are the *gauche gauche*

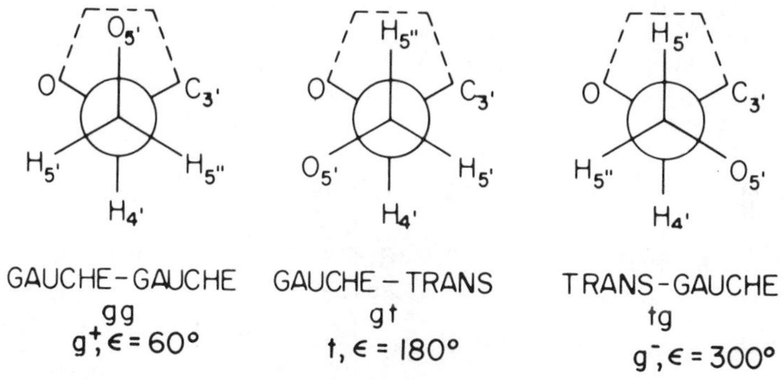

GAUCHE-GAUCHE   GAUCHE - TRANS   TRANS-GAUCHE
gg   gt   tg
$g^+, \varepsilon = 60°$ $t, \varepsilon = 180°$ $g^-, \varepsilon = 300°$

Figure 13. Classical staggered rotamers about the C4'-C5' bond.

(gg $\varepsilon = 60°$, g^+) gauche trans (gt $\varepsilon = 180°$, t) and *trans gauche (tg, $\varepsilon = 300°$, g^-)* conformers. In aqueous solution there is rapid interconversion among these conformers and the relative population is influenced by a large number of factors including the nature of the base and the value of χ, furanose ring puckering and solvent properties.[36-38] Usually, distribution of conformers about the C4'-C5' bond is calculated from the experimental sum $J_{4'5'} + J_{4'5''} \Sigma$ and this allows the determination of the population of *gg* conformers and the combined population of *gt* and *tg*. This approach is used because the evaluation of the individual contributions of *gt* and *tg* requires a knowledge of the absolute assignments of H5' and H5'', and this is not known. The use of the experimental sum is adequate for monitoring perturbations in the time averaged conformation about C4'-C5'. Any perturbation resulting in the increase of the *gg* rotamer populations will result in the decrease in the magnitude of $J_{4'5'} + J_{4'5''}$ where as an increase in *gt* and *tg* populations at the expense of *gg* rotamers will result in an increase in the observed sum.

The empirical equations developed earlier[39-41] for the calculation of rotamer distributions about C4'-C5' have been modified by Lee and Sarma[42] to reflect the more reasonable values of $J_t = 11.7$ Hz and $J_g = 2.0$ Hz. The modified expression is:

$$\text{percent } gg = (13.7 - \Sigma) 100/9.7 \qquad (2)$$

where $\Sigma = J_{4'5'} + J_{4'5''}$. The percentage $(gt + tg) = 100 -$ percent gg.

This modified expression has been recently used to evaluate the population distribution of conformers about the C4′-C5′ bond in 3′5′ribodinucleoside monophosphates,[43/44] 3′5′ deoxyribodinucleoside monophosphates,[45] 2′5′ ribodimers[46] and trinucleoside diphosphates.[16]

The torsion about C5′-O5′ bond is denoted by δ and the minimum energy conformers are shown in Figure 14. Population distribution of conformers about this

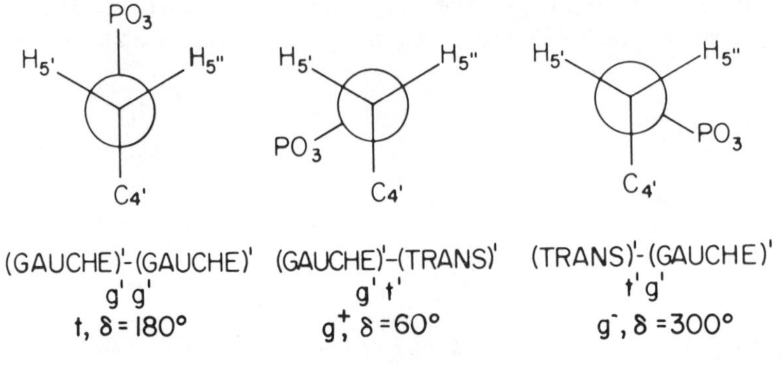

(GAUCHE)′-(GAUCHE)′   (GAUCHE)′-(TRANS)′   (TRANS)′-(GAUCHE)′
      g′ g′   g′ t′   t′ g′
   t, δ = 180°   g⁺, δ = 60°   g⁻, δ = 300°

Figure 14. Staggered rotamers about the C5′-O5′.

bond can be obtained from the magnitude of the vicinal coupling $J_{5'P}$ and $J_{5''P}$ and is given by the expression

$$\text{percent } g'g' = (25 - \Sigma') \, 100/20.8 \tag{3}$$

where Σ' is the sum $J_{5'P} + J_{5''P}$. It may be noted that when C4′-C5′ and C5′-O5′ rotamers exist in the *gauche gauche* arrangement the H4′ and P in a 5′-mononucleotide lie in an inplane 'W' path and this will result in a long-range four bond coupling.

Four bond J_{HCCOP} couplings of magnitude 2.4 - 2.7 Hz have been observed between the ³¹P and H4′ in many a nucleic acid systems. Their presence were discovered in this laboratory[47] by recording spectra in which phosphorus-31 was decoupled. Inspection of spectra in Figures 15a and c makes it unmistakably clear that H4′ is coupled to the phorphorus in 5′-AMP and the accurate value of the coupling can be obtained from computer simulation (Figure 15b and d). Significantly large coupling between H4′ and phosphorus atoms in nucleic acids results when the C4′-C5′ and C5′-O5′ are predominantly oriented in *gg* and *g′g′* conformation creating a zig-zag 'W' inplane path between H4′ an P (Figure 16). Hall and coworkers[48-50] have observed similar four bond coupling in many non-nucleic acid systems. Sarma, *et al.*[51] have observed linear relations between J_{HCCOP} and the *gg* and *g′g′* populations about the C4′-C5′ and C5′-O5′ bonds. The four bond coupling J_{HCCOP} has been shown to be a simultaneous measure of the orientation about the C4′-C5′ and C5′-O5′ bonds.

In Figure 17a is shown the plot of $\Sigma\,(J_{4'5'} + J_{4'5''})$ versus $J_{P4'}$ and in Figure 17b plot of $\Sigma'(J_{5'P} + J_{5''P})$ versus $J_{P4'}$. Also given along the y-axis are the fractional populations of gg (Pgg) and g'g' (Pg'g') computed from vicinal coupling expressions given earlier. Reasonably linear correlations exist between both Σ (Pgg) and Σ' (Pg'g') and $J_{P4'}$ with *increases* in $J_{P4'}$ accompanied by *decreases* in Σ and Σ'. The concurrent variations suggest an interdependence between rotational preferences about C4'-C5'

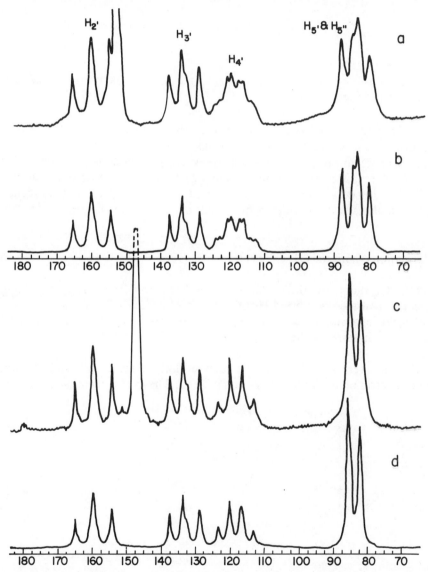

Figure 15. 100 MHz ^{1}H NMR spectra of 5'AMP under conditions in which phosphorus-31 was coupled (a) and decoupled (c). Compare the H4' and H5', H5'' regions in both spectra. The corresponding simulations are in b and d.

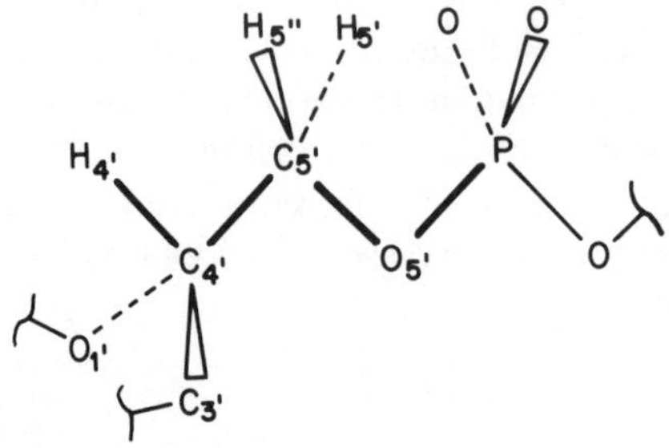

Figure 16. The inplane zig-zag 'W' path between H4′ and P when C4′-C5′ and C5′-O5′ are in gg and g′g′ conformations (see Figures 13, 14 and 22).

and C5′-O5′ and demonstrate that $J_{P4'}$ is a simultaneous measure of the conformational distribution about C4′-C5′ and C5′-O5′ for a β-5′ nucleotide. The plots in Figure 17 are of some practical value since they may provide some estimates of Σ and Σ′ (as well as Pgg and Pg′g′) from $J_{P4'}$ and may be useful in those situations where the 5′5″ regions defy analysis. In addition to these the vicinal coupling constant $J_{POC5'C4'}$ should also[52] provide information about rotational preferences about C5′-O5′.

Conformation About the C3′-O3′ (α) Bond

Theoretically, there are three staggered conformations possible about C3′-O3′ (Figure 18), and the three bond HCOP coupling between H3′ and the phosphorus should yield information about C3′-O3′ torsion. However, analysis of rotation about C3′-O3′ in terms of the three staggered forms leads to conflicting results.

The problem is rendered simple if one assumes that the *trans (α = 60°)* conformer does not contribute to the conformational blend. The experimental and theoretical

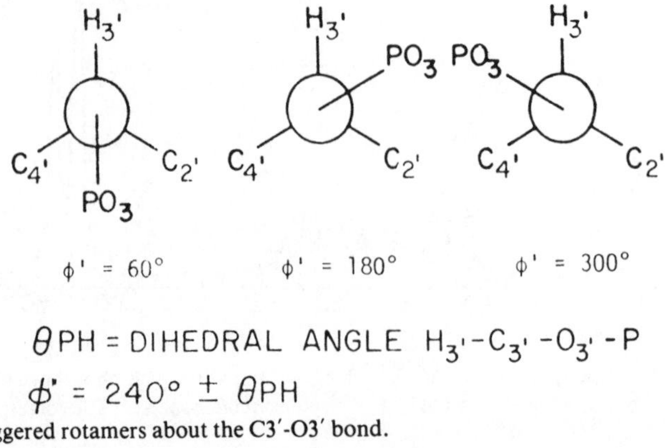

$$\theta PH = DIHEDRAL\ ANGLE\ H_{3'}-C_{3'}-O_{3'}-P$$

$$\phi' = 240° \pm \theta PH$$

Figure 18. Staggered rotamers about the C3′-O3′ bond.

studies which support this assumption are detailed elsewhere.[10] We have shown[10]

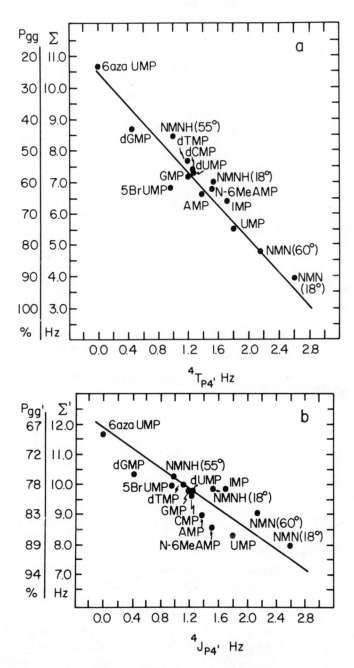

Figure 17. (a) Plot of Σ (P_{gg}) *vs* $J_{P4'}$ (b) Plot of Σ' ($P_{g'g'}$) *vs* $J_{P4'}$. Note that $\Sigma = J_{4'5'} + J_{4'5'}$ and $\Sigma' = J_{5'P} + J_{5''P}$.

that simple interpretation of the coupling constants in terms of a $\alpha^+ \rightleftharpoons \alpha^-$ equilibrium is a reasonable one. Here (+) and (-) denotes conformations in the range of 270° – 285° and 195° – 210°, respectively. The correlation between the vicinal HP coupling and the ϕPH is given by the expression

$$J_{HP} = 18.1 \cos^2\phi - 4.8 \cos\phi \qquad (4)$$

The value of ϕPH = 0 when α = 240°. A close correlation exists between the torsion about C3′-O3′ and the conformation of the ribose ring in nucleic acids and this correlation can be expressed as the following equilibrium

$$^2E\alpha^+ \rightleftharpoons \,^3E\alpha^- \qquad \text{(Figure 19)}$$

Destacking causes an increase in the magnitude of χ in the anti domain (*vide infra*) and this in turn causes a shift toward $^2E\alpha^+$ population and this in turn manifests in a long range four bond coupling between H2′ and P.

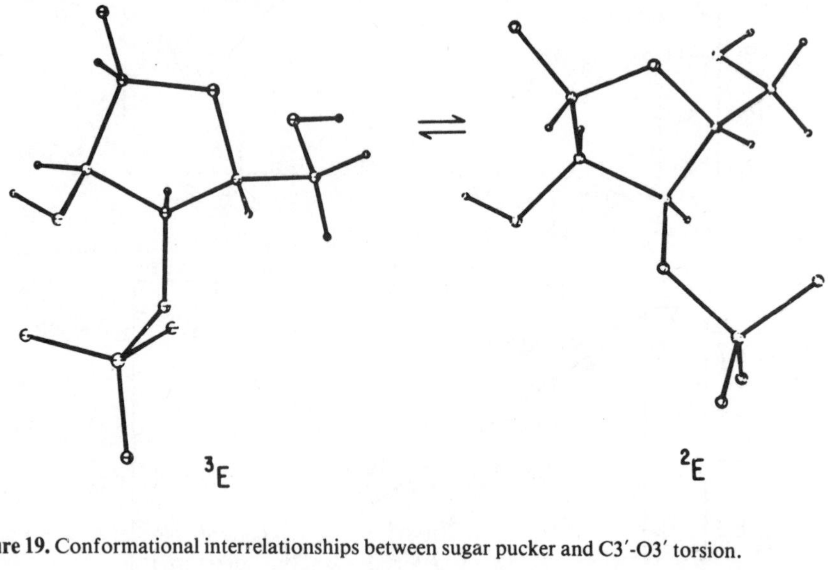

^3E ^2E

Figure 19. Conformational interrelationships between sugar pucker and C3′-O3′ torsion.

Four bond couplings between H2′ and the phosphorus atom have also been observed.[44/53] Inspection of Figure 19 indicates that the geometric relationship between H2′ and ^{31}P is an inplane 'W' in $^2E\alpha^+$ conformation. As is seen in Figure 20, elevation of temperature causes the appearance of fine structure in the H2′ region of Ap- of ApU clearly indicating $J_{H2'-P}$ couplings. This is due to a shift in the $^3E\alpha^- \rightleftharpoons \,^2E\alpha^+$ equilibrium toward the right with increasing temperature. Elevation of temperature causes unstacking of the bases, which results in an increase in the magnitude of χ_1 and a coupled shift of $^3E\alpha^-$ to $^2E\alpha^+$ conformation. These conformational interrelationships are described elsewhere.[43/44/53]

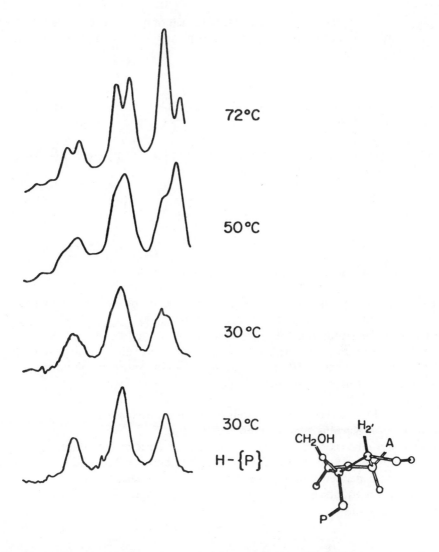

Figure 20. Temperature dependence of the H2′ signals of Ap- of ApU. The bottom spectrum is the phosphorus decoupled spectrum. Elevation of temperature causes the emergence of fine structure due to H2′-P four bond coupling.

Conformation About the Glycosidic Linkage (χ)

This torsion is denoted by χ and when $\chi = 0°$, it corresponds to the eclipsed conformation of bonds O4′-C1′ and N9-C8 (purines) and O4′-C1′ and N1-C6 (pyrimidines). Clockwise rotation of N9-C8/N1-C6 bonds relative to O4′-C1′ when looking along the C1′-N9/C1′-N6 results in positive values. The domain of $\chi \simeq 0 \pm 90°$ is *anti* and $\chi \simeq 180° \pm 90°$ is *syn* (Figure 21).

Employment of vicinal coupling constants to determine χ has been very difficult. This is because there is no observable proton-proton vicinal coupling that will provide information about χ.

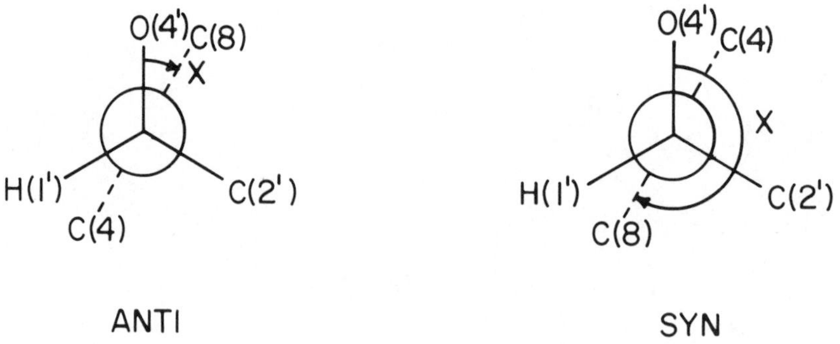

ANTI　　　　　　　　　　　　　　　SYN

Figure 21. Definition of sugar-base torsion (see also Figure 22).

Theoretically, the glycosidic torsion should manifest in the vicinal coupling constant $J_{H1'-C1'-N9-C8}$ and $J_{H1'-C1'-N9-C4}$ for purines and $J_{H1'-C1'-N1-C6}$ and $J_{H1'-C1'-N1-C2}$ for pyrimidines. Determination of these coupling constants have been very difficult because of the poor abundance of ^{13}C nuclei. The natural abundance proton coupled ^{13}C Fourier transform NMR spectra of suitably concentrated solutions has made possible the determination of such couplings.

To translate the observed vicinal coupling constants into conformational parameters one requires a Karplus type equation correlating J_{HCNC} with the dihedral angle. Lemieux and coworkers[54-56] from a study of ^{13}C enriched pyrimidine derivatives showed that the vicinal proton-carbon coupling between H1′ and C2 exhibited a Karplus type dependence in which the magnitude of J varied from 0 to 8 Hz and is governed by the equation:

$$J_{HC} = 6.7 \cos^2\phi + 1.3 \cos\phi \tag{5}$$

Application of this equation to the $J_{C4H1'}$ and $J_{C8H1'}$ in cyclic purine derivatives[57/58] gave unsatisfactory fitting. This may be due to the strain in the cyclo systems and it is possible that the above relation is applicable only for the pyrimidines.

Four bond couplings J_{HCCCH} between sugar H1′ and the base protons have been observed in a number of C-nucleosides, i.e., α-pseudouridine,[71] β-pseudouridine and its 3′ monophosphate,[72] showdomycin[73/74] and oxazinomycine.[75] These couplings will be useful in determining glycosidic torsion once a relationship between χ and the magnitude of these couplings is established.

Five bond couplings of the type J_{HCCNCH} have been used to investigate glycosidic torsion in certain pyrimidine and triazole nucleosides.[76/77] Evans and Sarma[78] have observed five bond coupling of about 0.5 Hz between H1′ and H7 in tubercidin 5′ phosphate which indicates detectable populations of conformers in which χ is in the anti-domain (Figure 22).

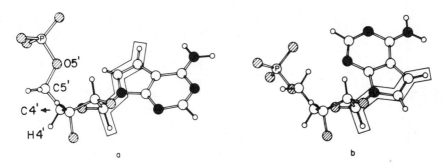

Figure 22. Perspective drawing of the *anti* (a) and *syn* (b) orientations about the glycosyl bond. In the *anti* conformation the bond system between H1′ and H7 has an in-plane zig-zag geometry (boxed regions). In the *syn* conformation, the zig-zag pattern is destroyed. and the C4′-C5′ is *gt*. In the *anti* conformation note the in-plane 'W' path between phosphorus and H4′ which is absent in the *syn* conformation.

Qualitative information about glycosidic torsion in nucleic acid structures can be derived from the changes in base proton chemical shifts due to the selective field effect of a charged phosphate group,[8/59/60] effect of substituents,[61-63] effect of metal cations,[9/64] anisotropic effect of the hetero atom and the ring current field effects of the heterocycle on the sugar proton chemical shifts,[36/37/64-68] NOE and T_2 measurements.

The Intramolecular Order in Single Stranded Nucleic Acids

From the foregoing it is clear that for nucleic acid fragments such as a dimer, NMR spectroscopy can provide reasonably accurate information about the nature of sugar pucker and all the torsion angles except the sugar base and the phosphodiester torsions. This is also true for higher oligomers where the situation is complicated by difficulty in analysis and assignments. One cannot obtain information about the intramolecular order without a knowledge of χ, β and γ (Figure 4). This can be seen from an inspection of Figure 23 which contains the perspectives for ApA in which, except for β and γ, all torsions are kept constant; one immediately notices that depending upon the magnitude of β and γ the molecule can adopt widely different intramolecular structures. In all these perspectives χ_1 and χ_2 were kept at 0° and variation in these can generate additional conformations. Then, at the level of higher oligomers, such as trimers, tetramers, etc. the possibilities of possible conformations continue to increase. Hence, NMR methods should be developed to determine sugar base and phosphodiester torsions.

Comparison of the base proton shift data between monomers and oligomers indicate that in a majority of cases the protons of the bases are shifted to higher fields in the

Figure continued, next page

Figure continued, next page

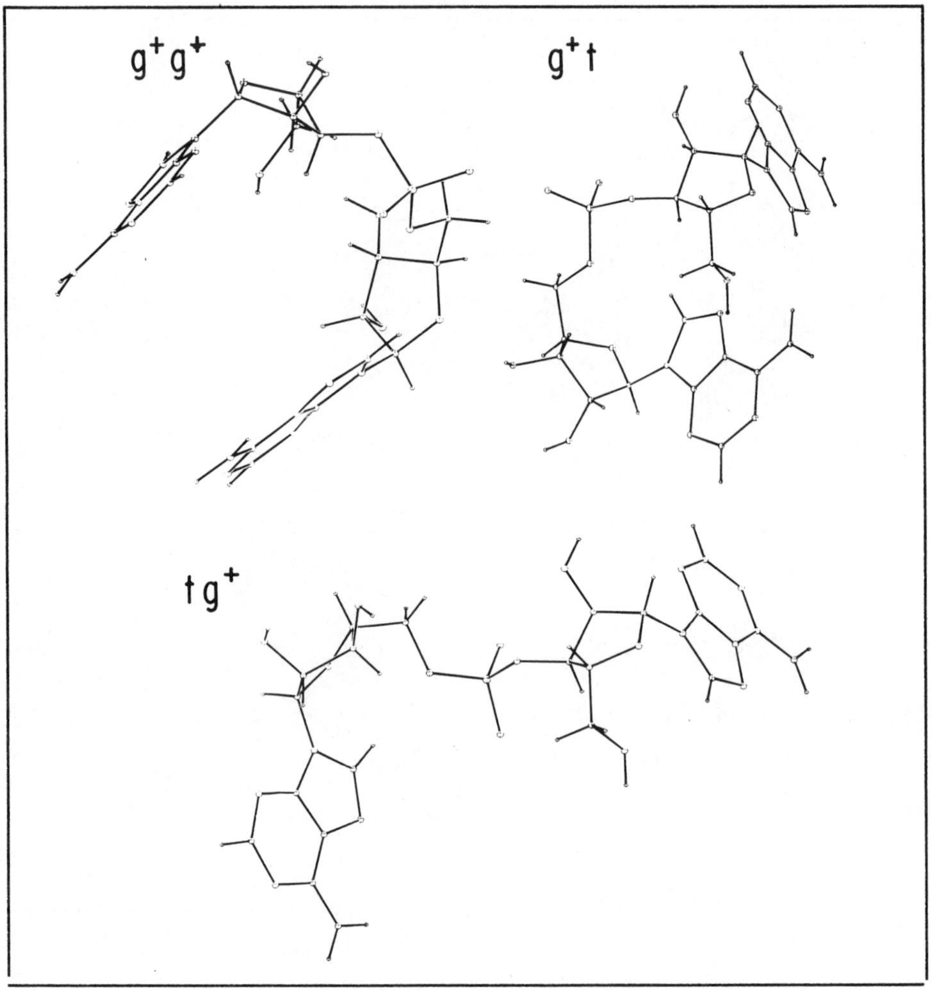

Figure 23. The nine possible conformations about the β, γ phosphodiester bonds. Note that in the g^-g^- and g^+g^+ arrays base stacking is possible. For all diagrams sugar pucker is ^3E, $\chi_1 = \chi_2 = 0°$, $\alpha = 206°$, $\delta = 180°$, ε and $\varepsilon' = 60°$.

oligomer suggesting intramolecular stacking interaction as in the g^-g^- arrangement, Figure 23. Further in the g^-g^- conformation the 2'OH of Xp- comes close to H5' of -pY causing a significant downfield shift[43/44] of the H5' of -pY in the dimer XpY. These qualitative approaches are useful but do not provide detailed information about intramolecular order.

Cheng and Sarma[45] have developed a method based on dimerization shift data (i.e., difference in chemical shifts between monomers and dimers) to determine χ_1, χ_2, β and γ based on ring current theory. Recently, the method was improved by Sarma, *et al.*[79] to take into consideration the contribution to shielding/deshielding from diamagnetic and paramagnetic anisotropy contributions. The method essentially involves a conformational search in the normally expected domains of χ_1, χ_2, β and γ to duplicate the experimentally observed dimerization data. The coupling constant data already provide information about the preferred torsional isomers about C3'-O3', O5'-C5', and C5'-C4' and about the sugar ring conformations. It is generally known that χ_1 and χ_2 are in the anti domain, and that the phosphodiester torsions (β,γ) may display any of the possible nine conformations (Figure 23). Inspection of these projections clearly reveals that in the g^-g^- conformation there is substantial base base stacking interactions. In fact, the dimerization data show that the base protons and a few protons of the sugar ring in, for example ApA, are shifted upfield compared to the monomers. This suggests the possible presence of g^-g^- arrays. It is important to emphasize that only in the g^-g^- stacked array in Figure 23 the *base overlap* and *base separation* permits the base protons to experience ring current upfield shifts, i.e., the observed upfield shifts of the base protons essentially reflect the presence of g^-g^- stacks. In the g^+g^+ loop stack, the base separation is too large to result in any ring-current shifts for the base protons. So the x, y, z coordinates for the hydrogens are generated for conformational arrays using experimentally observed torsion angles α, δ, ε, ε' and sugar pucker and varying β, γ, χ_1, χ_2 torsion angles initially at intervals of 20° and finally at 5° in the g^-g^-, and *anti* domains. From these coordinates contributions to shielding/deshielding from ring current fields, diamagnetic and paramagnetic anisotropy were determined using principles discussed in references 79 and 80 and references therein. The method essentially is an iterative procedure where the iterations are carried out on β, γ, χ_1 and χ_2 till a good

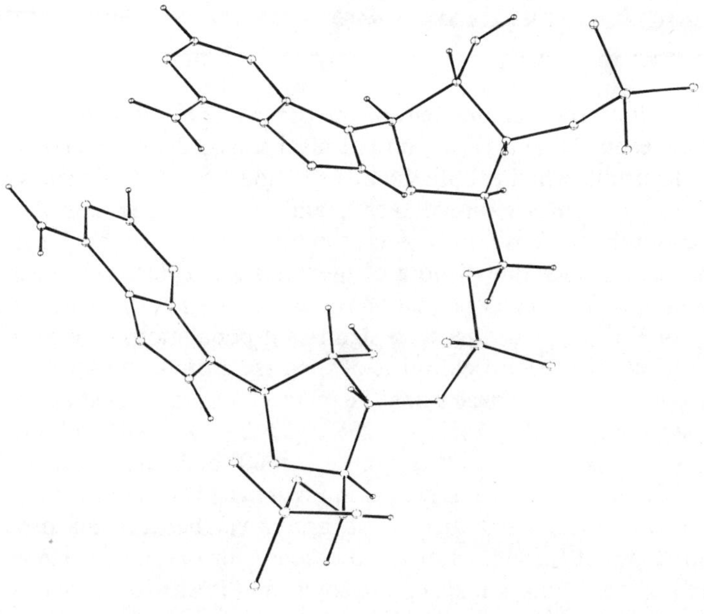

Figure 24. The preferred spatial configuration of ApA determined by the procedure outlined in the text.

fit is obtained between the observed dimerization shifts and those calculated for protons from NMR theory for a specific conformation. In Figure 24, the preferred spatial configuration of ApA so derived is shown. In Table II are listed the final torsion angles and the x, y, and z of the protons. Table II also contains the theoretically projected and experimentally observed dimerization data for the -pA segment of ApA. Except for H5′ and H3′, there is reasonable agreement between the calculated and observed values. The large deshielding of the H5′ of -pA originate from the juxtaposition between this H5′ and 2′OH of Ap- in the g^-g^- conformation (Figure 24, g^-g^-) and provide further evidence for the presence of g^-g^- arrays as has been discussed in extenso elsewhere.[43/44] The H3′ is affected because of the shift toward ^3E in the $^2E \rightleftharpoons {}^3E$ equilibrium upon dimerization.

Table II

Conformational Angles for the Derived Geometry
of ApA are $\alpha = 206°$, $\beta = 285°$, $\gamma = 290°$, $\delta = 180°$, $\varepsilon = 60°$, $\chi_1 = 25°$, $\chi_2 = 50°$, ^3E, ^3E.
Calculated Shieldings for the Effect of Ap- Residue on -pA Residue Protons.

Protons of -pA residue	x	y	z	Ring Current	Diamag Aniso	Paramag Aniso	δ Total	δ obs
H - 1′	-0.077	3.065	5.470	0.127	0.000	0.037	0.164	0.180
H - 2′	-1.730	4.921	4.633	0.118	-0.001	0.030	0.147	0.170
H - 3′	-2.797	3.709	2.991	0.093	-0.001	0.027	0.119	0.000
H - 4′	-2.735	1.323	4.690	0.018	-0.001	-0.003	0.014	0.015
H - 5′	-2.404	0.224	2.643	-0.013	-0.001	-0.012	-0.026	-0.245
H - 5″	-3.701	1.202	2.480	0.018	-0.001	0.005	0.024	-0.051
H - 2	3.917	5.287	5.789	0.090	0.002	0.055	0.147	0.201
H - 8	-0.797	3.701	1.855	0.262	-0.004	0.070	0.328	0.250

Even though an important component of the procedure elaborated above is the effect of intramolecular base stacking on chemical shifts, the same approach can be employed to determine whether base stacked and non-base stacked arrays contribute significantly to the conformational blend, and even to determine the extent of oscillations about the various bonds. A case in point is provided by the dinucleoside monophosphates in which one or both of the bases are replaced by the modified Y base εA (Figure 25). The dimerization data for εApεA indicated substantial deshielding[81] of H5′ of -pεA suggesting significant populations of g^-g^- arrays. This was further reflected in the upfield shifts of the base protons in εApεA. A search in the g^-g^- *anti* conformations space using the methodology discussed above indicated the best fit happens when $\beta = 285°$, $\gamma = 295°$, $\chi_1 = 25°$, $\chi_2 = 40°$ and the remaining torsion angles: $\alpha = 205°$, $\delta = 180°$, $\varepsilon = 60°$, $\varepsilon' = 60°$ and sugar puckers ^3E, ^3E. A perspective of this conformation is shown in Figure 26. The data for this conformation in Table III indicate that there is an agreement between the projected and observed shifts, except for H8. In order to account for the shift of H8 an intensive search in the g^-g^- conformation space was made and it was found that when β was changed from 285° to 300°, γ from 295° to 290° and χ_2 from 40° to 45°, the projection for H8 agreed with the observed value and under these slightly different torsion

Table III

The Projected Shieldings in ppm for the Base Protons of the εAp- Residue in the g^-g^- Conformation Space.

Also Given Are The Observed Shieldings

Conformation: $\beta = 285°$, $\chi_1 = 25°$, $\gamma = 295°$, $\chi_2 = 40°$

Base Protons	Projected Shieldings	Observed Shieldings
H2	0.39	0.409
H8	0.06	0.324
H10	0.45	0.449
H11	0.49	0.534

Conformation: $\beta = 300°$, $\chi_1 = 25°$, $\gamma = 290°$, $\chi_2 = 45°$

Base Protons	Projected Shieldings	Observed Shieldings
H2	0.08	0.409
H8	0.30	0.324
H10	0.05	0.449
H11	0.15	0.534

angles, the remaining base protons experience very little shielding (Table III). Thus the data clearly demonstrate that εApεA populate significantly in the g^-g^- conformation space and that the O3'-P torsion (γ) has a local flexible domain (285°-300°) in the g^-g^- array.

In many an εA containing dinucleoside monophosphates the H5' of the nucleotidyl unit at the 5' end (i.e., the Np- residue) undergoes substantial shielding in the range of 0.440 to 0.192 ppm.[81] This cannot originate from g^-g^- conformations. Calculations shown in Table IV indicate that the most important conformer which can influence the chemical shift of εAp- H5' is g^+t (Figure 26) and hence dinucleoside monophosphates which contain εA will have a significant proportion of g^+t conformers.

Table IV

The Projected Shieldings in ppm for the H5' of the εAp- Residue in Various Conformations

Conformation	Proton	Projected Shielding
$g^+g^+(\beta = 80°, \gamma = 80°)$	H5'	0.12
$g^+t (\beta = 110°, \gamma = 215°)$	H5'	0.56
$tg^+ (\beta = 180°, \gamma = 80°)$	H5'	0.03

Because of the large number of variables and observables our procedure may not give a unique solution for β, γ, χ_1, and χ_2 but definitely will predict the correct domains for these torsion angles. Further, in arriving at the preferred intramolecular order we have assumed that the local conformational properties are *persistent*; for example, in ApA the ribose is 60% ^3E, ε is 74% gg and we assume that a majority of ^3E pucker is associated with gg conformation about C4'-C5'.

The methodology can be extended to single-stranded higher oligomers. This is discussed later in this volume[82] in the cases of CpCpA, ApCpC, d-TpTpT and d-TpTpC. Sometimes the above methodology can be used along with computer modelling to detect the presence of small amounts of unusual conformations. A case in point is provided by d-TpTpA.[16]

It was found that d-TpTpA predominately exists in a conformation in which χ = an-

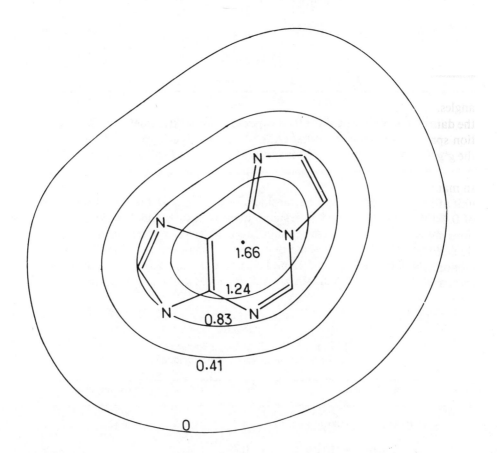

Figure 25. The structure of the modified Y base εA and its shielding values due to the sum of the ring current effect and of the atomic diamagnetic susceptibility anisotropy in a plane 3.4 Å distant from the molecular surface. From reference 81.

ti, sugar puckers ^2E, ε', ε_1, $\varepsilon_2 \simeq 60°$, δ_1, $\delta_2 \simeq 180°$, α_1, $\alpha_2 \simeq 199°$, $\beta_1\gamma_1 = tg^-$ and $\beta_2\gamma_2 = g^-g^-$ (Figure 27). However, the observed distant shieldings[83] of the terminal dTp-residue of d-TpTpA by the terminal adenine cannot be rationalized by the structure in Figure 27. The distant shieldings can originate from certain minor conformers which contribute to the conformational blend. Because the molecule is flexible, it

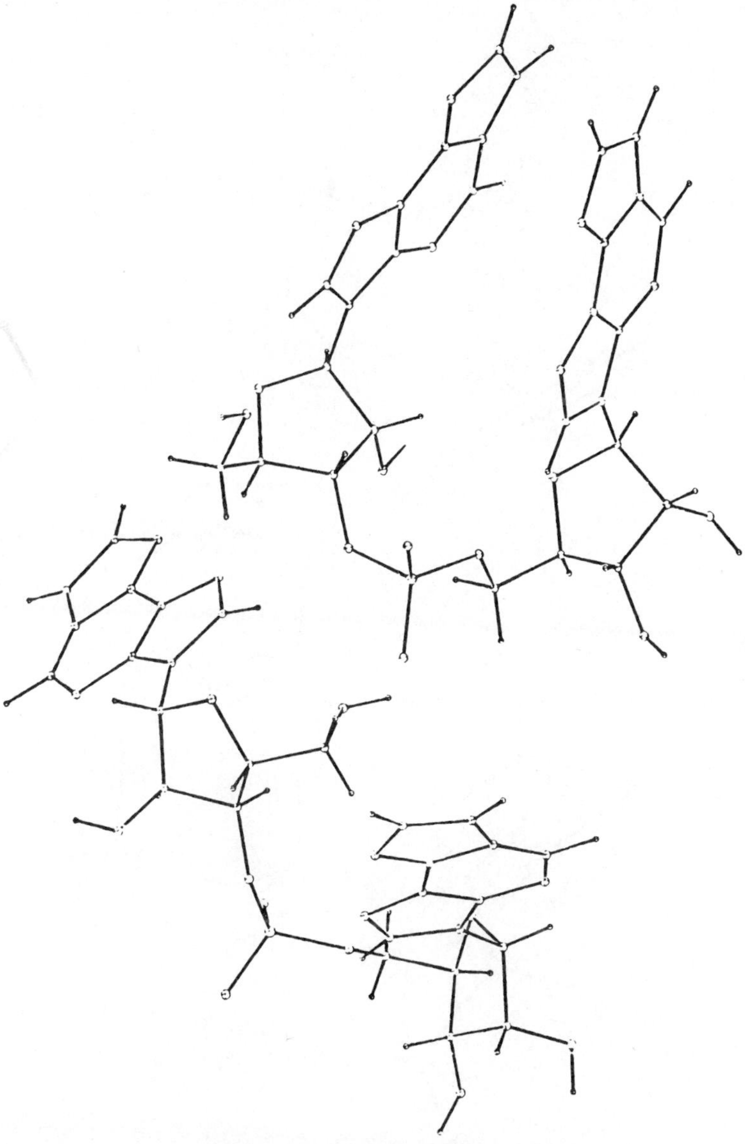

Figure 26. *Right* A perspective of εApεA in the g^-g^- conformation. The sugar pucker is ^3E and the torsion angles are: $\alpha = 205°$, $\beta = 285°$, $\gamma = 295°$, $\delta = 180°$, $\varepsilon = 60°$ $\chi_1 = 25°$ and $\chi_2 = 40°$. *Left* A perspective of εApεA in the g^+t conformation. Except for $\beta = 110°$, $\gamma = 215°$, $\chi_1 = 20°$, and $\chi_2 = 50°$, rest of the geometrical details same as the g^-g^- situation.

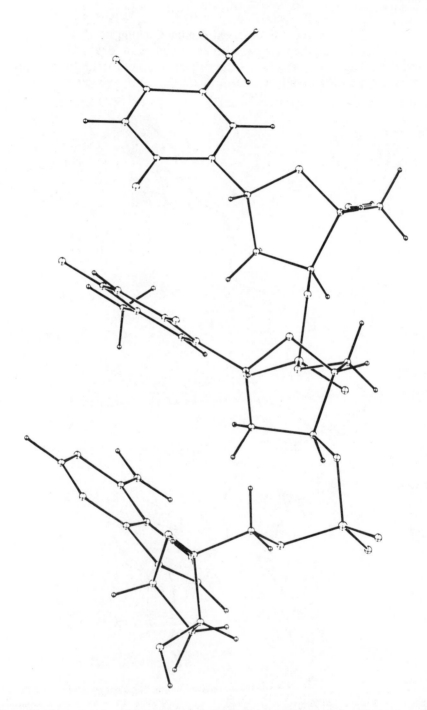

Figure 27. The preferred spatial configuration of d-TpTpA in aqueous solution. For torsion angles, see text.

may adopt different conformations in solution. The proclivity of base pairs such as purine-purine and pyrimidine-purine to interact and stack provide the driving force for the two end nucleotidyl units to undergo torsional motion about the phosphodiester bonds so that the bases can interact. In order to determine the spatial configuration of a trimer in which the two end nucleotidyl bases can interact and cause ring current shifts, we have examined a reasonable number of $\beta_1\gamma_1$ and $\beta_2\gamma_2$ combinations. The combinations used are | g⁺g⁻, g⁺g⁺|, |g⁺g⁻, g⁺g⁻|, |g⁺g⁺, g⁻g⁺|, |g⁺g⁻, g⁻g⁺| and |g⁺g⁺, g⁺g⁺|. The perspectives of some of these combinations are shown in Figure 28a, b, c, d. As is clear from Figure 28a, b, in the |g⁺g⁻, g⁺g⁻| and |g⁺g⁻, g⁺g⁺| combinations, the bases of the end nucleotidyl units can be parallel but the value of z (the perpendicular cylindrical coordinate) is too large to account for any distant shielding effects. In the combinations |g⁺g⁻, g⁻g⁺| and |g⁺g⁺, g⁻g⁺|, (Figure 28c, d) the bases cannot be parallel for reasonable values of χ_{CN}, and the distance between the base planes are too large to account for any distant shielding effects. The structure in which the phosphodiester bonds assume |g⁺g⁺, g⁺g⁺| conformation generates a spatial array in which the end nucleotidyl bases are approximately parallel and the value of z is reasonable to induce the ring current shifts. Hence in the cases of d-TpTpA and d-ApTpT the bases of the end nucleotidyl units can interact if $\beta_1\gamma_1$ occupies g⁺g⁺ and $\beta_2\gamma_2$ occupies g⁺g⁺ domains. A perspective of such a

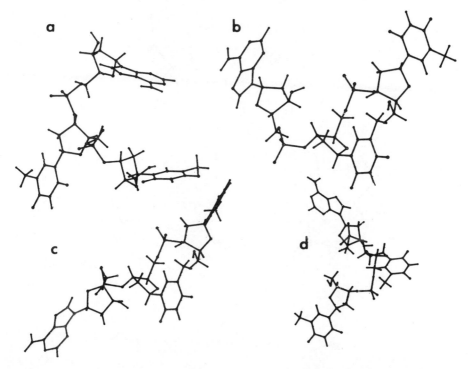

Figure 28. Perspectives of d-TpTpA in which $\beta_1\gamma_1$ and $\beta_2\gamma_1$ are in |g⁺g⁻, g⁺g⁻| (a) |g⁺g⁻, g⁺g⁺| (b) |g⁺g⁺, g⁻g⁺| (c) and |g⁺g⁻, g⁻g⁺| (d).

conformation for d-TpTpA is shown in Figure 29. It is clear from this figure and the corresponding cylindrical coordinates in Table V that in this structure, the adenine ring will shield the H1′, H6 and methyl protons of dTp- residue. The projected shieldings are larger than the observed ones because conformation in Figure 29 is only a minor contributor.

It is important to emphasize that in arriving at the bulged configuration in Figure 29, we have kept all torsion angles except χ_1, χ_3, $\beta_1\gamma_1$ $\beta_2\gamma_2$ same as the preferred ones. One indeed could alter any set of torsion angles and by computer manipulation may arrive at other bulged configuration. The purpose of the above exercise is to show how the methodology of NMR and computer modelling could be combined to arrive at molecular geometry.

Intramolecular Order of Double-Stranded Nucleic Acids

The methods of NMR spectroscopy as elaborated above based on coupling constants and chemical shifts and discussed for small nucleic acid fragments and single-stranded structures can be, in principle applied to delineate the intimate details of

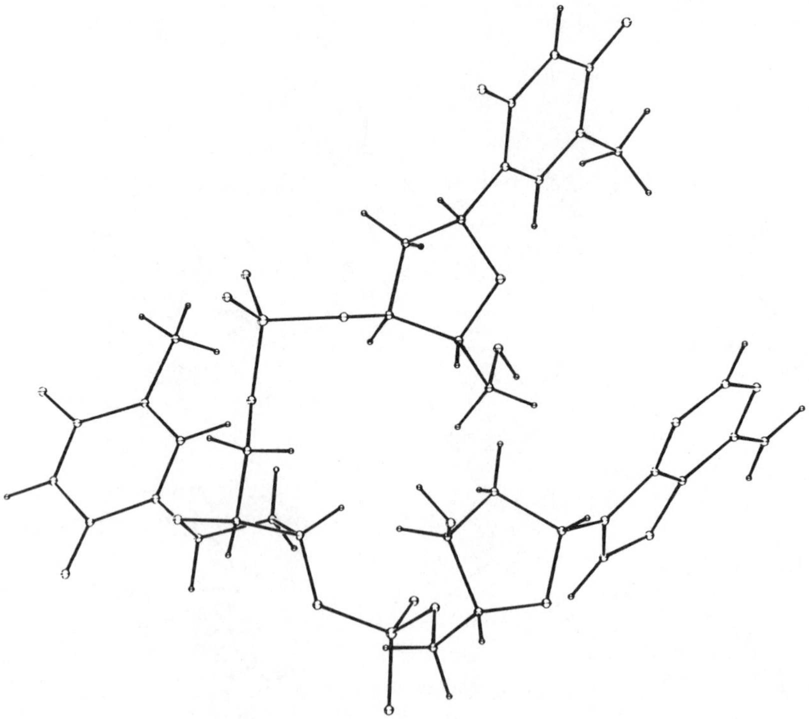

Figure 29. Perspective of d-TpTpA in the bulged configuration; $\beta_1\gamma_1 = g^+g^+$, $\beta_2\gamma_2 = g^+g^+$; rest of the torsion angles as in Figure 27, χ_1 and χ_3 were adjusted for base parallelism and distant shielding effects.

Table V
The Cylindrical Coordinates z, ϱ_5, ϱ_6 in Å of Tp- Residue With Respect to
Adenine Moiety in TpTpA (g⁺g⁺, g⁺g⁺, $\chi_1 = 60°$, $\chi_2 = 70°$, $\chi_3 = 30°$).
The projected shieldings are in ppm.*

Proton	z	ϱ_5	ϱ_6	$(\delta_5)^{cal}$	$(\delta_6)^{cal}$	$(\delta_5+\delta_6)^{cal}$	δ_{obs}**
H-1′	6.3	3.5	2.6	0.03	0.15	0.18	0.065
H-2′	7.8	1.9	2.0	0.02	0.05	0.07	
H-2″	7.0	0.9	1.2	0.06	0.12	0.18	
H-3′	6.2	1.4	3.2	0.10	0.10	0.20	
H-4′	3.9	2.4	3.1	0.15	0.13	0.28	
H-5′	3.0	0.3	1.9	0.8	0.50	1.30	
H-5″	3.9	1.2	3.3	0.4	0.13	0.53	
H-6	5.8	2.7	1.1	0.1	0.25	0.35	0.087
H-	6.7	4.8	3.5	0.0	0.10	0.10	
H- CH₃	6.4	6.1	4.4	0.0	0.05	0.05	0.015
H-	5.2	5.0	3.6	0.0	0.10	0.10	

*The conformation search was performed only in the anti domain and the values of χ_1 and χ_3 were adjusted for base parallelism and distant shielding effects. Rest of the torsion angles as in Figure 27.

**From reference 83.

the geometry of double-stranded structures. However, in most double-stranded systems the spectra are so complex that an accurate complete analysis of their NMR spectra are formidable and many times impossible at the frequency of the presently available superconducting systems. The one exception known to date is the case of a miniature double helix of an analog of ApU in which χ was fixed at $\simeq 120°$ by chemical modification. Complete solution of the molecular geometry of this system based on methodology discussed in this chapter is presented later.[84]

It is worthwhile to explore whether NMR spectroscopy can be used to obtain detailed geometry information from the spectra of oligomers and polymers where one could assign only a few chemical shifts. This is possible if one can work backwards, i.e., instead of using chemical shifts and coupling constants to arrive at the geometry, one may search for molecular geometries derived from X-ray which may be compatible with the observed limited number of shifts. For example, it is known that DNA may exist in various forms such as A[85], B[86], C[87], D[88], Z[89], β-kinked[90], propellar-twisted[91], alternating B[92], side by side[93/94] and so on. In most of the above cases, the x, y, z coordinates are available. One may assume that the double helix of a tetramer, hexamer or a polynucleotide in solution may exist in any of the above forms. Hence, one can theoretically derive from the appropriate x, y, z the expected shieldings for the various protons in the duplex state and compare with the experimentally observed changes as the duplex melts into the single strands. We illustrate this approach with an actual example below.

Suppose one wants to learn the geometry of the self-complementary duplex d-

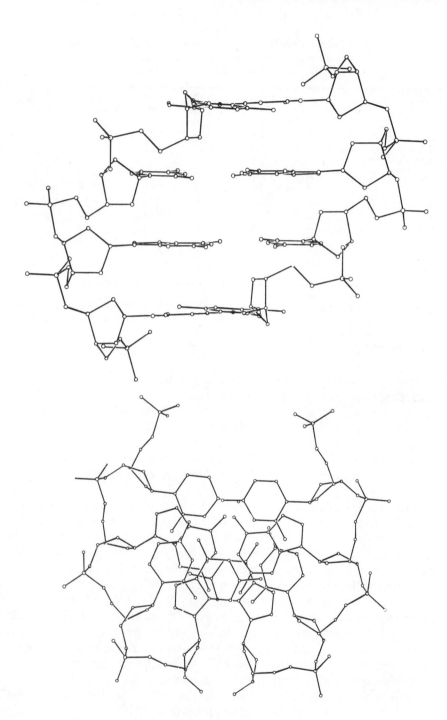

Figure 30. The B-DNA geometry of the duplex of d-pCpGpCpG. View perpendicular to the helical axis is on the top and the view along the helical axis is at the bottom.

CpGpCpG. Then, initially, one may assume that the duplex may exist in the B-DNA configuration. The B-DNA geometry for the duplex of d-CpGpCpG can be derived from Arnott's coordinates,[86] and this is shown in Figure 30. Then one has to calculate the effect on the chemical shift of all protons in a nucleotide unit by the remaining units. Thus, for example, the shift of H5 of the internal base marked C_{1a} (Figure 31) will be influenced by the remaining seven bases. One can calculate the ring current, diamagnetic and paramagnetic anisotropy contribution to the chemical shift of H5 of C_{1a} from the remaining seven bases from the x, y, z coordinates for the B-DNA configuration (Figure 30) using the principles outlined in Giessner-Prettre and Pullman,[80] and the results are presented in Table VI. Similar calculations can be

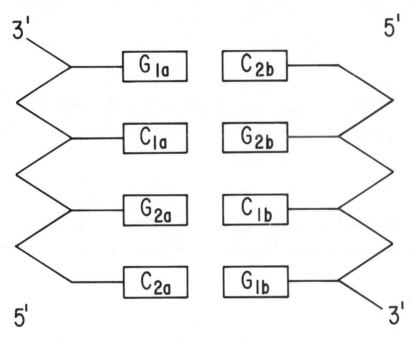

Figure 31. A schematic drawing of the duplex of d-CpGpCpG.

Table VI
The Various Contributions in ppm to the Chemical Shift
of H5 of C_{1a} from the Remaining Seven Bases (Figure 31).

| Nature of Effect | Individual Contribution From | | | | | | | |
	G_{1a}	G_{2a}	C_{2a}	G_{1b}	G_{2b}	C_{1b}	C_{2b}	Total
Ring Current	0.008	0.396	0.022	0.008	-0.037	-0.005	-0.007	0.385
Diamagnetic	0.001	-0.006	-0.001	-0.001	0.000	0.001	0.001	-0.005
Paramagnetic	0.015	0.187	0.029	0.010	-0.066	-0.016	-0.005	0.154
Total	0.024	0.577	0.050	0.017	-0.103	-0.021	-0.011	0.533

done for each proton in each nucleotide unit and the data like the ones in Table VI can be assembled.

The data of the type assembled in Table VI indicate the chemical shift change the H5 of C_{1a} will experience when it is moved from an isolated situation to the one in B-DNA as in Figure 32. As is evident from this figure, the calculation does not take in-

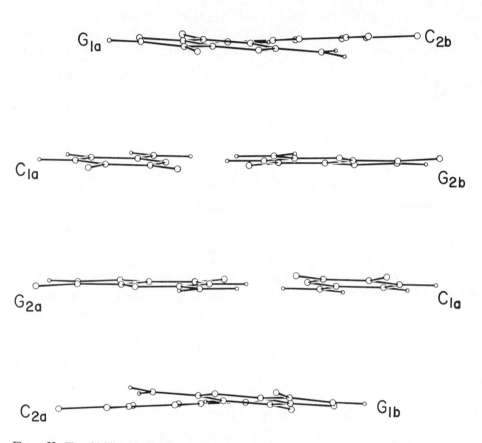

Figure 32. The shielding/deshielding environments in the B-DNA structure of the duplex of d-CpGpCpG, that are taken into consideration in the calculations.

to consideration the effect of the ribose phosphate backbone. Hence, experiments designed to measure the calculated data in Table VI should be such that the effects from the backbone should remain essentially constant. This is somewhat the situation, though not identical, when the duplex is melted to the strands. One can measure the chemical shifts in the above two states and compare with what is predicted from theory. However, there are complications. The chemical shifts in the individual strands should be measured at conditions at which no intramolecular base-base interactions exist; usually at fairly high temperatures the single strands approach this situation because they usually exist in random coil forms at these temperatures. In order to truly achieve the conditions of the calculation it is also necessary that the chemical shifts of the protons are affected *equally* by the backbone in the duplex and in the random coil forms. This is not the case. For example, in the transition base stacked duplex → unstacked random coil, the χ changes and this in turn will change the shifts of the base and sugar protons. Hence, a correction must be calculated and applied to the type of data in Table VI. How this can be done will become clear later.

In Table VII, we have reported the theoretically calculated and experimentally observed shift data for the formation of the B-DNA duplex of d-CpGpCpG from single strands. The data is given only for the base protons. Comparison of the calculated total and the experimentally observed values do not agree. In fact, out of the six sets of data, two show reasonable agreement, i.e., H6 terminal and H5 internal, 4 sets show disagreement, i.e., H6 internal, H5 terminal, H8 terminal and internal. This does not necessarily mean that the duplex of d-CpGpCpG does not display B-DNA structure. In the above calculation, we have assumed that χ is 82° in both B-DNA and in the single strands. If one assumes that B-DNA form of d-CpGpCpG is formed from single strands in which internal C (C_{1a} and C_{1b}, Figure 31) exists with χ of 55° and terminal and internal G (G_{1a}, G_{1b}, G_{2a}, G_{2b}, Figure 31) exist with a χ of 100° and upon any duplex formation they change to 82°, then the theoretically calculated data will agree with that reported experimentally. Under the above assumptions, then the only data set among the six that does not agree is that for H5 terminal which experimentally shows a shift of over 0.2 ppm and the calculation predicts little shift. This is a serious disagreement. Part of this observed high field shift is due to fraying at the ends (transient opening and closing at the ends)[84/96/97] because fraying removes the inplane deshielding of C_{2b} and C_{2a} by G_{1a} and G_{1b} respectively; part of the shift may come from aggregation and column formation and one can check this by conducting melting experiments at concentrations less than that used by Patel.[95] It is also possible that the idealized B-DNA model is not

Table VII

Theoretically Calculated and Experimentally Observed Shift
Data For The Formation of the B-DNA Duplex of
d-CpGpCpG From Single Strands

Base Proton	Position	Calculated Contributions in ppm from				Experimentally Observed[95] (ppm)
		Ring Current Term	Diamagnetic Term	Paramagnetic Term	Total Calculated	
H6	Terminal	0.020	0.000	-0.014	0.005	0.05
	Internal	0.105	-0.006	0.006	0.105	0.3
H5	Terminal	-0.001	-0.002	-0.019	-0.022	>0.2
	Internal	0.385	-0.005	0.154	0.533	0.45
H8	Terminal	0.081	-0.005	0.085	0.162	0.0 - 0.05
	Internal	0.052	-0.006	0.104	0.150	0.0 - 0.05

the correct model for the duplex of d-CpGpCpG. Hence, before any conclusions are made, the calculations should be extended to see whether the data will agree with other forms of DNA such as A, C, D, Z, etc., cited a while ago.

The purpose of this fairly involved section is not to attempt to arrive at the spatial configuration of the duplex of d-CpGpCpG, but rather to indicate that the theory of NMR spectroscopy can be used to arrive at the detailed solution geometry of complex systems. In Table VII we have presented calculations and results only for the base protons; in fact, data for the interior protons should also be examined; particularly crucial is the H1′ because of its sensitivity to stacking as well as to χ changes. These are not included in Table VII because no experimental data on these are available for oligomer duplexes. The assignment of the interior protons in their NMR spectra is still a problem. Progress in this direction is underway in this laboratory from the spectra of the duplexes of selectively deuterated oligomers discussed in the beginning of this chapter. The method that is presented here is the best that is known today because it calculates the NMR shielding values from the x, y, z coordinates taking into consideration the ring current, the diamagnetic and paramagnetic antisotropy terms; the extent of shielding of any proton in the duplex can be calculated, and the method can be applied to duplexes of oligomers and polymers and even to globular polynucleotides such as tRNA; in all these cases, the difficult problem is the experimental assignment of signals unambiguously and recording spectra at a very low concentration when molecules do not aggregate. The case of tRNA in this respect makes an exciting story.[2/98]

Acknowledgement

This research was supported by Grant CA12462 from the National Cancer Institute of NIH and Grant PCM-7822531 from National Science Foundation. This research was also supported by Grant 1-PO7-PR-PR00798 from the Division of Research Resources, NIH. The author is greatly indebted to Dr. C. Giessner-Prettre and Professor B. Pullman for discussions as well as for the various computer programs. The author expresses his thanks and appreciation to Drs. Dhingra and Mitra for their valuable help.

References and Footnotes

1. Szent-Gyorgyi, A., *International J. Quant. Chem. QBS 4*, 179 (1977).
2. Kearns, D.R. *Ann. Rev. Biophys. Bioeng. 6*, 477 (1977).
3. Sundaralingam, M. *International J. Quant. Chem. QBS 4*, 11 (1977)
4. Schweizer, M. P., Chan, S. I., Helmkamp, G. K., Ts′o, P. O. P., *J. Am. Chem. Soc. 86*, 696 (1966).
5. Bullock, F. J., and Jardetzky, O., *J. Org. Chem. 29*, 1988 (1964).
6. Akasaka, K., Imoto, T., and Hatano, H., *Chem. Phys. Lett. 21*, 398 (1973).
7. Ts′o, P. O. P., Kondo, N. S., Schweizer, M. P., and Hollis, D. P., *Biochemistry 8*, 997 (1969).
8. Schweizer, M. P., Broom, A. D., Ts′o, P. O. P., and Hollis, D. P., *J. Am. Chem. Soc. 90*, 1042 (1968).
9. Chan, S. I., and Nelson, J. H., *J. Am. Chem. Soc. 91*, 168 (1969).
10. Cheng, D. M., and Sarma, R. H., *Biopolymers 16*, 1687 (1977).
11. Remin, M., and Shugar, D., *Biochem. Biophys. Res. Commun. 48*, 636 (1972).

12. Lee, C. H., Evans, F. E., and Sarma, R. H., *FEBS Letters 51*, 73 (1975).
13. Evans, F. E., Lee, C. H., and Sarma, R. H., *Biochem. Biophys. Res. Commun. 63*, 106 (1975).
14. Kondo, N. S., Ezra, F., and Danyluk, S. S., *FEBS Letters 53*, 213 (1975).
15. Evans, F. E., and Sarma, R. H., *Nature 263*, 567 (1976).
16. Cheng, D. M., Dhingra, M. M., and Sarma, R. H., *Nucleic Acids Research 5*, 4393 (1978).
17. Karplus, M., *J. Chem. Phys. 33*, 1842 (1969).
18. Sternhell, S., *Q. Rev. Chem. Soc. 23*, 236 (1969).
19. Sternhell, S., *Rev. Pure Appl. Chem. 14*, 15 (1964).
20. Barfield, M., and Chakrabarti, B., *Chem. Rev. 69*, 757 (1969).
21. Hall, L. D., and Manville, J. F., *Carbohydr. Res. 8*, 295 (1968).
22. Blackburn, B. J., Lapper, R. D., and Smith, I. C. P., *J. Am. Chem. Soc. 95*, 2873 (1973).
23. Hruska, F. E., Mak, A., Singh, H., and Shugar, D., *Can. J. Chem. 51*, 1099 (1973).
24. Barfield, M., Spear, R. J., and Sternhell, S., *J. Am. Chem. Soc., 93*, 5322 (1971).
25. Barfield, M., McDonald, C. J., Peat, I. R., and Reynolds, W. F., *J. Am. Chem. Soc. 93*, 4195 (1971).
26. Davies D. B. in *Progress in Nuclear Magnetic Resonance Spectroscopy* (Emsley, J. W., Feeney, J., and Sutcliffe, L. H., Eds.), *12*, 135 (1978).
27. Altona, C., and Sundaralingam, M., *J. Am. Chem. Soc. 94*, 8205 (1972); *95*, 2333 (1973).
28. Sundaralingam, M., *Biopolymers 7*, 821 (1969).
29. Sundaralingam, M., in *Conformations of Biological Molecules and Polymers. The Jerusalem Symposium on Quantum Chemistry and Biochemistry V.* Editors Bergman, E., and Pullman, B., p. 417 (1973).
30. Jardetzky, C. D., *J. Am. Chem. Soc. 82*, 229 (1960).
31. Jardetzky, C. D., *J. Am. Chem. Soc., 84*, 62 (1962).
32. Sarma, R. H., and Mynott, R. J., *J. Am. Chem. Soc. 95*, 1641 (1973).
33. Evans, F. E., and Sarma, R. H., *J. Biol. Chem. 249*, 4754 (1974).
34. Sasisekharan, V., in *Conformations of Biological Molecules and Polymers. The Jerusalem Symposium on Quantum Chemistry and Biochemistry V.* Editors Bergman, E., and Pullman, B., p. 247 (1973).
35. Saran, A. Perahia, D., and Pullman, B., *Theor. Chim. Acta 30*, 31 (1973).
36. Lee, C. H., Evans, F. E., and Sarma, R. H., *J. Biol. Chem. 250*, 1290 (1975).
37. Sarma, R. H., Lee, C. H., Evans, F. E., Yathindra, N., and Sundaralingam, M., *J. Am. Chem. Soc. 96*, 7337 (1974).
38. Hruksa, F. E., in *Conformation of Biological Molecules and Polymers. The Jerusalem Symposium on Quantum Chemistry and Biochemistry V.* Editors Bergman, E., and Pullman, B., p. 345 (1973).
39. Hruska, F. E., Wood, D. J., Mynott, R. J. and Sarma, R. H., *FEBS Letters 31*, 153 (1973).
40. Wood, D. J., Mynott, R. J., Hruska, F. E., and Sarma, R. H., *FEBS Letters 34*, 323 (1973).
41. Wood, D. J., Hruska, F. E., Mynott, R. J., and Sarma, R. H., *Can. J. Chem. 51*, 2571 (1973).
42. Lee, C. H., and Sarma, R. H., *J. Am. Chem. Soc. 98*, 3541 (1976).
43. Lee, C. H., Ezra, F. S., Kondo, N. S., Sarma, R. H., and Danyluk, S. S., *Biochemistry 15*, 3627 (1976).
44. Ezra, F. S., Lee, C. H., Kondo, N. S., Danyluk, S. S., and Sarma, R. H., *Biochemistry 16*, 1977 (1977).
45. Cheng, D. M., and Sarma, R. H., *J. Am. Chem. Soc. 99*, 7333 (1977).
46. Dhingra, M. M., and Sarma, R. H., *Nature 272*, 798 (1978).
47. Sarma, R. H., Mynott, R. J., Hruska, F. E., and Wood, D. J., *Can. J. Chem. 51*, 1843 (1973).
48. Hall, L. D., and Malcolm, R. B., *Can. J. Chem. 50*, 2092 (1972).
49. Hall, L. D., and Malcolm, R. B., *Can. J. Chem. 50*, 2102 (1972).
50. Donaldson, B., and Hall, L. D., *Can. J. Chem. 50*, 2111 (1972).
51. Sarma, R. H., Mynott, R. J., Wood, D. J., and Hruska, F. E., *J. Am. Chem. Soc., 95*, 6457 (1973).
52. Schleich, T., Cross, B.P., Blackburn, B.J. and Smith, I.C.P. in *Structure and Conformation of Nucleic Acids and Protein-Nucleic Acid Interactions*, Ed., Sundaralingam, M. and Rao, S.T., University Park Press, Baltimore, p. 223 (1975).
53. Sarma, R. H., and Danyluk, S. S., *Int. Natl. J. Quant. Chem. QBS 4*, 269 (1977).
54. Lemieux, R. U., Nagbhushan, T. L., and Paul, B., *Can. J. Chem. 50*, 773 (1972).
55. Delbaere, L. T. J., James, M. N. G., and Lemieux, R. U., *J. Am. Chem. Soc. 95*, 7866 (1973).

56. Lemieux, R. U., *Ann. N. Y. Acad. Sci. 222*, 915 (1973).
57. Dea, P., Kreishman, G. P., Schweizer, M. P., and Witkowski, J. T., *Proceedings of the Ist Int. Conf. on Stable Isotopes in Chemistry, Biology and Medicine.* Editor, Klein, P. D., p. 84 (1973).
58. Schweizer, M. P., and Kreishnan, G. P., *J. Magn. Reson. 9*, 334 (1973).
59. Evans, F. E., and Sarma, R. H., *FEBS Lett 41*, 253 (1974).
60. Danyluk, S. S., and Hruska, F. E., *Biochemistry 7*, 1038 (1968).
61. Remin, M., and Shugar, D., *J. Am. Chem. Soc. 95*, 8146 (1973).
62. Remin, M., Darzynkiewicz, E., Dworak, A., and Shugar, D., *J. Am. Chem. Soc. 98*, 367 (1976).
63. Follman, H., and Gremels, G., *Eur. J. Biochem. 47*, 187 (1974).
64. Prestegard, J. H., and Chan, S. I., *J. Am. Chem. Soc. 91*, 2843 (1969).
65. Remin, M., Ekiel, I., and Shugar, D., *Eur. J. Biochem. 53*, 197 (1975).
66. Schweizer, M. P., Banta, E. B., Witkowski, J. T., and Robins, R. K., *J. Am. Chem. Soc. 95*, 3770 (1973).
67. Giessner-Prettre, C., and Pullman, B., *J. Theor. Biol. 65*, 171 (1977).
68. Giessner-Prettre, C., and Pullman, B., *J. Theor. Biol. 65*, 189 (1977).
69. Hart, P. A., and Davis, J. P., *J. Am. Chem. Soc. 93*, 753 (1971).
70. Imoto, T., Akasaka, K., and Hatano, H., *Chem. Letters 73* (1974).
71. Grey, A. A., Smith, I. C. P., and Hruska, F. E., *J. Am. Chem. Soc. 93*, 1765 (1971).
72. Schleich, T., Blackburn, B. J., Lapper, R. D., and Smith, I. C. P., *Biochemistry 11*, 137 (1972).
73. Darnall, K. R., Townsend, L. B., and Robins, R. K., *Proc. Natl. Acad. Sci. U. S. A. 57*, 548 (1967).
74. Dhingra, M. M. and Sarma, R. H. (unpublished data).
75. Haneishi, T., Okazaki, T., Hata, T., Tamura, C., Nomura, M., Naito, A., Seki, I., and Arai, M., *J. Antibiotics 24*, 797 (1971).
76. Dea, P., Schweize, M. P., and Kreishman, G. P., *Biochemistry 13*, 1862 (1974).
77. Hruska, F. E., *Can. J. Chem. 49*, 2111 (1971).
78. Evans, F. E., and Sarma, R. H., *Cancer Research 35*, 1458 (1975).
79. Sarma, R. H., Dhingra, M. M., and Feldman, R. J. in *Stereodynamics of Molecular Systems*, Sarma, R. H. (Ed.), Pergamon Press, Inc., New York, p. 3, 1979.
80. Giessner-Prettre, C., and Pullman, B., *Biochem. Biophys. Res. Commun. 70*, 578 (1976).
81. Dhingra, M. M., Sarma, R. H., Giessner-Prettre, C., and Pullman, B., *Biochemistry 17*, 5815 (1978).
82. Sarma, R.H. in *Nucleic Acid Geometry and Dynamics*, ed., Sarma, R.H., Pergamon Press, New York, Oxford, Frankfurt, Paris   p. 143 (1980).
83. Kan, L.S., Barrett, J.C., and Ts'o, P.O.P., *Biopolymers 12*, 2709 (1973).
84. Sarma, R.H., in *Nucleic Acid Geometry and Dynamics*, ed., Sarma, R.H., Pergamon Press, New York, Oxford, Frankfurt, Paris   p.83 (1980).
85. Arnott, S., Dover, S.D. and Wonacott, A.J., *Acta. Cryst. 25*, 2192 (1969).
86. Arnott, S., Hukins, D.W.L., Dover, S.D., Fuller, W. and Hodgson, A.R., *J. Mol. Biol. 81*, 107 (1973).
87. Marvin, D.A., Spencer, M., Wilkins, M.F.H., and Hamilton, L.D., *J. Mol. Biol. 3*, 547 (1961).
88. Arnott, S., Chandrasekaran, R., Hukins, D.W.J., Smith, R.S.C. and Watts, L., *J. Mol. Biol. 88*, 523 (1974).
89. Wang, A., Quigley, G.J., Kolpak, R., van Boom, J.H., and Rich, A., Summer Meeting of Crystallography, Boston, August, 1979.
90. Lozansky, E.D., Sobell, H.M., and Lessen, M. in *Stereodynamics of Molecular Systems*, Ed., Sarma, R.H., Pergamon Press, Oxford, New York, Frankfurt, Paris, p. 265 (1979) and references therein.
91. Levitt, M., *Proc. Natl. Acad. Sci. U.S.A. 75*, 640 (1978).
92. Klug, A., Jack, A. Viswamitra, M.A., Kennard, O., Shakked, Z., and Steitz, T.A., *J. Mol. Biol., 131*, 669-680 (1979).
93. Rodley, G.A., Scobie, R.S., Bates, R.H.T. and Lewitt, R.M., *Proc. Natl. Acad. Sci. USA 73*, 2959 (1976).
94. Sasisekharan, V., Pattabiraman, N. and Gupta, G., *Proc. Natl. Acad. Sci. USA 75*, 4092 (1978).
95. Patel, D.J. *Biopolymers 15*, 533 (1976).
96. Sarma, R.H. and Dhingra, M.M. in *Topics in Nucleic Acid Structure*, Ed. Neidle, S. McMillan (in press).

97. Patel, D.J. in *Nucleic Acid Geometry and Dynamics*, Ed. Sarma, R.H., Pergamon Press, New York, Oxford, Frankfurt, Paris   p. 185 (1980).
98. Reid, B.R. in *Topics in Nucleic Acid Structure*, Ed. Neidle, S. McMillan (in press).

Single Crystal Crystallography for Nucleic Acid Structural Studies

Nadrian C. Seeman
Institute of Biomolecular Stereodynamics
Center for Biological Macromolecules
and Department of Biological Sciences
State University of New York at Albany
Albany, New York 12222

Introduction

X-ray diffraction is the most powerful tool at our disposal for the elucidation of the three-dimensional structures of biological molecules. It has played a crucial role in the development of our understanding of the ways in which these molecules form discrete structures and the ways in which these structures interact with each other. There are two prominent types of X-ray diffraction procedures which have been used for this purpose: (1) single crystal analysis and (2) helical diffraction from oriented fibers and gels. While extensive single crystal analysis of peptides and proteins dates from the 1950's,[2/3] it is only within the last decade that investigators have been able to successfully determine the structures of oligonucleotides[4-17] and short macromolecular nucleic acids such as transfer RNA.[18-20] Prior to that time, all of our knowledge of nucleic acid structures larger than monomers relied on helical diffraction studies. Since the other chapters in this volume do not deal with helical diffraction extensively, this chapter is restricted to being an introduction to single crystal analysis. Those interested in the basics of helical diffraction are referred to the fine introduction written by Wilson,[21] and to the papers of Arnott[22] and his colleagues.

The result of a successful X-ray crystallographic structure determination is a statue of the substance being investigated, at the resolution to which the crystallographic data has been obtained. Distances, angles and particularly conformational parameters are therefore directly observable from the crystallographic study. It is important to realize that a structure derived by single crystal techniques is a minimally inferential depiction of the molecule; the bias introduced by the investigator is small, compared with those spectroscopic techniques which rely heavily upon a combination of indirect structural data with molecular model building. The crystallographic theory for relating the positions of atoms within the crystal to the intensities of diffracted rays is sufficiently well developed, that the quality of the crystals and the quality of the diffraction data are the limiting variables in the determination of nucleic acid structures. The following sections will be a description of that theory, emphasizing those parts of particular interest for non-crystallographers studying nucleic acid structure. For a more complete introduction to crystallography, the reader is referred to the excellent book by Glusker and Trueblood,[23] or the somewhat more advanced treatment of Stout and Jensen.[24] A good description of macromolecular crystallography may be found in Blundell and Johnson.[25]

A Crystal is a Periodic Array of Material

Those crystals that we will be discussing will be 3-dimensional arrays, but our illustrations will frequently be limited to two-dimensional examples for clarity. Figure 1a shows a simple motif, a set of dots. Figure 1b shows a lattice of points which establish a given periodicity in two dimensions. Figure 1c shows a periodic array, or crystal, of the structural motif shown in Figure 1a, with the periodicities of Figure 1b.

Figure 1. *Forming a crystal from a structural motif.* (a) A structural motif formed by a series of dots. (b) A repeating lattice. (c) A "crystal" of the dot pattern formed by placing the structure at each of the lattice points. This is analogous to a molecular crystal; each of the "molecules" is parallel to the others.

Figure 1c illustrates the importance of molecular crystals in the determination of molecular structure. A crystal is the only means which we have at present for holding a large number of molecules parallel to each other and allowing us to change their orientation (with respect to a laboratory reference frame) in unison. In solution, the orientation of molecules is random. Thus, any observation on a large number of molecules will yield a spherical average of their spatial properties. This is like trying to read the label on a phonograph record while it is playing. On the other hand, it should be remembered that crystal structures may be sensitive to small perturbations induced by the presence of the lattice. Furthermore, one conformation may crystallize under the conditions used to prepare the crystals, but others may exist in solution. Thus, crystallography should be used in conjunction with solution techniques in order to determine the complete range of structures available to the substance. The fundamental periodic unit of a crystal is called the *unit cell*. As shown in Figure 2, the three vectors which define the periodicity are termed **a**, **b**, and **c**. The angle between **b** and **c** is termed α, that between **a** and **c** is β, and the one between **a** and **b** is γ. Crystallographic coordinates are usually quoted as fractions of these fundamental repeat vectors. Note that there is no requirement for the repeat vectors to be equal in length, nor is there any requirement that they be perpendicular to each other. In the most general case, neither of these possibilities is fulfilled. However, the symmetry of a given unit cell's contents may impose such re-

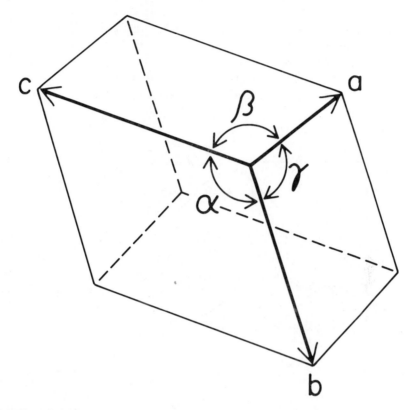

Figure 2. *The general unit cell showing the three vectors defining the periodicities in each of the three directions.* The three vectors are **a**, **b**, and **c**. The angle between **a** and **b** is γ, that between **a** and **c** is β and the one between **b** and **c** is α. Note that the lengths of **a**, **b**, and **c** are not equal, and that α, β and γ are neither equal to each other nor to 90°.

quirements.

The contents of a unit cell are frequently revealed to be a symmetric arrangement of molecules which extends throughout the lattice. Only a subset of crystallographically allowed symmetries are available to molecules which contain asymmetric carbon (or other) atoms, such as proteins and nucleic acids. These are the proper rotation symmetries involving 2-fold, 3-fold, 4-fold and 6-fold axes. The rotational symmetry may be combined with a translation parallel to it to yield a screw axis. A complete listing of crystallographic symmetries may be found in the International Tables for Crystallography, Vol. 1.[26] If symmetry does exist within a unit cell, the unique portion is termed the *asymmetric unit*.

X-Rays are Light Waves of Very Short Wavelength

The X-rays used in structure determination have wavelengths of about 1 Å. The most popular radiation is copper K_α, wavelength 1.5418 Å. Since we will be discussing the interaction of X-rays with matter, it is necessary that we understand the way in which the sinusoidally oscillating waves of which X-rays are composed interact with each other. A wave may be characterized for our purposes by its wavelength, λ, its amplitude, A, and its phase, ϕ, as shown in Figure 3. The wavelength is the distance between crests, the amplitude is just the height of the wave at the crest, and the phase is the distance of the crest from an arbitrary origin. It is convenient to describe

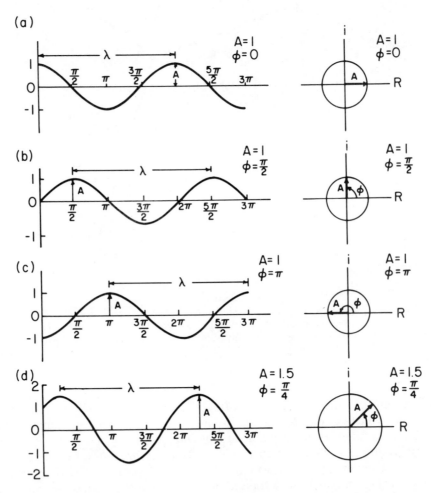

Figure 3. *Waves, amplitudes and phases.* The four parts of this figure illustrate waves both in their fully drawn form and by their representation in the complex plane. On the left of each part is a wave, whose amplitude is indicated as A, and whose phase, in radians is indicated by ϕ. The wavelength for each wave is the same, and is indicated by λ, the distance between the crests. The distance along the abscissa has been scaled in radians. On the right, the amplitudes are indicated by the length of the vector A, and the phase is indicated by the angle ϕ. The projection of each vector on the real axis of the complex plane is Acosϕ, and the projection of each vector on the imaginary axis is Asinϕ. The circles on the complex planes indicate the locus of the possible vectors corresponding to waves of the indicated amplitude. The actual wave illustrated is indicated by the vectors drawn in the complex plane. (a) A wave of unit amplitude and zero phase. (b) A wave of unit amplitude and phase of $\pi/2$, or 90°. (c) A wave of unit amplitude and phase of π, or 180°. (d) A wave of amplitude 1.5, and phase $\pi/4$, or 45°.

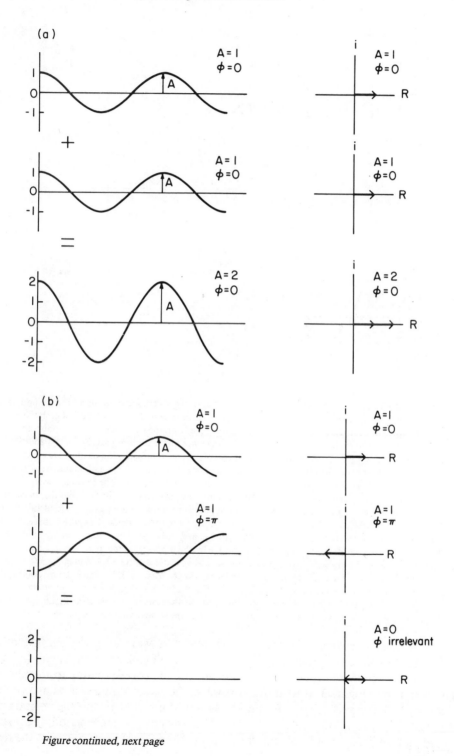

Figure continued, next page

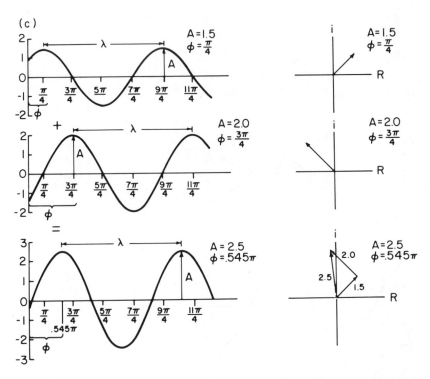

Figure 4. *Superposition of two waves.* The superpositions of waves are indicated in both wave and complex plane representations in this figure. Note that the waves represented on the left side of each figure are the same as those indicated by the vectors in the complex plane on the right side. Point by point addition of the two top waves in each part of the figure yields the wave at the bottom part. Similarly, addition of the two vectors in the top complex planes yields the complex plane representation of the wave which is the resultant. Note that the wavelengths of all waves are the same. Only amplitudes and phases are altered by the summation. (a) Addition of two unit amplitude waves of identical phase. The result is a wave of double amplitude with the same phase. Note that there is nothing special about the fact that the phase of 0 was chosen for this example. Whatever the phases of the two waves, so long as they are identical, the resultant wave will have that phase. (b) Addition of two unit amplitude waves of opposite phase. In this case, the two waves are exactly π radians (180°) out of phase, although their amplitudes are equal. The resultant wave is the complete cancellation of each wave by the other, resulting in a wave of 0 amplitude. (c) Addition of two arbitrary waves. In this case, the first wave is of amplitude 1.5, and phase $\pi/4$ radians (45°), while the second wave is of amplitude 2.0 and phase $3\pi/4$ radians (135°). The point by point addition of the two waves on the left yields the wave indicated at the bottom. This wave has amplitude 2.5 and phase 0.545 π radians (98.1°). Note that the addition of the two vectors representing the first two waves in the complex plane yields the vector which represents the resultant wave in the complex plane.

waves in polar coordinates in the complex plane. A wave at any given instant is represented as a vector from the origin of the complex plane, with length A, corresponding to the amplitude, and angular position ϕ, corresponding to the phase. This is shown in Figure 3. If we moved the wave, the vector would sweep out a circle, changing ϕ, but not A. Thus, the amplitude would not change, but the position of the crest from the origin would be altered. As it traversed a single period, the phase would go from 0 to 2π radians, and then repeat. Note that the projection on the real axis is Acos(ϕ) and the projection on the imaginary axis is Asin(ϕ). The wave is just the vectorial sum of these components, or A(cos(ϕ) + isin(ϕ)), which may be written as Ae$^{i\phi}$, by the Euler relationship.

When two waves of the same wavelength are superimposed, the resultant wave is merely the point by point addition of the two waves. As shown in Figure 4, this can also be represented by the vectorial sum of their complex plane representations. The resultant wave will have real axis component A$_1$(cos(ϕ_1)) + A$_2$(cos(ϕ_2)) and imaginary axis component A$_1$(sin(ϕ_1)) + A$_2$(sin(ϕ_2)). This may also be written A$_1$e$^{i\phi_1}$ + A$_2$e$^{i\phi_2}$.

The Scattering of X-rays by a Crystal Depends on The Superposition of Waves

Since no means has been found to focus X-rays, it is not possible to devise a lens which will give a direct picture of the contents of a crystal. Therefore we must make a more detailed analysis of the scattering process in order to interpret the pattern which is observed. Let us look at the general case, depicted in Figure 5. Our experimental setup consists of a source of X-rays at infinity, a scattering body in line with the source, and a detector at infinity. In order for scattering to be detected by the detector, it must be scattered through an angle at 2θ. For this experiment, we

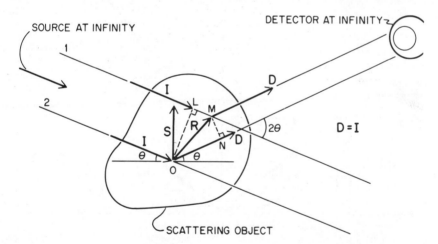

Figure 5. *Scattering of X-rays by an arbitrary object.* A single scattering experiment is depicted in this figure. The scattering object is being illuminated by a source which is at infinity, and we are considering scattering of X-rays by the object as detected by a device also at infinity. The deflection of the scattered beam being monitored from the incident beam is indicated by the angle 2θ. The 2θ angle is characteristic of the experimental setup. We consider the interference of two rays, one scattered from point O and the other scattered from point M. The vector from O to M is indicated by the vector **R**. **I** is a unit vector in the direction of the incident beam, while **D** is a unit vector in the direction of the scattered beam. The perpendicular from O, our origin, to incident ray 1 intersects that ray at point L, while the perpendicular from M to scattered ray 2 intersects that ray at point N. ON-LM is the path length difference between the two rays. **S** is the difference vector between **D** and **I**. **S** bisects the angle between **I** and **D**. Note that the perpendicular to **S** makes two equal angles θ with **I** and **D**.

have thus fixed the orientation of the object with respect to the source, detector and an arbitrary origin, O within it. We have furthermore fixed the scattering angle at 2θ. The question we are asking is, what will be observed at the detector under these conditions?

Let us begin by considering the scattering from two points within the object, our origin, O and another point, M, removed from O by the vector **R**. We shall only be concerned with the elastic scattering experiment, so the wavelengths of our incident and scattered beams will be equal in this analysis. In Figure 5, I represents a unit vector in the direction of the incident beam, and **D** represents a unit vector in the direction of the scattered beam. Clearly, the waves detected at infinity will be the superposition of the waves scattered from the points O and M. Since they are of the same wavelength, the path-length difference between them will determine their relative phases. The difference between the distance traveled by the second wave compared with the first, Δ, is just the length of $\overline{NO}$ less the length $\overline{LM}$; otherwise, they travel the same distances:

$$\Delta = \overline{NO} - \overline{LM}$$

The length of $\overline{NO}$ is just **D•R** (i.e., the projection of R on D). Similarly, the length

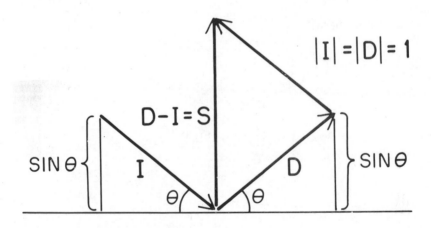

Figure 6. *Illustration of the magnitude of the* S *vector. The two components of the S vector are each* $\sin\theta$, since D and I are both unit vectors. Hence the magnitude of S is $2\sin\theta$.

of $\overline{LM}$ is just **I•R** (i.e., the projection of R on I). Thus,

$$\Delta = \mathbf{D \cdot R} - \mathbf{I \cdot R} = \mathbf{(D-I) \cdot R} = \mathbf{S \cdot R},$$

where S is the difference vector between **D** and **I**. Since the magnitudes of **D** and **I** are the same, S bisects the angle between them. If we construct the perpendicular to S through O, we can see that both I and D make an angle θ with this perpendicular.

The scattering angle, as you will recall, is 2θ. The length of S is $2\sin\theta$, since both I and D are unit vectors (see Figure 6). If we wish to convert th path length difference to a phase difference in radians, we must multiply it by $2\pi/\lambda$. This makes our phase difference $2\pi/\lambda$ S•R.

X-rays are scattered only by electrons.[27] The final result of a structure determination will therefore be the electron density map of the contents of unit cell, which will in turn be a statue of the molecules of which it is composed. Points of higher electron density will scatter X-rays more strongly than points of lower electron density. Thus, each of the diffracted waves originating at different points R will be weighted by a density term, $\varrho(R)$, and each scattered wave will be described by a term $\varrho(R)$ exp $(2\pi i/\lambda$ S•R). For this orientation of the scattering object with respect to the source and detector, the scattered wave observable at the detector, G(S), will simply be the sum of the interfering waves throughout the object:

$$G(S) = \int_{object}\varrho(R) \exp [(2\pi i/\lambda)(S\cdot R)]dR$$

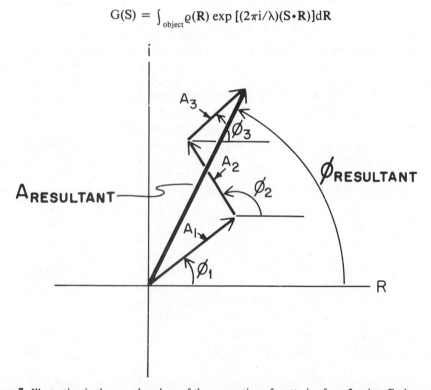

Figure 7. *Illustration in the complex plane of the summation of scattering from 3 points.* Each wave has an amplitude, A, indicated by the length of the vector, and a phase relative to an arbitrary origin, indicated by the phase angles ϕ. The resultant wave amplitude is indicated, with its resultant phase as well. This represents the contribution of each of 3 scattering points to the resultant wave observed at the detector. Clearly rotating the entire assemblage about the origin of the complex plane (changing the arbitrary origin to which phases are referred) would alter the resultant phase, but not the resultant amplitude. By adding more vectors, the scattering from each point in the body can be added into the summation to give a resultant for the entirely scattering object.

This G(S) term is just a complex number which characterizes the result of the scattering experiment described above. It is just the addition within the complex plane of all the scattering points within the object, weighted by the scattering density at those points and phased by their path length differences, as shown in Figure 7. In that figure, A_1, A_2 and A_3 are the amplitudes characteristic of scattering from three points in the object, and ϕ_1, ϕ_2, and ϕ_3 are the phases resulting from their path length differences from an arbitrary origin. The integral is the summation of all these vectors, of which we have shown three in the figure for simplicity. It is also clear from the figure that changing the origin would alter the phase, but not the amplitude of the resultant vector, since it would correspond to a rotation of the figure about the origin of the complex plane. The amplitude is determined by the relative distribution of scattering matter within the object.

The Total Scattering From An Object Is Defined Within
A Sphere Called Reciprocal Space

If we left the object in Figure 5 fixed, and rotated the souce and detection apparatus about the origin, keeping 2θ fixed, we would get a new scattering situation. (In practice, the object is usually rotated, rather than the experimental apparatus.) S would be of the same length, but would be pointed in a new direction, corresponding to the new orientation of the source and detector. This new vector S would still bisect the angle between I and D, but S•R would be different, for each R vector considered. The complete set of vectors S, corresponding to the scattering angle, 2θ, and all possible orientations of the experimental apparatus with respect to the scattering object defines a spherical shell of radius corresponding to the magnitude of S, namely $2\sin\theta$. At each of the positions on this shell, the scattered wave, G(S) is defined.

The scattering angle, 2θ, can clearly be varied, in the range from 0 to 180°. Thus, S $(=2\sin\theta)$ can range from 0 to 2. Thus, we have another shell of radius $2\sin\theta$ for each possible magnitude of S. This gives a solid sphere of radius 2 for the complete locus of the heads of vectors S. At each point within this sphere, the scattering function, G(S) may be observed. This sphere is called *reciprocal space*. It is important to remember that the G(S) function, characteristic of the scattering from the object, is tied to the object. If the object is rotated, G(S) rotates with it.

Scattering Factors for Individual Atoms
Characterize Molecular Scattering

It is very convenient to talk about the scattering arising from individual atoms, rather than from a continuum of scattering matter. For the resolutions with which we are dealing, we need not be concerned with the oriented aspects of atomic structure. Thus, the model used by crystallographers who deal with moderately large biological molecules (20 atoms or above) does not take into account the directionality of lone pairs or bonding electrons. The atom is treated as a spherically symmetric object whose oriented features have been eliminated in a spherical average. In deriving a scattering function f(S) (S now a scalar because of the spherical average), we

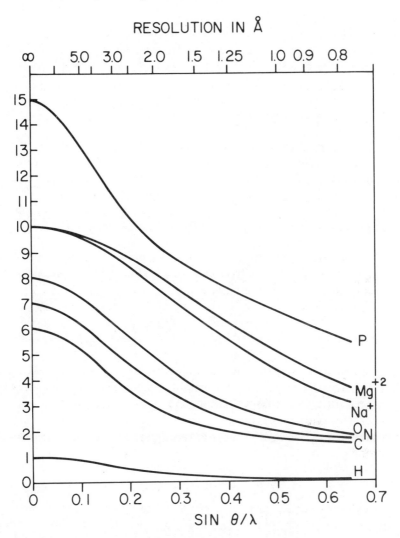

Figure 8. *Scattering curves for atoms found in nucleic acid structures.*[26] The scattering for each atom is indicated as a function of scattering angle, normalized for wavelength. Each curve is normalized for the number of electrons in the atom or·ion. The resolution is indicated at the top of the figure. Note that the curves are not all the same shape, and that the contributions of all atoms decline with increasing scattering angle.

assume that the electrons can be localized within atoms, that they scatter independently from each other, and that they scatter in the crystal just like they do when in isolated atoms. Representative scattering curves are shown in Figure 8. The only assumption which is not transparent to the treatment of data on nucleic acid fragments and tRNA is the independent scattering of electrons. In our treatment, we have assumed that the phase shift upon scattering (180°) is the same for all points in the scattering object. In certain cases, it is slightly different, resulting in *anomalous scattering.* This is a slight correction to the scattering from certain atoms (phosphorus is the only significant one amongst nucleic acids), resulting from some electrons scattering at a slightly different time from the others. Thus, the scattering factor f(S) is actually a sum of three terms,

$$f(S) = f_o(S) + f'(S) + if''(S),$$

where f_o is the normal scattering factor and f' is a real (and usually negative) correction term, while f'' (= .43 electrons for P with CuK_α radiation) is an imaginary correction term. Use of anomalous dispersion is extremely important in determining the structures of both nucleic acid fragments[12] and tRNA.[18]

Scattering From A Crystal Samples The Complete Scattering Pattern

With this assumption of atomic scattering, we can now describe the scattering just by adding up the contributions from each of the atoms in the molecule or molecules which we are observing with our apparatus:

$$G(S) = \sum_{j=1}^{N \text{ atoms}} f_j(S) \exp[(2\pi i/\lambda)(S\cdot R_j)]$$

To make a crystal of our molecule, all we have to do is place our molecules next to one another, separated by the fundamental periodicities **a**, **b**, and **c**, and add up the contributions for each of the unit cells in the crystals:

$$G(S) = \sum_{N_1=1}^{N_a} \sum_{N_2=1}^{N_b} \sum_{N_3=1}^{N_c} \sum_{j=1}^{N \text{ atom}} f_j(S) \exp[(2\pi i/\lambda)S\cdot(R_{j+}N_1a+N_2b+N_3c)]$$

where N_a, N_b and N_c are the number of unit cells in each of the **a**, **b**, and **c** directions, respectively. Thus, with a given value of the indices N_1, N_2, N_3 and j, we are accounting for the contributions to the scattering from the j'th atom in the N_1th unit cell in the **a** direction, N_2th unit cell in the **b** direction and N_3th unit cell in the **c** direction. It should be noted that we are seeing the spatial average of the crystal's content. If a mixture of structures exists within the crystal (e.g., two different solvent structures) the weighted average structure will be seen. This expression may be factored to give:

$$G(S) = (\sum_{j=1}^{N \text{ atom}} f_j(S) \exp[(2\pi i/\lambda)(S\cdot R_j)]) (\sum_{N_1=1}^{N_a} \exp[(2\pi i/\lambda)(S\cdot N_1a)]) \times$$

$$(\sum_{N_2=1}^{N_b} \exp[(2\pi i/\lambda)(S\cdot N_2b)]) (\sum_{N_3=1}^{N_c} \exp[(2\pi i/\lambda)(S\cdot N_3c)]).$$

The first term is just the molecular scattering term from an isolated molecule. The other terms are *fringe functions,* which are related to the periodicities of the lattice. Figure 9 demonstrates the effect of making a crystal out of a given structural motif in two dimensions. The boxes at the bottom of the figure are structures whose scat-

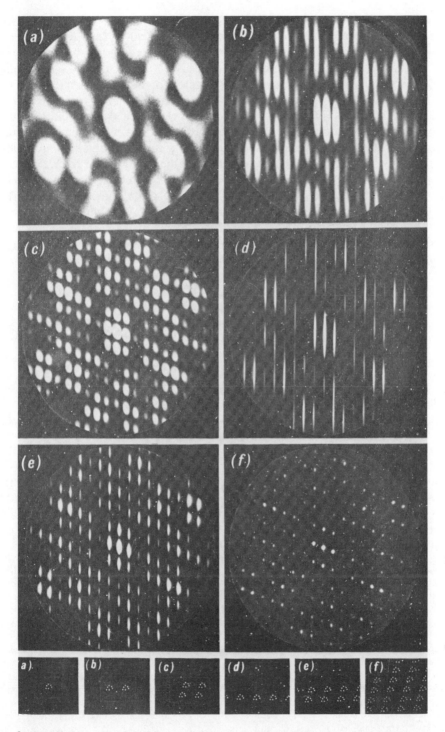

Figure 9. *The effect on scattering of incorporating a structural motif into a crystal.* The small blocks at the bottom of the figure are the objects whose scattering functions are shown in the large blocks with corresponding letters at the top. This is an emulation of the X-ray scattering case by using masks and optical diffraction tecniques.[46] (a) The structural motif and its scattering function. In this two dimensional exam-

tering has been demonstrated by optical diffraction techniques. In Figure 9a, we see the individual array and above it is its scattering function. This function is a continuum. In 9b, we have started to make a one dimensional crystal out of it, by placing it adjacent to another version of the same array. This results in a set of broad fringes across the transform. In 9d, an infinite (for our purposes) 1-dimensional crystal has been formed, and the fringes are very fine, although they have the same separation as in 9b. The function has a value of effectively 0 at all points off the fringe. By squinting at the figure, it may be readily seen that at the points where the fringes allow the function to come through, it looks the same as in 9a. Figures 9b, 9e and 9f show the same principles with respect to the addition of a second dimension of periodicity. The spacings of the fringes are inversely related to the periodicity. If the periodicity in Figure 9d is d, then the separation of the fringes will be λ/d. Thus, in Figure 9d, the central fringe runs through the origin of reciprocal space, and the first one over is removed from it by λ/d, the second one by $2\lambda/d$, the third by $3\lambda/d$ and so on. These separations may be rewritten as λ/d, $\lambda/(d/2)$, $\lambda/(d/3)$ ... $\lambda/(d/n)$, indicating that they contain information about the structure of resolutions d, d/2, d/3, ..., d/n, respectively. Thus,

$$\lambda/(d/n) = 2\sin\theta.$$

This expression may be rearranged to yield Bragg's law,

$$n\lambda = 2d\sin\theta.$$

which is the Bragg condition for diffraction by a crystal.

The reader should examine Figure 9f in detail and realize that even in the case of the two-dimensional crystal (and of course in the three-dimensional case as well), the parts of the scattering function which do come through the fringe function have the same intensity as they do in 9a. Thus, by putting the molecule in a crystal, and fixing many parallel copies of it so that we can manipulate it macroscopically, we find that we can only *sample* its scattering function, rather than being able to observe it con-

ple, reciprocal space is the contents of a circle, rather than a sphere. Each point within the circle corresponds to an **S** vector, from the origin at the center. The alteration in intensity corresponds to G(**S**), the scattering function characteristic of the structure. (b) A simple crystal has been constructed by juxtaposing two of the structures seen in (a). The effect on the scattering function is to overlay it with a fringe, caused by the interference between the two parallel structures. Because the crystal is very short, the fringe is very broad. Note that where the function is visible, however, it appears to have the same intensity as the scattering function seen in (a). (c) The crystallinity has been extended to a second dimension, resulting in a second set of fringes reflecting the newly added periodicity. Again the fringes are broad because of the limited nature of the crystal. (d) An infinite crystal of the structure in one dimension. The fringes are now narrow because of the infinite character of the crystal. (e) An infinite crystal in the horizontal direction but only two lattice rows in the other direction. Now the fringes are narrow in the horizontal direction, but are still broad in the other direction. (f) An infinite crystal in both directions. The fringes are sharp in both directions. Note that by squinting at the picture you can notice that the intensity distribution in (f) is the same, where sampled, as in (a). The complete optical masks for (d), (e), and (f) are not shown.

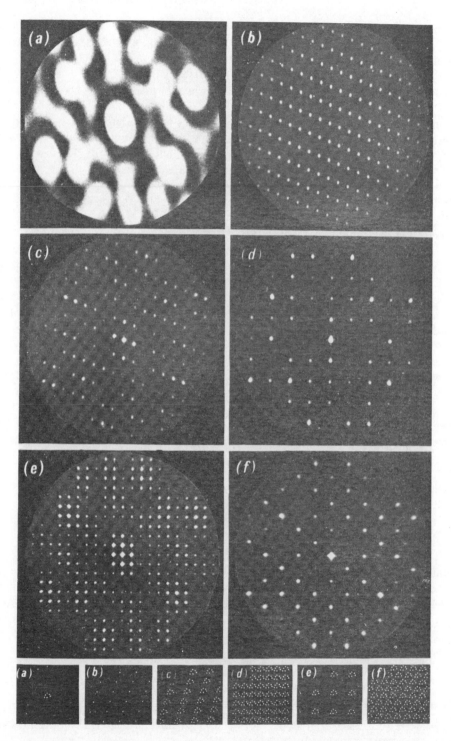

Figure 10. *The effect of changing the lattice on the scattering function.* The same protocol is used in this figure as in Figure 9. (a) The structural motif of Figure 9 (a) and its scattering function. (b) The lattice of Figure 9 (f) and its scattering function. Since there is only one point per unit cell, just the fringes are seen, due to the effect of the lattice. (c) The structural motif of (a) incorporated into the lattice of (b). This is the same pattern as seen in Figure 9 (f). (d) Incorporation of the structural motif into a lattice with a very

tinuously. There is no scattering between the fringes for a crystal since the number of unit cells is so high. Because of the value of being able to deal with oriented molecules, this is a relatively small price to pay.

Scattering From A Crystal Defines The Reciprocal Lattice

This array of intersections of the fringe functions forms a lattice in its own right, a *reciprocal lattice*. The fundamental lattice repeats in the reciprocal lattice are called a*, b*, and c*, and the angles between them are α^*, β^*, and γ^*. Thus, $a^* \cdot a = \lambda$, $b^* \cdot b = \lambda$ and $c^* \cdot c = \lambda$. These reciprocal vectors are defined as

$$a^* = \lambda b \otimes c / V, \ b^* = \lambda c \otimes a / V \text{ and } c^* = \lambda a \otimes b / V,$$

where V is the volume of the unit cell. The different points in the reciprocal lattice (termed, naturally enough, reciprocal lattice points) are indexed from the origin of reciprocal space in terms of positive or negative whole numbers. These indices are called *Miller indices*, h,k and l. Thus, the point which is 3rd along a*, 5th in the opposite direction from b* and 4th along c* in the three dimensional lattice would have miller indices h = 3, k = -5, and l = 4. This represents a vector in reciprocal space with components 3a*, -5b*, 4c*.

Figure 10 shows the same structural motif as in Figure 9 (and Figure 1), but this time assembled into lattices of different periodicities. The reader should note that the intensities are the same at points with the same S value in reciprocal space, regardless of the lattice which has been used. The other important fact to note is the demonstration of the reciprocal natures of the direct and reciprocal lattices. In 9d the unit cell is small, but the reciprocal lattice spacings are large; in 9e, the reverse is true: the unit cell is large, and the spacing in reciprocal space is small.

Now that we can index the points in our diffraction pattern, we no longer need to talk only about a G(S) function which is continuous,

$$G(S) = \sum_{j=1}^{N \text{ atoms}} f_j(S) \exp [(2\pi i / \lambda)(S \cdot R_j)].$$

Rather, we can use the Miller indices to refer to a given point in reciprocal space and fractions of a unit cell edge to refer to a point in direct space. Thus, the components of R_j will be $x_j a$, $y_j b$, $z_j c$ and those of S will be ha*, kb*, lc*. Thus, we may rewrite the scattering from the crystal at the points of interest, namely reciprocal lattice points, as

small unit cell. In this case, because of the inverse relationship between direct lattice vectors and reciprocal lattice vectors, the reciprocal lattice spacings are very large. (e) Incorporation of the structural motif into a lattice with a very large unit cell. This results in a very fine sampling of reciprocal space. Note that since the periodicities in both (d) and (e) are orthogonal, the reciprocal lattices are also orthogonal. (f) A small unit cell with a skewed periodicity. Note that in (c), (d), (e), and (f) the intensities of the sampled points are the same as they are in (a). Only the positions available for sampling are different. The complete optical masks for (b), (c), (d), (e), and (f) are not shown.

$$F(h,k,l) = \sum_{j=1}^{N\,atoms} f_j(S) \exp\left[(2\pi i/\lambda)(ha^* \cdot ax_j + kb^* \cdot by_j + lc^* \cdot cz_j)\right)$$

Remembering that $a^* \cdot a = b^* \cdot b = c^* \cdot c = \lambda$, this expression may be simplified to be

$$F(h,k,l) = \sum_{j=1}^{N\,atoms} f_j(S) \exp\left[(2\pi i)(hx_j + ky_j + lz_j)\right].$$

The complex number F is called the *structure factor* associated with the reciprocal lattice point h,k,l.

The Scattering From A Crystal Corresponds To The Fourier Transform Of The Unit Cell Contents

At this point, we must ask the meaning of this array of structure factors, F(h,k,l). In order to do this, let us leave discretely scattering atoms for a moment. Following the treatment of Stout and Jensen,[24]

$$F(h,k,l) = \int_{\text{volume of unit cell}} \varrho(x,y,z) \exp\left[(2\pi i)(hx + ky + lz)\right]dxdydz,$$

where $\varrho(x,y,z)$ is the electron density at point x,y,z within the unit cell. Since ϱ is a periodic function, it must be possible, by the Fourier theorem, to expand it in a Fourier series,

$$\varrho(x,y,z) = \sum_{h'=-\infty}^{\infty} \sum_{k'=-\infty}^{\infty} \sum_{l'=-\infty}^{\infty} C_{h'k'l'} \exp\left[(2\pi i)(h'x + k'y + l'z)\right],$$

where the $C_{h',k',l'}$ are the Fourier coefficients of $\varrho(x,y,z)$. Plugging this into our expression for F(h,k,l) we get:

$$F(h,k,l) = \int_{\substack{\text{volume of unit}\\\text{cell}}} \sum_{h'=-\infty}^{\infty} \sum_{k'=-\infty}^{\infty} \sum_{l'=-\infty}^{\infty} C_{h'k'l'} \exp\left[(2\pi i)((h+h')x + (k+k')y + (l+l')z)\right]dxdydz$$

Since trigonometric functions are orthogonal, when we carry through the integration, all terms will be 0, unless $h' = -h$, $k' = -k$ and $l' = -l$. If this is the case, the integral will just equal the volume of the unit cell, so

$$F(hkl) = VC_{-h,-k,-l}$$

Thus

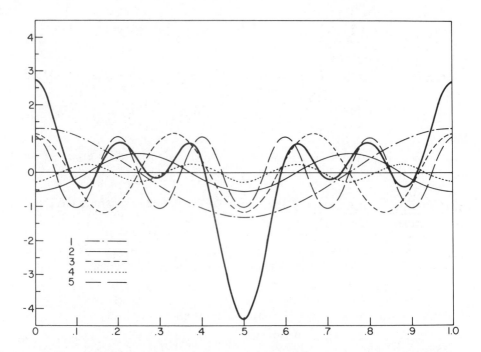

Figure 11. *A one-dimensional Fourier synthesis.* This drawing shows the summation (indicated by the dark line) of the first five terms of a Fourier synthesis. The structure corresponds to 6 unit-scattering atoms placed at fractional positions 0.05, 0.20, 0.35, 0.65, 0.80 and 0.95 in a unit cell 10Å long. The waves of periodicity 1,2,3,4 and 5 per unit cell are shown. Thus the resolution of this synthesis is 2Å. The wave of periodicity 1 has an amplitude of 1.311 and a phase of 0. The wave with periodicity 2 has an amplitude of 0.559 and a phase of π. The wave of periodicity 3 has an amplitude of 1.166 and a phase of 0. The wave of periodicity 4 has an amplitude of 0.256 and a phase of π. The wave of periodicity 5 has an amplitude of 1.071 and a phase of 0. It can be seen that those waves of phase 0 start at the origin of the unit cell on their crests, and those waves of phase π start at the origin of the unit cell at their minimum values. The presence of the mirror plane at 0.5 in this structure forces the phases to be only 0 or π in this example, although this is generally not true. A wave of 0 periodicity (just a constant value) of amplitude 6 has been omitted. Had it been included, all values in the resultant synthesis would be 6 greater than they are shown to be. Note that the atoms at 0.20, 0.35, 0.65 and 0.80 are fairly well resolved from each other, but those at 0.05 and 0.95 (which corresponds to -0.05) are not resolved, because they are separated by only 1Å, and this synthesis only includes terms of 2Å spacing or greater. The dip in the middle of the cell is indicative of series termination error. Addition of higher order terms (6,7,8...) would increase the resolution and make the image sharper, since their spacings are finer.

$$\varrho(x,y,z) = 1/V \sum_{h,k,l=-\infty}^{\infty} F_{(h,k,l)} \exp[(-2\pi i)(hx + ky + lz)].$$

Recalling our earlier equation,

$$F(h,k,l) = \int_{\text{volume of unit cell}} \varrho(x,y,z) \exp[(2\pi i)(hx + ky + lz)]dxdydz,$$

we can see that the F and ϱ functions are Fourier transforms of each other. This

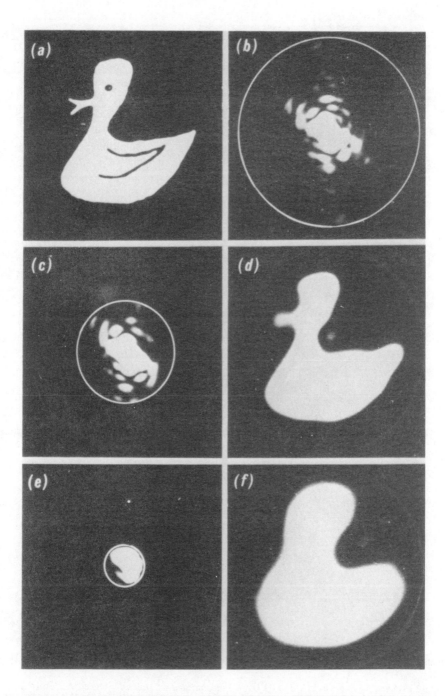

Figure 12. *Illustration of the effect of decreasing resolution on reconstructing an image from its transform.* (a) An image of a duck. (b) Transform of (a). (c) Part of the transform shown in (b) is covered up. (d) The reconstruction of the image shown in (a) using only that part of the transform shown in (c). Note that the fine details are lost, because those parts of the transform containing high resolution information have been suppressed. The transform of the fuzzy image shown in (d) would yield (c), since the outer parts of the transform can only arise if there are fine details, rather than fuzzy images. (e) and (f) are more extreme examples of (c) and (d) respectively.

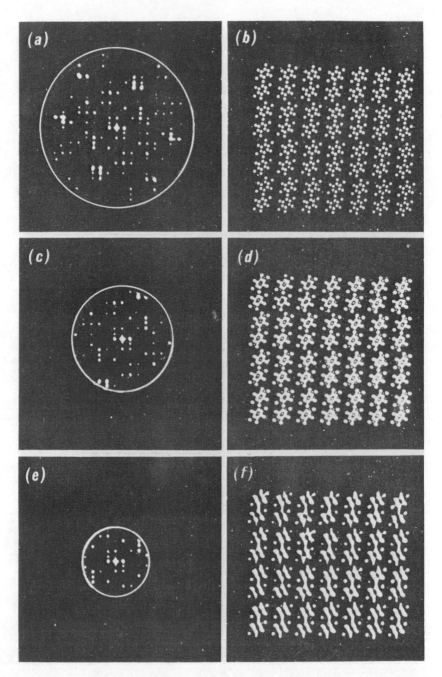

Figure 13. *Simulation of the effect of decreasing resolution on a molecular crystal.* (a) A diffraction pattern from the molecular crystal pattern shown in (b). The pattern shown in (b) has been obtained from the diffraction pattern shown in (a). This corresponds, on an atomic scale, to approximately 1Å resolution. In the pair (c) and (d), the resolution has been decreased to about 1.5Å resolution, and in the pair (e) and (f), the resolution has been further decreased to about 2.2Å resolution. Note the smeariness of the images reconstructed from the limited data sets. Note further, that if the *spatial average* of the images in (d) and (f) looked like any one of the reconstructed images (e.g., through slight misalignments), the patterns in (c) and (e), respectively, would result.

means that we can get from one to the other directly. Given the complete set of structure factors, F, we can reconstruct the electron density function ϱ, through some trivial arithmetic in the computer. Figure 11 shows how this is done for a few terms in a one dimensional case. In the multi-dimensional case, the waves are pointed along the reciprocal lattice vectors H(=h,k,l), and the wavelength corresponds to the d-spacing (d = $2\sin\theta/n\lambda$) characteristic of that vector H.

The Quality of the Structure Depends on the Extent of the Data

Since we are talking about a finite representation of our structure, we must ask what the effect of a limited data set will be. You will recall that information about finer and finer details of the structure is found in reciprocal lattice points further and further from the origin of reciprocal space. Let us see what happens when data from the outer portions of the Fourier transform of the structure is omitted. Figure 12a shows an image of a duck and Figure 12b shows it Fourier transform. By the direct or inverse Fourier transformation procedure (here done optically), either the transform or the picture of the duck may be produced, one from the other. Figure 12d shows the image of the duck when only the portion of the transform shown in 12c is used to produce the image of the duck. It is clearly fuzzier than the image shown in 12a. This is because the portion of the transform which contains information about the fine details of the structure has been excluded. In like fashion, the fuzzy image in 12d would only produce a transform out to the indicated circle in 12c: There is no higher order (finer detail) information in the image, so its transform will be blank beyond the circle. A more extreme case is shown in Figures 12e and 12f. Here the image is only marginally recognizable.

Figure 13 shows the same thing, this time with respect to a molecular crystal. Figure

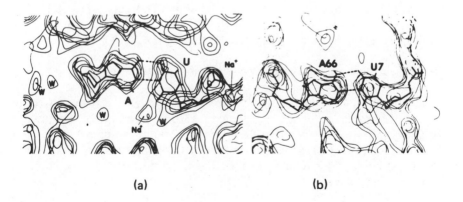

(a)   (b)

Figure 14. *Adenine-Uracil base pairs at 3Å resolution.* Base pairs are indicated by the molecular structures indicated. The electron density levels are indicated by the contouring. (a) A base pair from the structure of the dinucleoside phosphate, ApU.[9] This structure was determined to a resolution of 0.8Å, but only those terms to 3Å resolution have been used in this synthesis. (b) Base pair A66-U7 from the 3Å multiple isomorphous derivative map of Yeast tRNA[phe].[47]

13a shows the diffraction pattern of the crystal simulation shown in 13b. On an atomic scale, this would correspond to an approximate resolution of about 1Å. The atoms are clearly resolved from one another and the structure is readily apparent. With a more restricted dataset, shown in Figure 13c, the poorer quality image is seen in Figure 13d. This corresponds to approximately 1.5Å resolution—note that the individual atoms are no longer resolved from each other. The bottom pair in the figure, 13e and 13f represent a structure and its transform at about 2.2Å resolution. Fig. 14 is a picture of base pairs at 3Å resolution.

Why, one might ask, don't crystallographers always use data to 1Å resolution or above? Certainly, it is more work to get higher resolution data, since the number of data points increases with the cube of the reciprocal of the resolution. However, the answer lies not in professional slothfulness, but rather in dealing with the realities of nature. The data from a crystal are only so good as the crystal. The fine details of the structure, namely the data far from the origin of reciprocal space, may be lost through a number of circumstances. Thermal motion within the crystal is a prime cause of this problem for smaller structures. As the atoms in the molecule vibrate, they smear the time-averaged image that the crystallographer sees over the course of the scattering experiment. Besides that, crystal imperfection also contributes greatly to loss of data. If one unit cell is tilted slightly with respect to the next, data will be lost through the space averaging over all unit cells in the crystal. A typical crystal of an oligonucleotide will diffract to about 1Å resolution; tRNA crystals at best yield data to 2.5-2.2Å resolution, even when the temperature is lowered to reduce thermal motion. In order to interpret the electron density map from tRNA with a molecular model, it was necessary to understand the basic conformations which nucleic acids could assume from higher resolution studies of small fragments.[18] As is clear from Figure 14, a 3Å density map bears the same relationship to the underlying molecular structure that Figure 12f bears to a duck: If you know what can be there, the map is interpretable. It is therefore critical that crystallographers choose crystals which diffract to adequate resolution to answer the scientific questions in which they are interested, since a lot of time will be wasted if they do not.

The Phase Problem Complicates Crystallography Enormously

If the structure factors were directly observable through the X-ray diffraction experiment, the mere computation of structures would be all that was required of the crystallographer. Unfortunately, this is not the case. The phases of the complex structure factors are lost in the course of the diffraction experiment. The intensity of the diffracted ray is measurable, but this is proportional (when corrected for geometric and physical factors affecting the experiment) to the square of the amplitude of the structure factor: $I \propto F_{hkl} \cdot F^*_{hkl} = F_{hkl}\, e^{i\phi} \cdot F_{hkl}\, e^{-i\phi} = |F_{hkl}|^2$. The phase information is totally lost. This loss is termed the phase problem of crystallography. Without the phases, the structure is not directly knowable, and it must be solved by numerical and experimental techniques.

From the discovery of X-ray diffraction before World War I, until the middle

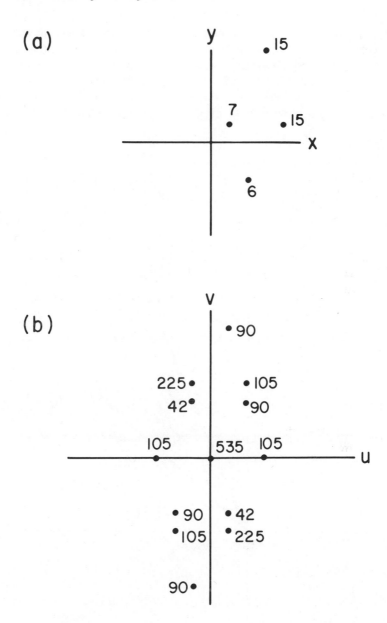

(a)

(b)

Figure 15. *The Patterson function of a simple distribution.* In (a) is shown a simple distribution of points, and in (b) its Patterson function is shown. Note that each point in the Patterson function corresponds to a vector between two points in the point structure, and is weighted by the product of the weights assigned to the points in (a). Note further, that the self-vectors pile up at the origin, and that for every vector from one point to another, the negative vector, from the second point to the first, also exists. This implies that both the structure and its enantiomorph may be derived from the same Patterson function. Note also that the same Patterson map could be constructed by placing each point of the structure on the origin of the Patterson function and redrawing the structure weighted by the weight of the point at the origin.

1930's, crystallographers were in the position of having to guess the structure, and then checking their guess against the diffracted intensities. This situation changed in the 1930's when Patterson asked if there were any meaning in the Fourier transform of the intensity distribution itself. Let this function be called $P(u,v,w)$:

$$P(u,v,w) = \sum_{hkl} |F_{hkl}|^2 \exp[(-2\pi i)(hu + kv + lw)]$$

Now,

$$|F_{hkl}|^2 = F_{hkl}F_{hkl}^*,$$

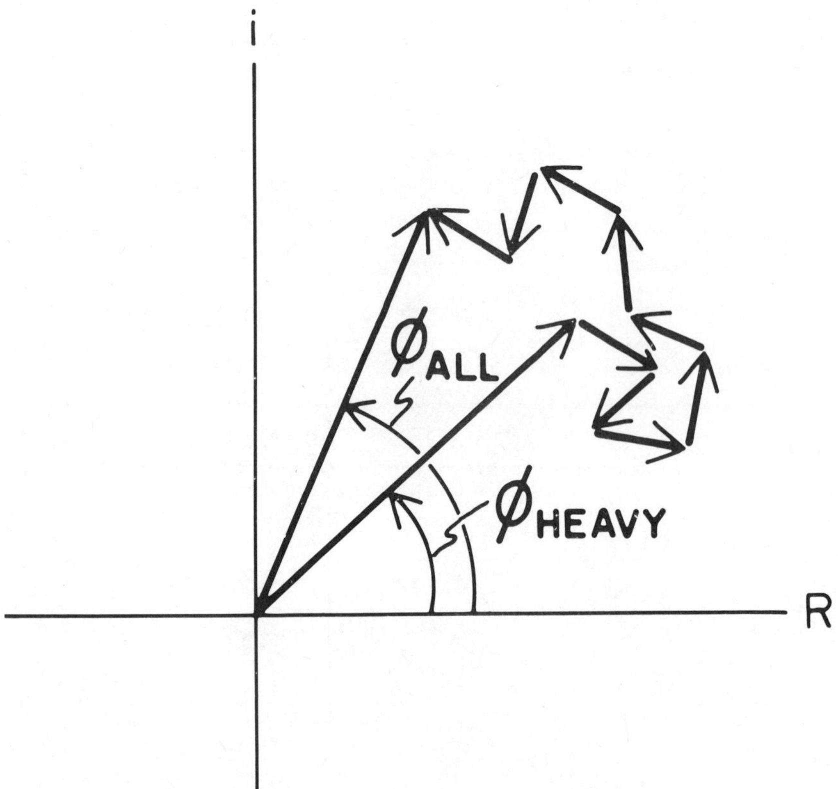

Figure 16. *The domination of the phase of a structure factor by the phase of a heavy atom.* The contribution of a heavy atom to the complete structure factor is illustrated in the complex plane. If the distribution of the contributions of the other lighter atoms, which have smaller scattering factors, is random as indicated, the phase of the heavy atom alone will be a reasonable approximation to the phase of structure factor.

and

$$F^*_{hkl} = \int \varrho(x,y,z) \exp [(-2\pi i)(hx + hy + lz)]dxdydz.$$

Thus,

$$P(u,v,w) = \sum_{hkl} F_{hkl} \int \varrho(x,y,z) \exp [(-2\pi i)(h(x + u) + k(y + v) + + l(z + w))]dxdydz$$

Reversing the integration and the summation yield

$$P(u,v,w) = \int \varrho(x,y,z) [\sum_{h,k,l} F_{hkl} \exp (-2\pi i(h(x + u) + k(y + v) + l(z + w)))]dxdydz$$

Ignoring scale factors, the term in brackets is simply $\varrho(x + u, y + v, z + w)$, thus yielding

$$P(u,v,w) = \int \varrho(x,y,z)\varrho(x + u, y + v, z + w)dxdydz.$$

We must now ask what this expression means. The Patterson function, as this synthesis is known, is a map of all the vectors in the structure. If we ignore the integral in the above equation, for a moment, and consider atomic point scatterers, we can see that if two atoms are separated by the vector (u,v,w) then there will be a peak in the Patterson function at (u,v,w) whose height will be proportional to the product of their electron densities. That is illustrated for a simple 4-point structure in Figure 15. The integral merely means that if there is more than one such pair of atoms in the structure those other products will also be added to the value of the Patterson function at u,v,w, as well. The number of vectors in the Patterson function increases with the square of the number of atoms, N, in the cell. N (self-vectors) of these will be concentrated in the origin of the function, however, leaving $N(N-1)+1$ total vectors. Determining a structure from the Patterson function is not trivial, but at least this interpretation of the intensity distribution gave crystallographers a means of determining a structure directly from the X-ray diffraction data.

It should be obvious from the expression for the Patterson function that if the structure contains a pair of atoms which are markedly denser (which correlates fairly well with higher atomic number) than the others, the vector between them should stand out in the map. Quite frequently, these may be the same atom in two different molecules related by symmetry. This has been useful in the phasing of a number of dinucleoside phosphate structures which contained derivatized bases or heavy cations.[14/15/28] Once one has identified the relative locations of the heavy atoms, it is possible to calculate a Fourier synthesis using the observed structure amplitudes and the phases derived merely from the identification of the sites of the heavy atoms. As shown in Figure 16, this will frequently be a good approximation to the actual phase of the structure factor because the phase is likely to be dominated by the heavy atom with the other atoms contributing randomly. When new atoms found in this synthesis have been added to the model, another Fourier synthesis is calculated, including the contributions of these atoms, and the procedure is iterated. This procedure is called Fourier refinement. Such a synthesis will frequently reveal much of the rest of the structure. Although the quality of the structure will not be as good as in the absence of the heavy atom, this technique will frequently solve the phase problem.

Short Nucleic Acid Fragments May Be Solved Directly From the Intensity Data

Since nucleic acids come with their own somewhat heavy atom, phosphorus, it is not necessary to derivatize them in order to solve their structures. Although phosphorus-phosphorus vectors will not be markedly visible in the Patterson function, due to the unusually heavy amount of overlap characteristic of Patterson functions (i.e., the summation over a lot of light atom-light atom vectors being greater than the product of two heavy atom vectors), it is possible to locate their position through the use of special techniques, such as resolution difference[4/9] and anomalous dispersion[12] procedures. Although the contribution of the phosphorus atom is not usually a dominant part of the structure factor, it is not too hard for the experienced investigator to determine the structure starting with an initial synthesis calculated with phosphorus phases. Generally several more cycles of Fourier refinement are necessary to deduce the rest of the structure in the absence of a very heavy atom. Such a procedure is akin to Patterson superposition procedures, for the details of which the reader is referred to references 29-31.

Another phasing technique which has found some use in the solution of oligonucleotide structures is termed *direct methods*. These procedures are based primarily on the non-negativity of the electron density distribution. If one invokes this constraint, strengthened by the localization of the electron density in a relatively small number of discrete scattering centers (atoms), it is possible to derive a very powerful formula relating the phases of the structure factors, the *tangent formula*.[32]

$$\tan(\phi_H) = \frac{\sum_{H'} |E_{H-H'}||E_{H'}|\sin(\phi_{H-H'} + \phi_{H'})}{\sum_{H'} |E_{H-H'}||E_{H'}|\cos(\phi_{H-H'} + \phi_{H'})}$$

In this formula, ϕ_H is the phase of the reflection $H (= h,k,l)$, $\phi_{H-H'}$ and $\phi_{H'}$ phases of two reflections (h-h', k-k', l-l' and h',k',l') whose indices vectorially add up to **H**. $E_{H-H'}$ and $E_{H'}$ are normalized structure amplitudes related to the ordinary structure amplitudes of $F_{H-H'}$ and $F_{H'}$. If one can assign or guess a few initial phases, it is possible to iteratively bootstrap oneself to the complete structure using this formula and associated probability relationships in the same way that one can bootstrap oneself to the complete structure by knowing the location of the phosphorus atoms.

Macromolecules Must Be Derivatized To Be Solved

Neither direct methods nor Patterson methods are directly applicable to the *de novo* solution of macromolecular structures. This is primarily due to the dependence of these procedures on the ability to resolve individual atoms. The theoretical basis of the tangent formula and the practical application of most Patterson procedures are greatly weakened if individual atoms are not resolvable.[33] Since macromolecular crystals rarely diffract beyond 2Å resolution, an alternative phasing means is necessary. This procedure, known as isomorphous replacement, had its first macromolecular application to protein crystals,[36] but it is quite general, and was used for the solution of Yeast tRNA^{phe}.[18/19/37]

Macromolecular crystals are different from crystals of small molecules, in that they tend to be about 50-70 percent solvent, where small molecule crystals usually range from 0 percent to 20 percent solvent. Because of the large amount of solvent in macromolecular crystals, much of it is in the liquid state, rather than tightly bound to the molecule. This enables the investigator to diffuse heavy atoms into the crystal. If these bind in a few places, without perturbing the original lattice, then one has prepared an *isomorphous derivative* of the macromolecule. How is this useful? If we look at Figure 17a, we see a triangle in the complex plane formed from the vectors corresponding to the structure factor for the native macromolecule, F_M, the structure factor for the derivitized macromolecule, F_{M+H} and the contribution to the derivatized macromolecules from the heavy atom derivative alone, F_H. Clearly, the following vector equations are valid:

$$F_{M+H} = F_M + F_H$$

$$F_M = F_{M+H} - F_H$$

If we draw a circle of radius $|F_M|$ about the origin of the complex plane, and draw another circle of radius $|F_{M+H}|$ about the point ($-F_H$), they will intersect at two sites, both of which satisfy the above equations (Figure 17c). Both intersection points on the $|F_M|$ circle are possible phases for F_M. The ambiguity can be broken by using a second derivative, whose coefficients are F_{M+K} (Figure 17b) and following the same procedure (Figure 17c). This procedure must be repeated for all coefficients F_M. In practice, several derivatives are usually used, and the phases must be extensively refined before they can be used.[38] You will note that we have been using the amplitudes of the native macromolecules and the derivatized macromolecule (F_M and F_{M+H} or F_{M+K}, respectively), but we have been using the amplitude and *phase* of the derivative itself (F_H or F_K). One might reasonably ask where this phase and amplitude come from. What must be done is to solve the structure of the derivative by itself. One must use difference coefficients between the native and derivatized macromolecule, and solve this difference structure, with somewhat shaky amplitudes, in the same way that high resolution small molecule structures are solved.

Another technique used in conjunction with the multiple isomorphous replacement

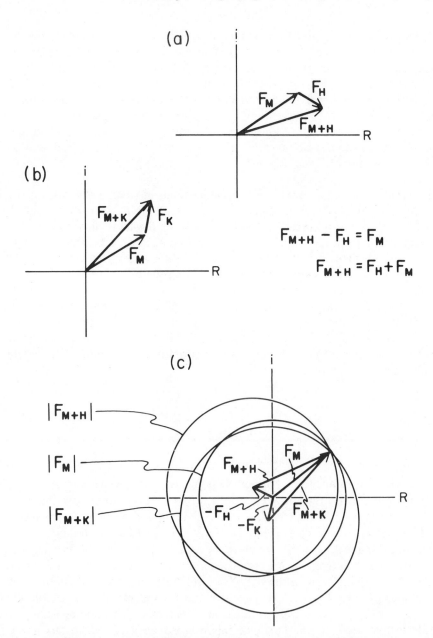

Figure 17. *The principle of multiple isomorphous replacement.* (a) The representation of the resultant structure factor when a macromolecule is derivatized. F_M represents the structure factor of the macromolecule, whose amplitude is known, but whose phase is sought. F_H represents the contribution of a heavy atom derivative. F_{M+H} is the resultant structure factor for the derivatized macromolecule. (b) The same structure factor, F_M, but this time combined with a different derivative, whose contribution F_K results in the structure factor F_{M+K} for the derivative. (c) Construction indicating how the phase of F_M can be derived from knowledge of the *amplitudes* of F_M, F_{M+H}, and F_{M+K} which are directly observable

procedure described above involves the use of anomalous dispersion. Let us consider the two structure factors F_{hkl} and the one on the opposite side of reciprocal space from it, $F_{-h,-k,-l}$.

$$F_{hkl} = \sum_{j=1}^{N\,atom} f_j \exp [2\pi i\,(hx_j + ky_j + lz_j)]$$

$$F_{-h,-k,-l} = \sum_{j=1}^{N\,atom} f_j \exp [2\pi i((-h)x_j + (-k)y_j + (-l)z_j)] = \sum_{j=1}^{N\,atom} f_j \exp [-2\pi i(hx_j + ky_j + lz_j)].$$

These structure factors are clearly equal in amplitude, but of opposite phase. This is known as Friedel's Law, and the two reflections are called a Friedel pair. Friedel's Law only holds in the case of real scatterers, however, and breaks down in the case of complex, or anomalous, scattering, as shown in Figure 18. If one has an isomorphous derivative which is an anomalous scatterer for the wavelength of radiation used, one may then treat the data from Friedel pairs like the data from two separate derivatives, as shown in Figure 19.

It should be pointed out that phases derived from isomorphous replacement with or without anomalous dispersion are dependent on the small differences observed between large numbers. As a result, they are likely to be off by an average of 35° in the best cases, and more typically 50° or worse. With the map derived from these phases, at the best resolution obtainable, it is the crystallographer's job to fit a molecular model to the electron density; this is much like fitting the duck in Figure 12a to the density in Figure 12f. This is the point at which there is an interpretive aspect to macromolecular crystallography. Although the solution of the Patterson function or phasing via direct methods may require large amounts of chemical and crystallographic intuition, the crystallographer using these techniques at high resolution can quickly determine whether or not the structure is correct. This can be done by comparing structure factors calculated from the model, F_c, with those experimentally determined, F_o. A useful index in this respect is known as the R-factor,

$$R = \frac{\sum ||F_o| - |F_c||}{\sum |F_o|}$$

For a high resolution structure, before refinement, an R-value of .2 to .3 is in-

and the *amplitudes* and *phases* of F_H and F_K, which may be obtained by solving the derivative structure. A circle of radius $|F_M|$ can be constructed about the origin of the complex plane. A circle of radius $|F_{M+H}|$ can be constructed with its center at the point $(-F_H)$. This will intersect the F_M circle at two points, one of which corresponds to the correct phase. The ambiguity may be broken by constructing a second circle of radius $|F_{M+K}|$ with its center at the point $(-F_K)$. As can be seen from the figure, the intersection points obey the vector addition laws for the summation of structure factors.

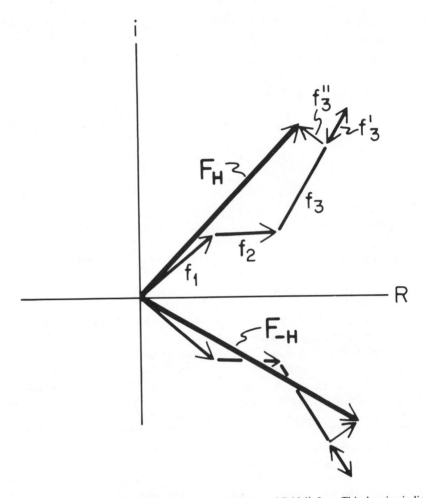

Figure 18. *Anomalous dispersion in the complex plane; violation of Fridel's Law.* This drawing indicates the contributions of three atoms to the structure factors F_H and F_{-H} which are on opposite sides of the origin of reciprocal space from each other. The first 2 atoms are both real scatterers, so that they contribute equally to the scattering of each reflection, but on opposite sides of the real axis; thus, the amplitudes of the resultant waves will be equal, but their phases will be opposite. The third scatterer is an anomalous scatterer which has a complex scattering factor. The f', or real portion of the anomalous scattering is taken to be negative here. The f'', or imaginary, portion of the anomalous scattering is indicated perpendicular to the real part. It can be seen that the resultant amplitudes for F_H and F_{-H} will not be equal, and their phases will also not be negatives of each other, since the imaginary component does not obey the mirror symmetry about the real axis.

dicative of a probably correct structure, with 0.59 being the expected result for a random array of atoms.[39] Correct macromolecular structures usually give R-values of 0.4 to 0.5, before refinement. Thus, the macromolecular crystallographer has much less feedback from the data about the correctness of the interpretation. Therefore, much more reliance must be placed on data from sequence, chemical modification and genetic studies, as well as high resolution crystallography of oligomers in order to make a meaningful interpretation of the map.

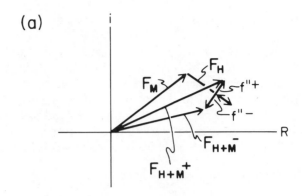

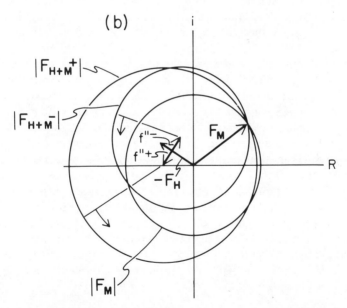

Figure 19. *Use of anomalous dispersion data in phasing.* (a) Vector diagram indicating the vector relationships between the structure factor for a macromolecule, F_M and the structure factors for an anomalously scattering heavy atom derivative, F_H, combining to give Friedel paired reflections F_{H+M}^+ and F_{H+M}^-. *The reader should note that the portion of the diagram corresponding to F_{H+M}^- has been reflected through the real axis to clarify the relationships.* The imaginary components, f'', are indicated. (b) In a similar manner to Figure 17(c), three circles have been constructed. One of radius $|F_M|$, as before, has its center at the origin of the complex plane. The complex scatterer, F_H, is drawn, and again, the sign of the vector is reversed to give ($-F_H$). The imaginary and real portions of the anomalous scattering are then taken into account, giving the two indicated points on either side of the $-F_H$ vector. One point corresponds to the scattering contribution for F_H to F_{H+M}^+, and the other corresponds to the scattering contribution for F_H to F_{H+M}^-. Circles of these radii are constructed about these two points, respectively, and they can be seen to intersect at the phase of F_M on the F_M circle.

Refinement Is The Acid Test For A Structure

Once the crystallographer derives a model for the structure from the diffraction data, it must be refined so that the best fit to the data is obtained. The most common techniques for refinement involve least squares procedures, although others exist. The reader is referred to the treatment by Stout and Jensen[24] for the details and derivations of these methods. With small molecules, one initially refines positional coordinates for each atom, as well as an isotropic thermal parameter. During this phase of the refinement, one usually calculates difference Fourier syntheses,

$$\Delta\varrho(x,y,z) = \sum_{hkl} \; (|F_o| - |F_c|) \; e^{i\phi_c} \exp\left[-2\pi i(hx + ky + lz)\right],$$

where ϕ_c is the calculated model phase associated with $|F_c|$.

These syntheses reveal small errors in the structure and are very useful in filling in the solvent region of the unit cell if this has not been done previously. In oligonucleotide crystals, there is frequently a large amount of solvent, which must be interpreted correctly. It is frequently statistically disordered (namely, it might not be the same in all unit cells), and great care must be taken to insure that any solvent molecules added to the model are readily interpretable in a chemical sense: the solvent molecules in all ordered and disordered sites must form reasonable hydrogen bonded or non-bonded contacts with their putative neighbors. Once the solvent is filled in and isotropic convergence has been obtained (usually with an R-factor between 0.10 and 0.20), anisotropic temperature factors may be employed. The anisotropic thermal motion model for atomic motion increases the number of parameters per atom from 4 to 9, and thereby decreases the overdetermination of the number of data points to parameters varied, by more than a factor of two. Agreement between data and model will automatically improve by doing this, and care must be taken to see that the thermal motion is meaningful and that the improvement in agreement is greater than that expected just by increasing the number of parameters varied.[40] Figure 20 illustrates the anisotropic thermal motion derived in the dinucleoside phosphate ApU. At the end of the refinement, the R-factor will typically be about .1 or less.

In large high-resolution structures, such as oligonucleotides, the uninterpretablity of the solvent region combines with the poor scattering from hydrogen atoms to make these atoms very hard to find in the electron density map. As can be seen from Figure 8, hydrogen atoms will be very small contributors to the diffraction pattern, and their density in the Fourier synthesis will suffer from errors in the data and in other parts of the structure. The positions of some hydrogens are generable from the stereochemistry of the atoms to which they are bonded. Accurate positioning of hydrogen atoms is best accomplished by performing a neutron scattering experiment. Unfortunately, nobody has grown a crystal of an oligomeric nucleic acid fragment large enough to do this yet.

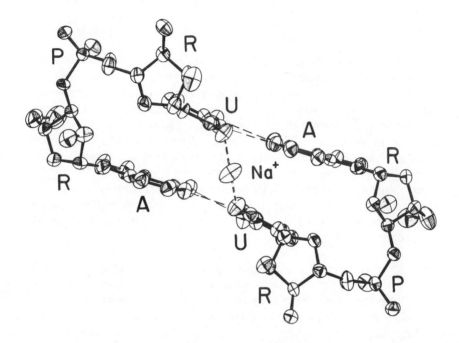

Figure 20. *Anisotropic thermal motion of the atoms in a miniature double helix.* This is an ORTEP plot[48] of the molecular structure of ApU,[9] indicating the anisotropic thermal motion of the individual atoms in the bimolecular complex. This is a view from the minor groove of the double helix. The principal components of the thermal motion are parallel to the major axes of the ellipsoids depicted, and the relative r.m.s. amplitudes of vibration in these directions are proportional to the relative lengths of the major axes. The sodium ion indicated at the center of the figure was found coordinated to the two uracil O2 atoms as indicated in the minor groove. The motion of the ion can be seen to be almost perpendicular to the liganding direction.

The refinement of macromolecules is more difficult, because the resolution is poorer and the overdetermination of observations to parameters is much less. The refinement process also has much more of a combined "solution and refinement" character than with smaller molecules studied at higher resolution, since a larger number of solvent molecules are found in the course of the refinement. Furthermore, portions of the structure may have to be slightly reinterpreted as the refinement proceeds. In order to insure the chemical integrity of the structure, bond distances, bond angles and conjugated ring planarities are restrained to values determined by high resolution crystallography of smaller systems.[41] R-values between .1 and .20 have been reported for macromolecules nearing the end of their refinement. The R-value for Yeast tRNA[phe] is now 0.19.[42] The R-value by itself is not meaningful, however, and may be dramatically decreased if attention is not paid to retaining good stereochemistry throughout the course of the refinement. For refinement, in general, and macromolecular refinement in particular, the chemical reasonableness of the model is the *sine qua non*, with R values being of secondary importance.

Concluding Remarks

I have tried to present the reader with a brief introduction to X-ray crystallography, particularly that of nucleic acid fragments and macromolecules. Space has not permitted discussion of experimental matters, such as the generation and properties of X-rays, and the collection of diffraction intensity data. The reader is referred to the books by Stout and Jensen,[24] Arndt and Willis,[43] and Arndt and Wonacott[44] for a discussion of these matters. The growth of suitable crystals (a major stumbling block to many nucleic acid studies) is not yet systematized, and the reader is referred to original papers in the area and the review by McPherson.[45] Many areas of crystallography have only been briefly mentioned, in particular symmetry, structure solution procedures and refinement. I feel that the material presented above will allow the reader to read the structural papers in this volume critically, thereby removing some of the mystique from this field.

Nucleic acid crystallography is an exciting field which is still in its infancy. A relatively small number of oligonucleotide structures have been determined, and very few tRNA molecules have been solved. Because of the great importance of understanding the structural aspects of nucleic acids, I am sure that we will see many more structures in the near future. With those structures will come a greater appreciation for the molecular-structural aspects of life.

Acknowledgements

Figures 1, 9, 10, 12 and 13 are reprinted from G. Harburn, C. A. Taylor and T. R. Welberry: *Atlas of Optical Transforms.* Copyright © 1975 by G. Bell and Sons, Lts. Used by permission of the publishers, Cornell University Press.

This work was aided by a Basil O'Connor Starter Research Grant from The National Foundation-March of Dimes, by University Award Number 7421 from the Research Foundation of the State University of New York and by Grant GM26467 from the National Institute of Health.

References and Footnotes

1. Notation: The amplitudes of complex numbers and the lengths of vectors are represented by vertical bars, e.g., $|F|$ is the amplitude of the complex number F and $|D|$ is the length of the vector **D**. Vectors are represented by bold face type, e.g., **D**. The length of a line segment between two points A and B is indicated by $\overline{AB}$. The complex conjugate of a complex number F will be denoted F*. Structure factors will be denoted F_{hkl} when referring to the indices of interest, or will be denoted F_M or F_{M+H} when referring to Fourier coefficients from different crystals. Sometimes the triple indices (h,k,l) will be abbreviated as the reciprocal lattice vector **H**.
2. Kendrew, J. C., Bodo, G., Dintzis, H. M., Parrish, R. E., Wyckoff, H. and Phillips, D. C., *Nature 181*, 662 (1958).
3. Corey, R. B. and Pauling, L., *Rc. Ist. Lomb. Sci. Lett. 89*, 10 (1955).
4. Seeman, N. C., Sussman, J. L., Berman, H. M. and Kim, S. H. *Nature New Biology 233*, 90 (1971).
5. Sussman, J. L., Seeman, N. C., Kim, S. H. and Berman, H. M., *J. Mol. Biol. 66*, 403 (1972).
6. Rubin, J., Brennan, T. and Sundaralingam, M., *Biochemistry 11*, 361 (1972).

7. Camerman, N., Fawcett, J. K., and Camerman, A., *Sceienc 182*, 1142 (1973).
8. Rosenberg, J. M., Seeman, N. C., Kim, J. J. P., Suddath, F. L., Nicholas, H. B. and Rich, A., *Nature 243*, 150 (1973).
9. Seeman, N. C., Rosenberg, J. M., Suddath, F. L., Kim, J. J. P., and Rich, A., *J. Mol. Biol. 104*, 109 (1976).
10. Day, R. O., Seeman, N. C., Rosenberg, J. M., and Rich, A., *Proc. Nat. Acad. Sci. (USA) 70*, 849 (1973).
11. Rosenberg, J. M., Seeman, N. C., Day, R. O., and Rich, A., *J. Mol. Biol. 104*, 145 (1976).
12. Seeman, N. C., Day, R. O., and Rich, A., *Nature, 253*, 324 (1975).
13. Suck, D., Manor, P., Germain, G., Schwalbe, C. H., Weimann, G., and Saenger, W., *Nature New Biology 246*, 161 (1973).
14. Tsai, C. C., Jain, S. C., and Sobell, H. M., *Proc. Nat. Acad. Sci. (USA) 72*, 2626 (1975).
15. Sakore, T. D., Jain, S. C., Tsai, C. C., and Sobell, H. M., *Proc. Nat. Acad. Sci. (USA) 74*, 188 (1977).
16. Neidle, S., Achari, A., Taylor, G. L., Berman, H. M., Carrell, H. L., Glusker, J. P., and Stallings, W. C., *Nature 269*, 304 (1977).
17. Viswamitra, M. A., Kennard, O., Jones, P. G., Sheldrick, G. M., Salisbury, S., Falvello, L., and Shakked, Z., *Nature 273*, 687 (1978).
18. Kim, S. H. Suddath, F. L., Quigley, C., McPherson, A., Sussman, J. L., Wang, A. H. J., Seeman, N. C., and Rich, A., *Science 185*, 435 (1974).
19. Robertus, J. D., Ladner, J. E., Finch, J. T., Rhodes, D., Brown, R. S., Clark, B. F. C., and Klug, A., *Nature 250*, 546 (1974).
20. Schevitz, R. W., Podjarny, A. D., Krishnamachari, N., Hughes, J. J., Sigler, P. B., and Sussman, J. L., *Nature 278*, 188 (1979).
21. Wilson, H. R., *Diffraction of X-rays by Proteins, Nucleic Acids and Viruses*, London, Edward Arnold Publishers, Ltd. (1966).
22. Arnott, S. and Hukins, D. W. L., *J. Mol. Biol. 81*, 93 (1973) and references therein.
23. Glusker, J. P. and Trueblood, K. N., *Crystal Structure Analysis: A Primer*, New York, Oxford University Press (1972).
24. Stout, G. M. and Jensen, L. M., *X-ray Structure Determination: A Practical Guide*, New York, The Macmillan Company (1968).
25. Blundell, T. L. and Johnson, L. N., *Protein Crystallography*, New York, Academic Press (1976).
26. International Tables for X-ray Crystallography, Vol. 1-3, Birmingham, England, Kynoch Press (1969); Vol. 4 (1974).
27. The scattering is inversely proportional to the mass of the scattering particles, thus making electrons the only constituent of molecules to give a noticeable signal.
28. Wang, A. H. J., Nathans, J., van der Marel, G., Van Boom, J. H., and Rich, A., *Nature 276*, 471 (1978).
29. Buerger, M. J., *Vector Space*, New York, Wiley (1959).
30. Seeman, N. C., in *Crystallographic Computing*, F. R. Ahmed, ed, Copenhagen, Munksgaard (1970), pp. 87-89.
31. Seeman, N. C., Ph.D. Thesis, University of Pittsburgh (1970).
32. Karle, J. and Karle, I. L., *Acta Cryst. 21*, 849 (1966).
33. One important exception to this statement involves Patterson search procedures. If a crystal is expected to contain a well characterized electron density distribution, such as an adenine ring or a RNA double helix, it is possible to search the observed Patterson function for this distribution of vectors. This technique has been applied to the crystal structure of *E. coli* tRNA^{Metf} (34). While its applicability is very powerful, and it is not dependent on atomic resolution, it is not a *de novo* crystallographic phasing technique, since one must know what electron density distribution is being sought. For more information on this technique, see Rossmann.[35]
34. Woo, N. H., and Rich, A., Priv. comm. (1979).
35. Rossmann, M. G., *The Molecular Replacement Method*, New York, Gordon and Breach (1972).
36. Green, D. W., Ingram, V. M., and Perutz, M. F., *Proc. Roy. Soc. A 225*, 287 (1954).
37. Kim, S. H., Quigley, G. J., Suddath, F. L., McPherson, A., Sneden, D., Kim, J. J., Weinzierl, J., and Rich, A., *Science 179*, 285 (1973).

38. Blow, D. M. and Crick, F. H. C., *Acta Cryst. 12*, 794 (1959).
39. Wilson, A. J. C., *Acta Cryst. 3*, 347 (1950).
40. Hamilton, W. C., *Statistics in Physical Science*, New York, Ronald Press (1964).
41. Konnert, J. J., *Acta Cryst. A32*, 614 (1976).
42. Kim, S. H., Priv. comm. (1979).
43. Arndt, U. W. and Willis, B. T. M. *Single Crystal Diffractometry*, Cambridge University Press, Cambridge (1966).
44. Arndt, U. W. and Wonacott, A. S., *The Rotation Method in Crystallography*, Amsterdam, North-Holland Publishing Co. (1977).
45. McPherson, A., *Methods of Biochemical Analysis, 23*, 249 (1976).
46. Taylor, C. A. and Lipson, H., *Optical Transforms*, Ithaca, New York, Cornell University Press (1964).
47. Quigley, G. J., Wang, A. H. J., Seeman, N. C., Suddath, F. L., Rich, A., Sussman, J. L., and Kim, S. H., *Proc. Nat. Acad. Sci. (USA) 72*, 4866 (1975).
48. Johnson, C. K., *ORTEP*, ORNL-3794, Oak Ridge National Laboratory, Oak Ridge, Tennessee (1965).

Is DNA Really a Double Helix?
Its Diverse Spatial Configurations and the
Evidence for a Vertically Stabilized Double Helix

Ramaswamy H. Sarma
Institute of Biomolecular Stereodynamics
State University of New York at Albany
Albany, New York 12222

Somewhere Near The Beginning of the Race

Friday evening, 6:30 p.m., May 2, 1952, King's College, London, England. Rosalind Franklin mounted a good DNA fiber on the X-ray camera and precisely controlled the relative humidity by bubbling hydrogen into the chamber through a solution of saturated ammonium sulfate. 7:30 p.m., she started the exposure by turning on the X-ray tube. After about 62 hours, the exposure was stopped. The picture she obtained was vivid. "The overall pattern was a huge blurry diamond. The top and bottom points of the diamond were capped by heavily exposed, dark arcs. From the bull's eye, a striking arrangement of short horizontal smears stepped out along the diagonals in the shape of an X or a maltese cross. The pattern shouted helix."[1] It is this picture—the celebrated #51—it has been alleged that Watson saw 10 months later and that this propelled him and Crick to build the model for DNA[2]—a right-handed double helix with the two antiparallel chains wound plectonemically round a common axis, the chains being connected by Watson-Crick hydrogen bonds (Plates 2 and 6).

Is DNA Really a Double Helix?

A number of recent reports[3-7] suggest that the two strands of DNA do not coil round one another but lie side-by-side, thus abolishing the difficulty of separating the two chains during replication. The new structure retains the antiparallel complementary base paired chains of the Watson-Crick arrangement and the 10-base pair repeat each turn. Mutual coiling is precluded in this structure by a sequence of five base pairs having a right-handed twist followed by five with a left-handed twist, and so on, indefinitely (Figure 1). "The resulting model is analogous to a warped zipper."[8]

Crick, *et al.*[9] have indicated several points against the side-by-side structure; it is inelegant; the reversal of the hand of the screw at every fifth residue is arbitrary; in the crystal structures of tRNA (Plate 1) the helices are right-handed; the right-handed screw is not exactly equal to the left-handed one and hence the chains of the proposed structure do coil around one another, though much more slowly than in the classical double helix, one turn taking about 100 base pairs, rather than 10. These are weak arguments. However, Crick *et al.*[9] provide compelling evidence which leaves any doubt that the side-by-side model is wrong. This has to do with the linking number, i.e., the net number of times each chain is coiled around the other.[10-12] For a circular DNA of 5000 base pairs for a classical double helix, the linking number would be[13] +500 (i.e., N/10 where N is the number of base pairs). For a true side-by-side structure, it will be zero and for the proposed side-by-side structure, it will be around +50. Crick, *et al.*,[9] from various experiments in which circular DNA is unwound by known amounts, show that the linking number for relaxed DNA is roughly equal to N/10 as expected from the classical double helix. They[9] have further shown that supercoiled DNA molecules in gel electrophoresis migrate in a series of discrete bands—all the molecules in one band have the same value of linking number and that this value differs by unity from that of the adjacent bands. This observation can be rationalized only on the basis of the classical double helical configuration for DNA.

Arnott[8] has provided crystallographic evidence that nucleic acid duplexes are regular helices rather than asymmetric zippers. He indicates that in the side-by-side structure the phosphate groups lie more or less on the surface of a cylinder with radius about 9Å and the base pairs are stacked 3.4Å apart and hence the calculated diffraction pattern would be similar to that observed from DNA fibers. However, Arnott[8] wonders whether the crystallographic expectations, i.e., molecules with fourfold

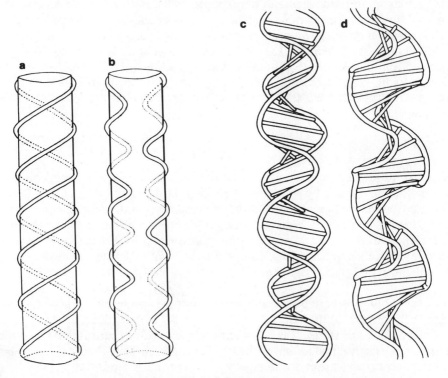

Figure 1. Idealized drawings of (a) an elementary Watson-Crick model, (b) an elementary side-by-side structure; (c) and (d) are sketches of two views of the side-by-side model. From Reference 3.

screw symmetry crystallize with tetragonal symmetry, 12 fold helices of A'-RNA crystallize with rhombohedral symmetry, etc., would be realized so accurately if polynucleotide duplexes in general and B-DNA in particular were not close to being perfectly regular helices. Based on these, we will proceed on the assumption that DNA is really a double helix.

The Diverse Spatial Configurations of DNA

More than a quarter of a century has elapsed since the double helix was propounded by Watson and Crick. During this period, remarkable progress has been made in arriving at the intimate details of the spatial configuration of many a nucleic acid system by crystallography; many of these are summarized in this volume.[14-17] An important achievement of this decade in this regard is the determination of the detailed geometry of yeast phenylalanine tRNA.[18-21] An artistic rendition of this structure by Irving Geis is depicted in Plate I. Recently single crystal studies have revealed the details of the initiator tRNA.[22]

The A, B, C, and D Forms of DNA

Fiber diffraction investigations have shown that DNA may assume different shapes depending upon the mode of preparation and the extent of humidity content. It was Rosalind Franklin who made the fundamental discovery that DNA fibers yield two different types of diffraction patterns depending upon whether they are in "crystalline" or "wet" state; they were designated by her as the A and B forms,[23/24] a designation which we still use. In addition to the A and B forms, it is thought that it may exist in forms called C and D. The C form occurs in LiDNA fibers[28] and in NaDNA fibers under conditions of low hydration and salt contents intermediate between those appropriate for A-DNA and B-DNA.[27] The eight fold D-DNA has been reported to be the structure of DNAs containing alternating purine pyrimidine sequences.[27] In Plates 2 through 5 are depicted the computer generated stereoperspectives of these forms. These were generated using the coordinate of Arnott, *et al.*.[25-27]

Table I
Backbone and Glycosidic Torsion Angles for Various Forms of DNA

Molecule	Sugar Pucker	Torsion Angles In Degrees						Ref.
		$\alpha(\phi')$	$\beta(\omega')$	$\gamma(\omega)$	$\delta(\phi)$	$\epsilon(\psi)$	χ	
A-DNA	³E	178	313	285	208	45	26	25
B-DNA	²E	155	264	314	214	36	82	26
C-DNA	²E	211	212	315	143	48	73	28
D-DNA	²E	141	260	298	208	69	84	27
Levitt	⁰⁴E-₁E*	178	275	295	170	65	47	29
Levitt Superhelix	⁰⁴E-₁E*	177	278	301	169	60	48	29

*pucker in between O4'-*endo* and C1'-*exo*.

Plates 6 and 7 are the artistic renderings of B and A forms by Irving Geis. Even though there are differences in the backbone torsion angles (Table I) and base pair overlap geometries among the four forms (Plates 10-13) one may consider the B, C, and D to belong to the B-family. The principal differences between the A and B type of structures center around the sugar pucker and the magnitude of glycosidic torsion (Table I). In the A form sugar pucker is ^3E and $\chi_{CN} = 26°$ and in the B family sugar pucker is ^2E and $\chi_{CN} \simeq 80°$. In the A form there is a vacant inner core of 6-7Å in diameter and this is absent in the B family of structures (see Seeman[14] in this volume). The dense packing of the base pairs with no vacant inner core in the B form is evident in the bottom of Plate 9. The helical parameters for A-DNA and B-DNA are summarized in Table II. In the B form both the major and minor grooves (Plates

Table II
Helical Parameters for A-DNA and B-DNA*

	A-DNA	B-DNA
Translation (Å)	2.59	3.38
Rotation (°)	32.7	36.0
Pitch (Å)	28.5	33.8
Residue/turn	11.0	10.0
Base tilt (°)	19.0	6.0

*From Arnott, S., and Hukins, D.W.L., *Biochem. Biophys. Res. Commun. 47*, 1504 (1972).

6 and 2) are approximately of equal depth. In the A form the minor groove is nearly flat while the major groove is narrow and deep (Plates 7 and 3). In all these forms the structure is a right-handed double helix with the two chains wound plectonemically round a common axis.

The Propeller Twist Model

In addition to the A, B, C and D forms of DNA discussed, several different spatial configurations have been advocated based on theoretical calculations, drug intercalation studies and single crystal studies of oligonucleotides. All these structures are double helical but differ in intimate details and have been advocated in many instances to rationalize certain biological and physical behavior of DNA. For example, Levitt[29] has modified the regular B-DNA structure by empirical energy calculation so that it can be bent smoothly and uniformly into a superhelix with a small enough radius (45Å) to fit the dimensions of chromatin. The Levitt model differs principally from the regular B form in its sugar pucker and χ_{CN}. The glycosidic torsion has changed from 80° in B-DNA to 47° in the Levitt model; the sugar pucker is intermediate between $04'$-*endo* and $C1'$-*exo* instead of $C2'$-*endo*. The various torsion angles are summarized in Table I. The base normals which initially tilt 6.3° from the helix axis now tilt 17.6° and a consequence of this is the *propeller-like twist* of the base pairs of 28.1° (Figure 2b). This 10 fold DNA was further refined by Levitt to smoothly deform into a superhelix of radius 45Å and pitch of 55Å and the resultant

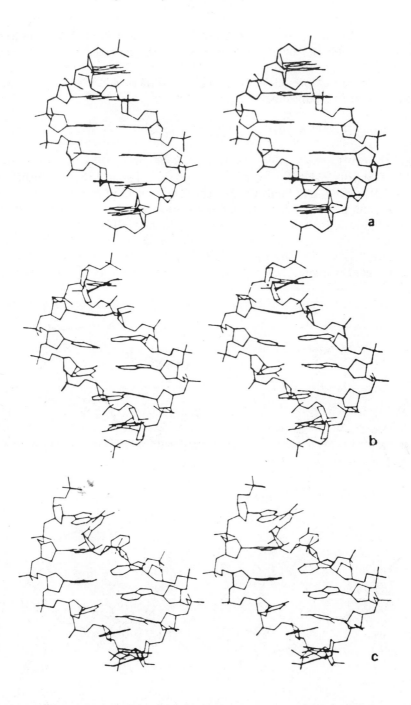

Figure 2. Stereoscopic views of about one-half turn of DNA double helix (a) B-DNA; (b) energy refined straight 10-fold propeller twist DNA; (c) energy refined 10.4-fold propeller twist super helical DNA. From reference 29.

conformation had 10.4 base pairs per turn and several torsion angles underwent minor changes (Table I). The propeller-like twist of the base pairs of the superhelical DNA is slightly larger than for straight DNA (Figure 2c). Crothers, *et al.* in this volume and elsewhere[30] provide electric dichroism data on DNA which have been interpreted as supporting this model.

The Alternating B-DNA Model

Klug and coworkers[31] have advocated what they call *alternating-B* structure for the poly(dA-dT)•poly(dA-dT) tracts of DNA. It is their thesis that poly(dA-dT)•poly(dA-dT) binds the *lac* repressor protein of *Escherichia coli* about 100 to 1000 times more stronglly than does bulk calf thymus DNA because the spatial configuration of the deoxyribose-phosphodiester backbone is altered as a consequence of the regular alternation of the bases in sequence. This idea, the authors state,[31] is inspired by the single crystal structural studies on (dA-dT)₂ by Viswamitra, *et al.*[32] summarized by Seeman[14] elsewhere in this volume. In the alter-

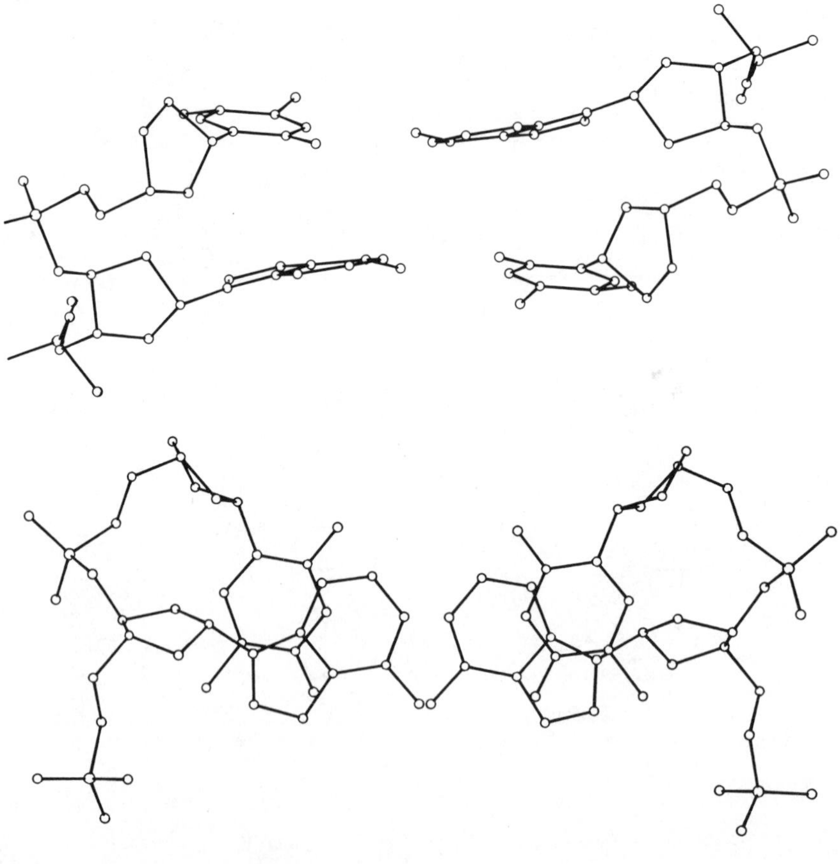

Figure 3. The side and top views of a repeating unit in the alternating B-DNA structure of poly(dA-dT)•poly(dA-dT). The figure was generated by Mitra and Sarma from coordinates in Reference 31.

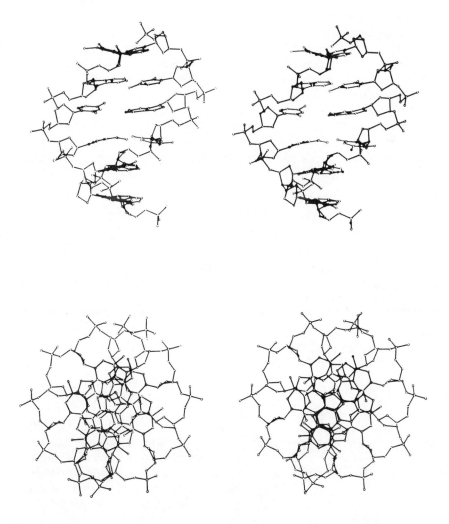

Figure 4. The side and top views in stereo of a six base pair segment of the alternating B-DNA structure of poly(dA-dT)•poly(dA-dT),. The figure was generated from coordinates of Figure 3 by applying a right-handed screw rotation of 36° and a displacement of 3.4Å along the Z axis as per instructions in Reference 31. We have no explanation why the bases appear bent, except that in the coordinates[31] some of the atoms attached to the bases are out of plane.

nating B structure, every second phosphophate-diester linkage has a conformation different from that of the normal B form. Two perspectives of the repeating unit is shown in Figure 3. Figure 4 shows the side and top views of the alternating B structure for a six base pair segment. This structure is arrived at by the authors[31] by energy minimization. The main difference between the normal and alternating structures are in the phosphodiester torsion angles β and γ. In the normal B form $\beta =$ 264° and $\gamma =$ 314° and they are the same throughout; **in the alternating structure $\beta = \gamma = 300°$ between A and the T, but $\beta = 240°$ and $\gamma = 310°$ between T and A.** The sugar ring conformation in B-DNA is ^2E, but in the alternating B structure A residue has its sugar in ^3E form and the T in the ^2E form. The authors[31] state that an energetically viable alternating B form may be generated in which all the sugar rings have the same conformation. In arriving at this structure the authors[31] have provided a reinterpretation of the NMR data of Patel and coworkers on dA-dT oligomers and polymers. Patel[33] later in this volume interprets his NMR data on poly(dG-dC)•poly(dG-dC) at high salt along the lines of an alternating B structure. However, a thorough analysis of the NMR data as discussed in the beginning[34] from x, y, z coordinates is desirable before NMR evidence is presented to support the alternating B structure.

The β-kinked B-DNA

Sobell from his extensive studies of nucleic acid drug complexes, summarized elsewhere in this volume[15] has advanced a β-*kinked DNA* which contains a dinucleotide with the mixed **C3′ endo (3′-5′) C2′ endo** sugar puckering as the asymmetric unit, and which has helical parameters that lie midway between B and A-DNA. Plates 16 through 20 show several aspects of β-kinked DNA. Insofar as this volume contains an extensive chapter by Sobell where the role β-kinked DNA plays in kinetic and structural aspects of drug binding, DNA dynamics and breathing, B ⇌ A transition and protein-DNA interaction, has been thoroughly discussed, the reader is directed to read this opus by Sobell.

Z-DNA

The structures described above have been proposed based on fiber diffraction studies in which atoms are not resolved, theoretical calculations, single crystal studies of drug intercalated dinucleoside monophosphate complexes along with rich imagination and intuition. These approaches cannot provide the unequivocal *detailed* molecular configuration of DNA. This can come only from the solution of the three dimensional structure of *double helical DNA fragments of reasonable length at atomic resolution.* To date, this has been accomplished only in one case,[36] i.e., the self complementary hexanucleoside pentaphosphate d-CpGpCpGpCpG. The hexamer was synthesized by van Boom and van der Marel, University of Leiden and the structure was solved to atomic resolution of 0.9Å by Alexander Rich and his coworkers Wang, Quigley, Kolpak, and Crawford at MIT.[36] The authors, based on the hexamer duplex data and on the disposition of the molecules within the crystal lattice, have proposed a structure for DNA—the Z-DNA.[36] Insofar as the structure they discovered is very novel and totally unexpected, and because this author believes that this discovery is an important event in the human endeavours to comprehend the molecular biology of living systems, a detailed description is provided below. This is entirely based on a preprint[36] kindly supplied to this author by Alexander Rich on November 27, 1979.

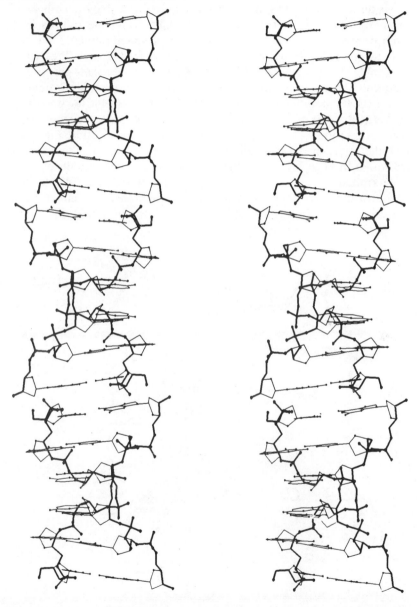

Figure 5. Stereodiagram of Z-DNA. Three of the double helices of d-CpGpCpGpCpG in the crystal lattice as they are stacked along the *c* axis in what looks like a continuous double helix. A phosphate residue is absent every sixth residue along the chain. Diagram supplied by Alexander Rich from Reference 36.

In Figure 5 is reproduced a stereo view of Z-DNA from Wang, *et al.*[36] This is a view of the continuous double helix as it appears within the crystal lattice because the symmetry related hexamers stack one upon another; obviously the helix is interrupted by the lack of a phosphate group on the terminal dG residue. What is immediately clear is that it is a **left handed double helix**. It contains twelve base pairs per turn, i.e., six dimer residues and the bases are aligned parallel to each other. It is apparent from the diagram that the sugar-phosphate backbone follows a zig-zag course. In a perspective of one-half of the duplex in the asymmetric unit with the residues C1 through G6, Wang, *et al.*[36] illustrate this dramatically (Figure 6). Note that in the van der Waal's diagram (Figure 6, *Right*), the phosphate groups pursue a zig-zag pattern. The sugar residues have an alternating orientation in that the dG residues have the O4′ pointing up while the dC residues have the O4′ pointing down (Figure 6, *Left*). Because of this alternation, the authors[36] note, the repeating unit in the helix is not a mononucleotide as in B-DNA, but a dinucleotide. Figure 6, *Left* also shows the stereochemical details of the residues. The guanine is *syn* and is associated with a ³E sugar pucker for the two dG residues in the middle of the strand while the terminal dG has a ²E sugar conformation. The C4′-C5′ bond is *gauche-trans* for the dG residues. The dC residues are in the *anti* conformation with ²E sugar pucker and *gauche-gauche* C4′-C5′ torsion—similar to B-DNA.

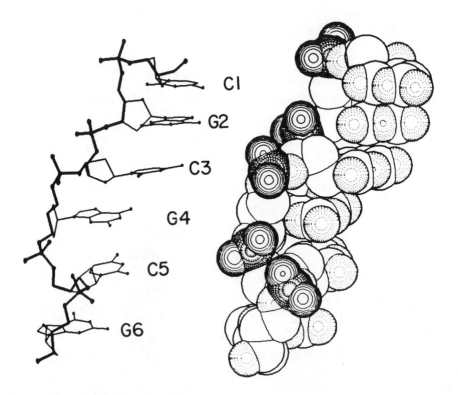

Figure 6. The skeletal and van der Waals diagram of one-half of the duplex of d-CpGpCpGpCpG in the asymmetric unit. Diagram supplied by Alexander Rich from Reference 36.

Wang, *et al.*[36] provide an extensive discussion of the base stacking pattern in Z-DNA, *viz-a-viz* B-DNA. In Figure 7, reproduced from their preprint,[36] is illustrated the base stacking pattern in the d-CpG and d-GpC segments in Z and B-DNA. The d-CpG base pairs in Z-DNA (Figure 7, *Left*) are not stacked upon each other directly but are associated with a lateral translation of about 7Å relative to each other so that they present a sheared appearance with the cytosine of the upper base pair partially stacked upon the cytosine of the lower base pair, i.e., interstrand base stacking instead of intrastrand base stacking similar to the vertical double helices described later. The authors[36] specifically draw attention to the fact that in the d-CpG base stacking, the guanines are stacked upon the O4′ oxygen atoms of the preceding deoxyribose residues (this is very clear in Figure 6, *Left*), and these patterns are remarkably different from that in B-DNA. However, Rich and coworkers[36] show that in the d-GpC residues (Figure 7, *Right*) there is a great deal of stacking overlap

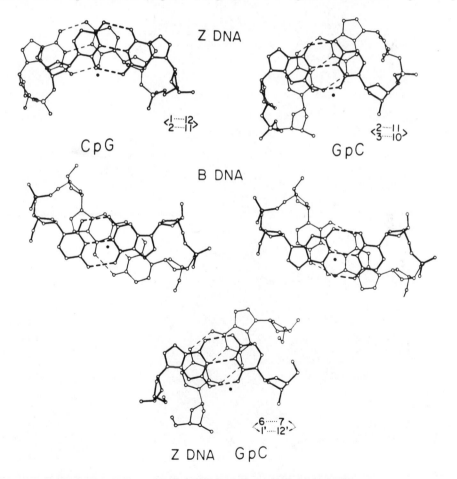

Figure 7. Stacking pattern in Z and B-DNA. Note that the minor groove in B-DNA is found at the top of the B-DNA diagram but at the bottom of the Z-DNA diagram. The helix axis is shown by a solid circle. Diagram supplied by Alexander Rich from Reference 36.

of the bases upon each other in a manner somewhat similar to that of B-DNA. The distance between the phosphate groups across the helix in the d-GpC segment is 15Å, close to that in B-DNA. The corresponding distance in d-CpG residues is 12.5Å.

Wang, *et al.*[36] graphically depict the reason for the hexamer units in the crystal lattice to appear as a continuous double helix and this is shown in Figure 7, bottom. Here the position of the base pair G6-C7 at the bottom of one helix, is shown

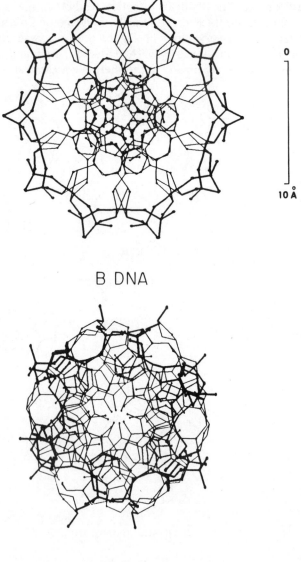

B DNA

Z DNA

Figure 8. End views of B-DNA and Z-DNA. Diagram supplied by Alexander Rich from Reference 36.

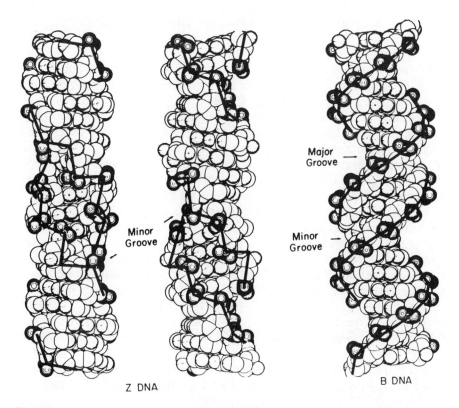

Minor Groove

Major Groove

Minor Groove

Z DNA

B DNA

Figure 9. Van der Waals side views of Z-DNA and B-DNA. Two views of Z-DNA are shown which are 30° apart in orientation about the helix axis. Diagram supplied by Alexander Rich from Reference 36.

relative to the position of the base pair C1′-G12′ of the helix immediately below it in the crystal lattice. It is remarkable that this stacking pattern is virtually identical to the internal stacking of the d-GpC residues!

An end view of Z-DNA compared to B-DNA is reproduced from reference 36 in Figure 8. The authors[36] draw attention to the following: In both projections one complete turn of the helix is shown. This consists of ten base pairs of B-DNA with an alternating G-C sequence and twelve base pairs of Z-DNA. In the Z-DNA, there is an approximate six-fold symmetry with *the guanine bases largely found on the outer part of the molecule* in B-DNA there is five-fold symmetry with *the base clustered near the center of the molecule.* The Z-DNA is a slightly thinner helix with a diameter of 18Å compared to the 20Å diameter of B-DNA. The Z-DNA model in Figure 8 does not have the regularity of the B-DNA model due in part to the fact that two phosphate groups are missing from the helix. The center of the Z form has an elliptical oxygen tunnel (cytosine 02). The ellipticity is due to the fact that the two segments containing the six base pairs are not oriented in a straight line but are tilted relative to each other by a very small angle (<2°). In addition to being somewhat thinner, the Z form is somewhat more extended than B-DNA. Twelve base pairs are found in 44.6Å, whereas B-DNA will accommodate thirteen base pairs in that same distance.

Reproduced from Wang, *et al.*[36] in Figure 9 is a van der Waal's side view of Z-DNA and B-DNA. The authors[36] emphasize the alternating positions of the phosphate group by drawing a heavy black line from phosphate to phosphate residue along the chain. "These lines are smooth, continuous and right-handed in B-DNA, but follow a zig-zag, discontinuous left-handed course in Z-DNA."[36] "Z-DNA looks like a cylinder and has less of a grooved appearance than B-DNA due to the fact that it has only one narrow groove."[36] The minor groove of Z-DNA is considerably deeper than the minor groove of B-DNA. "The major groove is absent and is more or less inverted relative to B-DNA. In B-DNA, the major groove is a concave surface with the base pairs at the bottom, while the analogous part of Z-DNA is a rounded convex surface made up of the base pairs with the imidazole residues of guanine prominently exposed on the surface."[36] In the Z-DNA, the deep minor groove "extends all the way down to the axis of the molecule. The phosphate groups are 8.5Å apart between the two edges and the groove itself is approximately 9Å deep as it extends to the axis of the 18Å diameter molecule."[36] The position of the helix axis in Z-DNA and B-DNA is shown by a solid circle in Figure 7. In that figure, note that the minor groove in B-DNA is found at the top of the B-DNA diagram but it is at the bottom of the Z-DNA diagram and the helix axis in this passes through the minor groove.

What is clear is that Z-DNA is substantially different from B-DNA, the only common feature being the antiparallel double helical arrangement with Watson-Crick

Table III
Structural Parameters of B-DNA and Z-DNA†

	B-DNA	Z-DNA*
Helix Sense	Right-handed	Left-handed
Residues/Turn	10	12 (6 dimers)
Diameter	~20Å	~18Å
Rise/Residue	3.4Å	3.7Å
Helix Pitch	34Å	45Å
Base Pair Tilt	6°	7°
Rotation/Residue	36°	-60° (per dimer)
Glycosidic Torsion Angle		
Deoxyguanosine	Anti	Syn
Deoxycytidin	Anti	Anti
Sugar Pucker		
Deoxyguanosine	C2′ endo	C3′ endo
Deoxycytidine	C2′ endo	C2′ endo
Distance of P from Axis		
d-GpC	9.0Å	8.0Å
d-CpG	9.0Å	6.9Å

†Table supplied by Alexander Rich from Reference 36.
*Average values are given and end effects are excluded.

base pairing. The important differences are summarized by the authors[36] in tabular form and this is reproduced in Table III.

At the end of their[36] elegant opus, opulent with van der Waals and stereodiagrams, the authors provide answers to some of the obvious and provocative questions such as: Is their Z-DNA identical to the high salt form of poly(dG-dC)•poly(dG-dC) in solution? What is the mechanism of B ⇌ Z interconversion? Do Z-DNA segments occur in natural DNA? Is the Z-DNA structure strictly limited to alternating guanine and cytosine sequences or could it accommodate any alternating purine and pyrimidine sequences? What is the relevance of Z-DNA structure to that of RNA? What is the biological relevance of Z-DNA in gene expression, regulation and recognition? We urge the students and scientists to read the original article[36] for the possible answers.

Spatial Configuration of High Anti Polynucleotides
The Vertically Stabilized Double Helices

There exists a theoretical controversy about the spatial configuration of high anti polynucleotides. It has been shown experimentally[35] that when χ is fixed at a high anti domain ($\simeq 120°$) the sense of *the base stack* is left-handed. However, the effect of a high anti arrangement on the *sense of the helical organization* of the backbone of a polynucleotide is not known with any certainty, and two diametrically opposed views have been reported about its helix handedness. This is taken up in detail next.

The Controversy

The observation that the cotton effect in the spectra of high anti polynucleotides was opposite to that of normal nucleic acid counterparts[37/38] suggested the possibility that their helical organization may be left handed. Fujii and Tomita[39] carried out conformation energy calculations and concluded that energetically these molecules prefer a left handed helical backbone. Recently, Yathindra and Sundaralingam[40/41] using the "rigid nucleotide concept," a concept whose validity has been challenged,[42/43] reported the helical parameters n and h where n is the number of nucleotide residues per turn and h is the residue height along the helix axis for high anti polynucleotides and concluded that the helical sense of the backbone in both single and double stranded systems is left handed and that the sense of base stack is also left handed. They have shown a model of the left handed duplex in reference 40.

However, Olson and Dasika[44] have advanced an alternative model for high anti polynucleotides. From extensive theoretical studies and model building these authors demontrated that the right handed helical backbone conformation with the bases in a left handed stack and with the base planes oriented parallel to the helix axis are energetically more favored than any of the alternative structures. They argued that the left handed base stack would explain the inverse cotton effect and maintained that the helical sense of the backbone remained right handed. At the double stranded level such a structure generates a right handed vertical double

helix[45] in which the base planes are parallel to the helix axis and the duplex is stabilized by vertical base stacking and familiar Watson-Crick hydrogen bonding. This is
absolutely an out of the ordinary and a novel double helix and vastly different in
gross morphology from the one advocated by Sundaralingam and Yathindra.[41] A
computer generated stereo space filling perspective of this vertical double helix is
shown in Plate 8. A striking feature of the vertical double helix is the large vacant inner core ($\simeq$ 36Å diameter) which is absent in B-DNA and is about 6-7Å in A-DNA
(Plate 9). This paper describes our experimental attempts to settle the controversy
between Olson[44/45] on the one hand and Sundaralingam and Yathindra[40/41] on the
other hand about the helical organization of the single and double stranded high anti
polynucleotides.

The Resolution of the Controversy

In the absence of single crystal x-ray data on short mini helices of high anti nucleic
acid structures we have attempted to resolve the controversy by NMR studies of such
systems. The high lights of our findings are summarized below. Details will be
presented elsewhere.[46]

Experimental

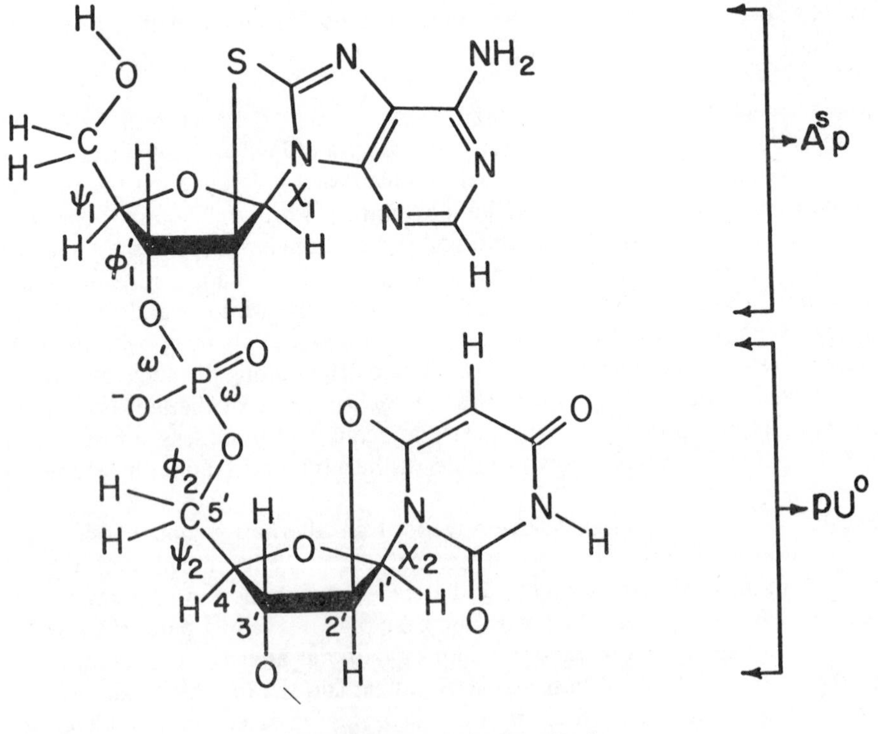

Figure 10. The structure of $A^s pU^o$.

The most difficult part of this investigation was the procurement of a nucleic acid constituent (i) in which the sugar base torsion can be chemically engineered to adopt a fixed value of $\simeq 120°$ (ii) in which a potentiality exists for the formation of double helices and (iii) whose NMR signals, particularly those from the base protons can be unambiguously assigned. All the above requirements are easily met by the cyclodinucleoside monophosphate A^{s}pU^o whose structure is shown in Figure 10. The A^{s}pU^o and the constituent mononucleotides A^{s}p and pU^o were synthesized in the laboratory of Ikehara and Uesugi.[46] r-ApU and d-pGpC and the constituent mononucleotides were purchased commercially. Pulsed ^{1}H NMR measurements in the time domain were performed using the Bruker 270 MHz FT system at the Southern New England High Field Facility, New Haven and the data were Fourier transformed to the frequency domain using a BNC 12 data system located at the laboratory in Albany. ^{31}P FT NMR measurements were made at 109.3 MHz at National Magnet Laboratory, MIT.

Evidence For the Formation of a Self-Complementary
Miniature Double Helix by A^{s}pU^o

The chemical shift data at 5°C and 20 mM indicated that the base proton H2 of the

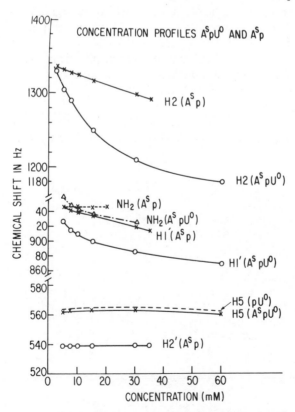

Figure 11. Concentration dependence of the chemical shifts of A^{s}pU^o and the mononucleotide A^{s}p.

A^{s}p- segment of A^{s}pU^o has moved markedly upfield (0.32 ppm) relative to the corresponding proton in the mononucleotide A^{s}p, while the chemical shift of the uridine H5 was little affected (Figure 11). This kind of a shift pattern for PUPY (PU = purine, PY = pyrimidine) dinucleoside monophosphates can be described as bizarre because such a situation has never been encountered in our detailed study of all the naturally occurring systems, i.e., ApC, ApU, GpC, GpU and the four corresponding deoxyribose systems.[47-53] In the naturally occurring PYPU and PUPY dinucleoside monophosphates one invariably observes the base protons of the PY residue shifting significantly to higher fields with very small upfield shifts for the base protons of th PU residue. These observations have been rationalized on the ground that PYPU and PUPY molecules show a general preference to exist in intramolecularly stacked arrays and in such an array the strong ring current shielding abilities of the purine causes significant upfield shifts of the pyrimidine protons and that the purine protons do move to higher fields only to a small extent because of the poor ring current effects of the pyrimidine systems. For example, in ApU, the H5 and H6 of -pU are shifted to higher fields by 0.374 and 0.263 ppm respectively whereas the adenine H2 and H8 of Ap- are affected in the range of 0.005 and 0.021 ppm respectively[48/53] but the situation has reversed in A^{s}pU^o.

The only way to rationalize the observed upfield shifts of H2 of A^{s}p- of A^{s}pU^o is to invoke adenine-adenine interactions. There are two ways in which this could happen: (i) aggregation of A^{s}pU^o molecules like all the nucleic acid constituents resulting in base stacking and column formation, (ii) formation of a Watson- Crick hydrogen bonded miniature double helix in which adenine of molecule 1 shields the adenine of molecule 2 and vice versa (Figure 12).

It has been shown[54] that in the adenine nucleotide systems intermolecular aggregations become dominant and cause marked upfield shifts only beond 50 mM concen-

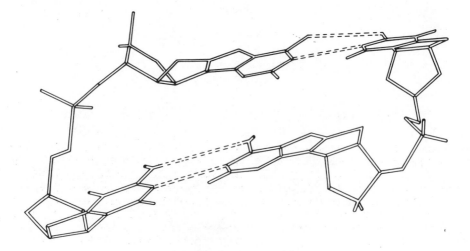

Figure 12. Self-complementary miniature double helix of A^{s}pU^o.

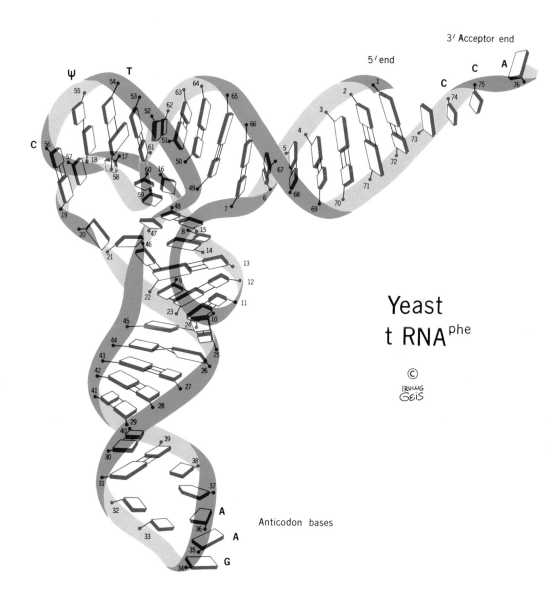

Plate 1. An illustration of the crystal structure of yeast phenylalanine transfer RNA. The phosphate ribose backbone is represented as a continuous blue ribbon with the bases attached to the light side of the ribbon. Bases are shown in red with purines as longer boards, pyrimidines as shorter boards and the highly modified base 37 as a pentagonal board. Hydrogen bonds between bases are shown with thin black lines. All the base pairs are of the Watson-Crick type with the following exceptions: 4-69, 8-14, 9-23, 10-45, 15-48, 18-55, 22-46, 26-44, and 54-58. Note especially the three way base pairing of 46-22-13, 9-23-12; the "propellor" twist of 44-25, and the "wobble" pair 4-69. Illustration by Irving Geis in collaboration with Sung-Hou Kim. Copyright 1979 by Irving Geis, 4700 Broadway, New York, New York 10040.

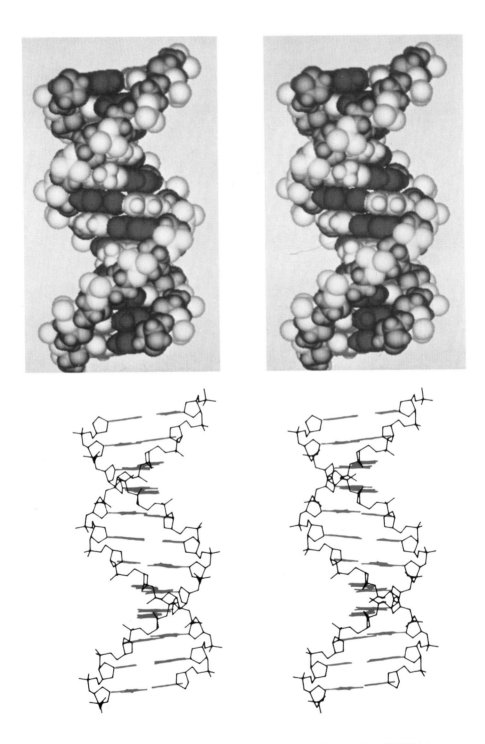

Plate 2. Computer generated space filling and the corresponding line drawing of B-DNA in stereo.

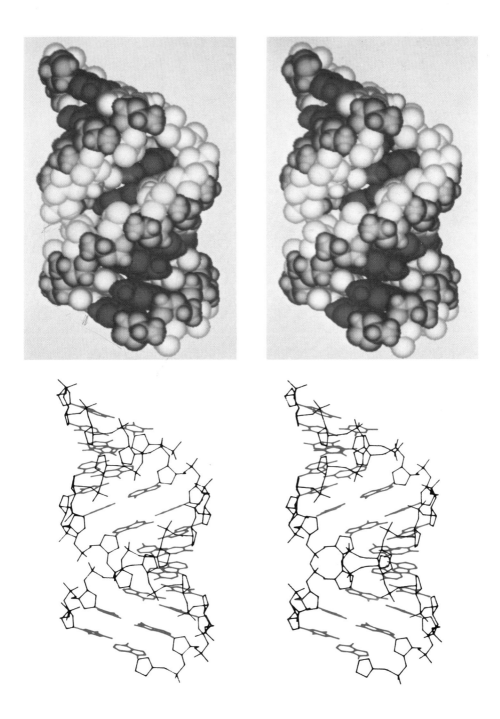

Plate 3. Computer generated space filling and the corresponding line drawing of A-DNA in stereo.

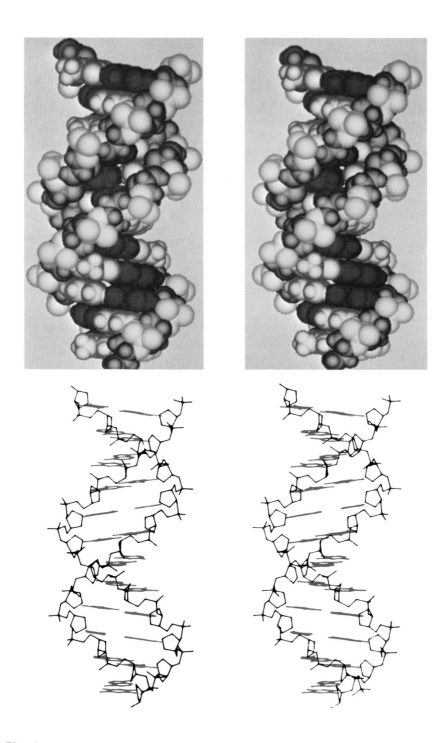

Plate 4. Computer generated space filling and the corresponding line drawing of C-DNA in stereo.

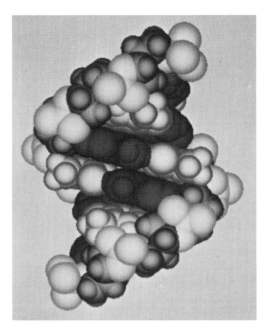
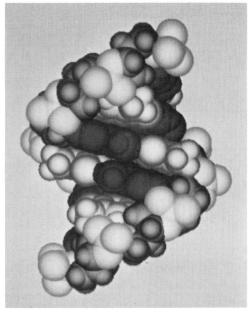
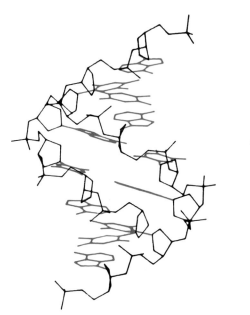
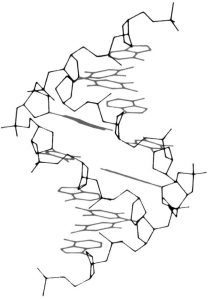

Plate 5. Computer generated space filling and the corresponding line drawing of D-DNA in stereo.

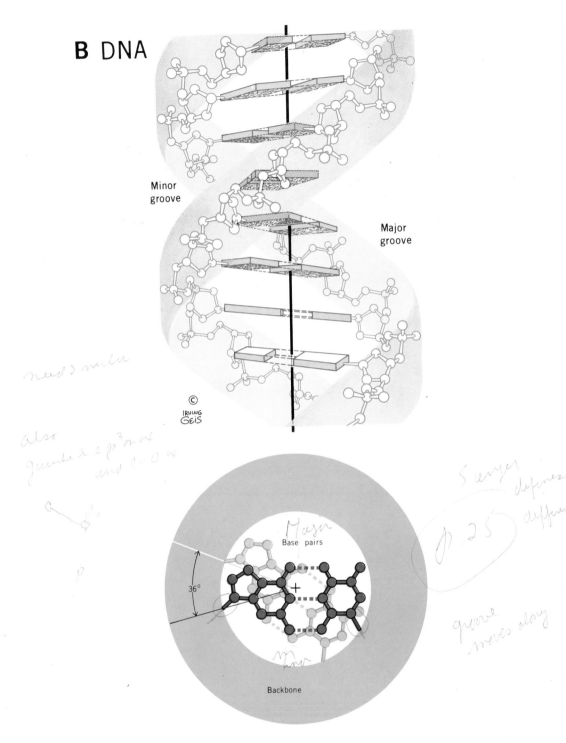

B DNA

Minor groove

Major groove

Base pairs

36°

+

Backbone

Plate 6. Double helical model of B-DNA, viewed perpendicular to the helical axis (top illustration). Base pairs are drawn in red and they are almost perpendicular to the helical axis. Ten base pairs make one complete turn. The phosphate deoxyribose backbone is enclosed in a blue ribbon. View along the helical axis (bottom illustration). Only two consecutive base pairs are shown. Note that the planes of base pairs intersect the helical axis.

A DNA

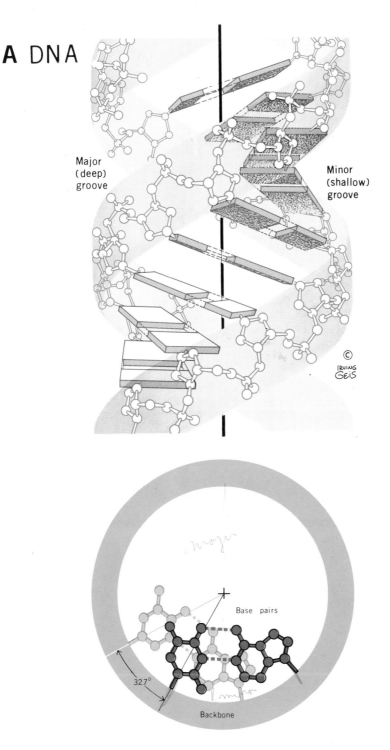

Major
(deep)
groove

Minor
(shallow)
groove

IRVING
GEIS

Base pairs

327°

Backbone

Plate 7. Double helical model of A-DNA, viewed perpendicular to the helical axis. (top illustration) Base pairs are shown in red and they are inclined to the helical axis. Eleven base pairs make one complete turn. As in the B-form, the phosphate deoxyribose backbone is enclosed in a blue ribbon. View along the helical axis (bottom illustration). Note that the base pairs go around the helical axis leaving the vertical axis open. Double helical A-RNA is very similar in conformation to A-DNA. Illustrations in these two pages, copyright by Irving Geis.

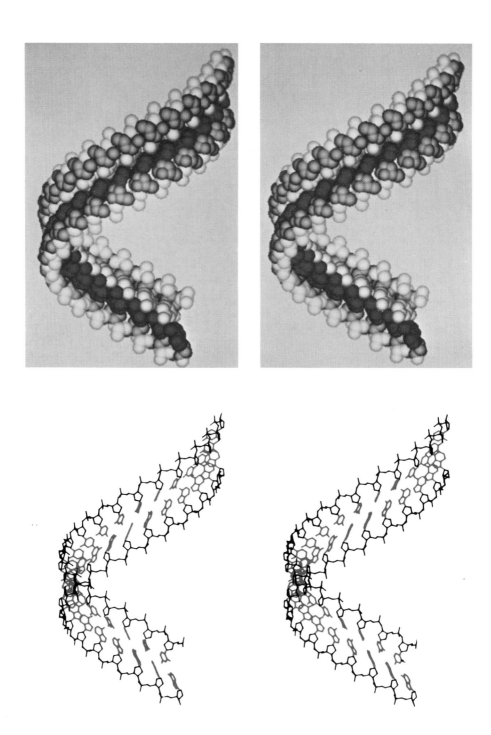

Plate 8. Computer generated space filling and the corresponding line drawing of the vertical double helix.

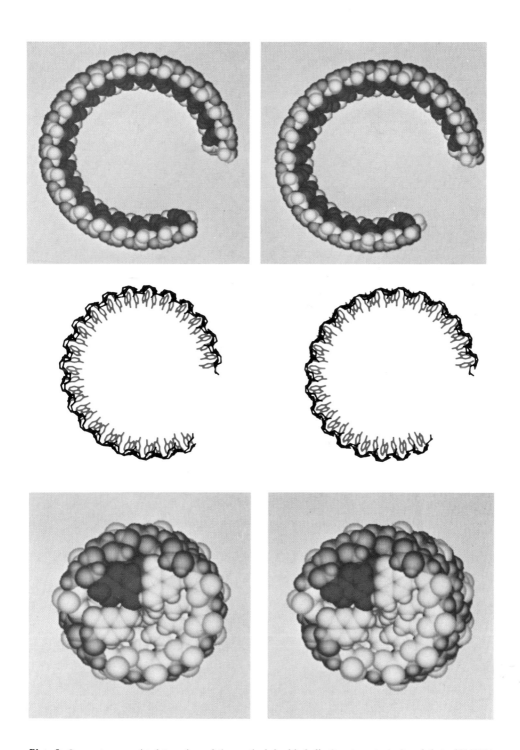

Plate 9. Computer generated top view of the vertical double helix (top two stereos) and that of B-DNA (bottom). Notice the large vacant inner core in the former and none in the latter.

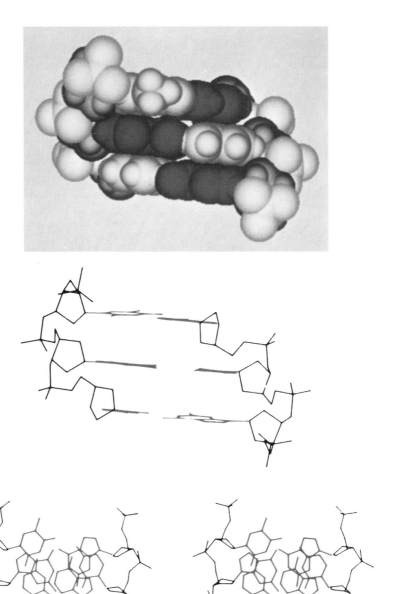

Plate 10. Computer generated geometrical relationships among the base pairs in a trimer duplex of B-DNA. The top two views are along the base planes and the bottom one is top view in stereo.

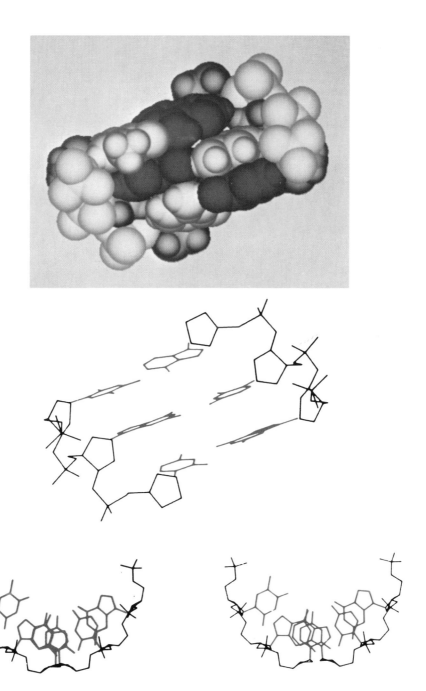

Plate 11. Computer generated geometrical relationships among the base pairs in a trimer duplex of A-DNA. Details same as in Plate 10.

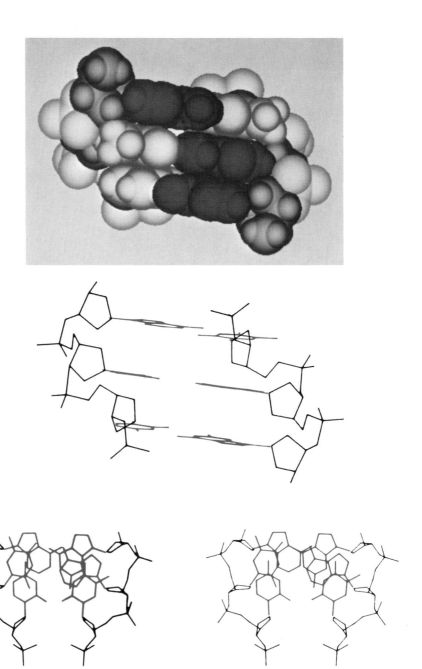

Plate 12. Computer generated geometrical relationships among the base pairs in a trimer duplex of C-DNA. Details same as in Plate 10.

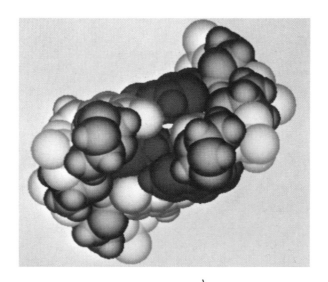

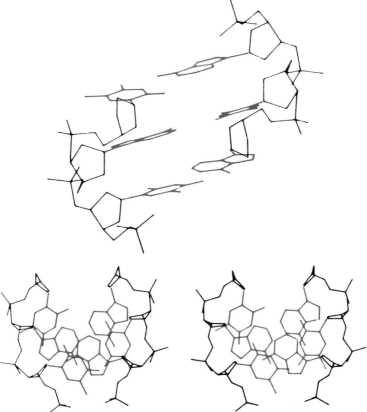

Plate 13. Computer generated geometrical relationships among the base pairs in a trimer duplex of D-DNA. Details same as in Plate 10.

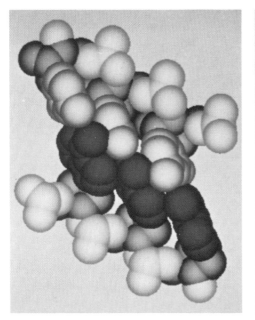
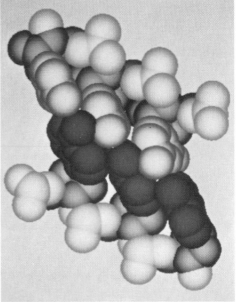
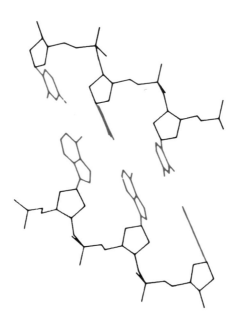

Plate 14. Computer generated duplex segment of Olson's vertical double helix.

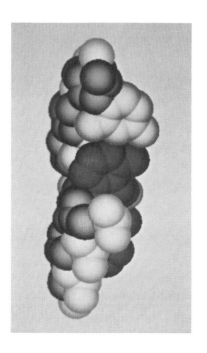
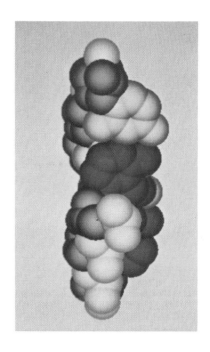

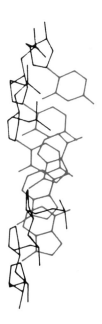
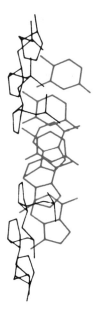

Plate 15. Same segment as in Plate 14, but the view more clearly reveals geometric relationships among the bases.

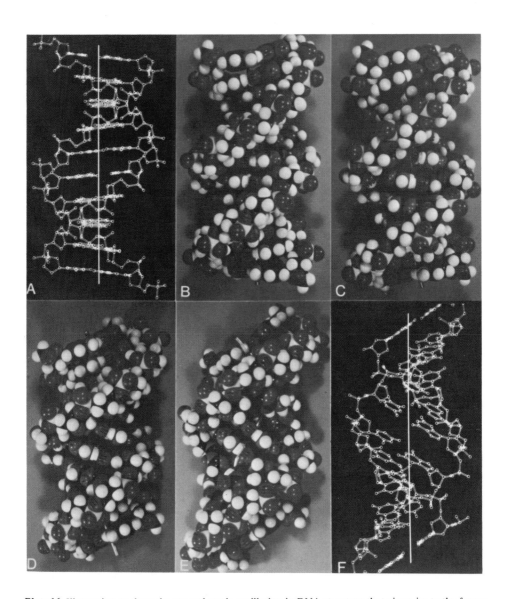

Plate 16. Illustration to show the normal mode oscillation in DNA structure that gives rise to the formation of β kinked DNA. A. B DNA, drawn by computer graphics. B. B DNA, as visualized with the Corey-Pauling-Koltun (CPK) space filling molecular models. C. Low amplitude normal mode oscillation. D. Higher amplitude normal mode oscillation. E. β kinked DNA structure. F. β kinked DNA, drawn by computer graphics. See text by Lozansky, Sobell and Lessen for detailed discussion.

trations. The concentration dependence of the chemical shifts of A^{s}pU^o and the mononucleotide A^{s}p (Figure 11) show that at even as low as 15 mM, the H2 of A^{s}pU^o is shifted to higher fields by as much as 66 Hz compared to the corresponding proton in the mononucleotide A^{s}p. In the concentration range of 5-30 mM at 5°C the H2 and H1' of the mononucleotide show a linear dependence and undergo an upfield shift of 30-35 Hz. In the case of A^{s}pU^o, the H2 and H1' of A^{s}p- part show a nonlinear dependence and undergo shifts to higher fields by 120 Hz (!) and 41 Hz respectively. The H5 of -pU^o of A^{s}pU^o is unaffected in the same concentration range. The concentration study indeed shows that weak and strong adenine- adenine interactions exist in the mononucleotide A^{s}p and the dinucleoside monophosphate A^{s}pU^o respectively and molecular topologies that cause these interactions are most likely different in the two systems. We interpret the upfield shifts with increasing concentration from 5 to 30 mM for the mononucleotide essentially reflect increasing aggregations, whereas the marked high field shift for A^{s}pU^o comes mostly from the increasing formation of adenine-adenine overlapping Watson-Crick base paired miniature double helices with increasing concentration from 5 to 30 mM. Krugh, *et al.*[55] and Young and Krugh,[56] by monitoring the shift value of the Watson-Crick hydrogen bonding protons, have demonstrated that at low temperatures self-complementary systems such as GpC, CpG, etc. form increasing amounts of double helices as the concentration is increased from 5 to 30 mM. Cross and Crothers[57] and Patel and coworkers[58-60] have shown that in systems which readily form double helical arrays, such as tetra-, penta-, and hexanucleotides, formation of double helical structures always results in large upfield shifts for the base protons as we have observed for A^{s}pU^o. The only difference is that our system is just a dinucleoside monophosphate. It appears likely that the rigid high anti χ_{CN} of $\simeq$ 120° in A^{s}pU^o renders the easy formation of a double helix stabilized by adenine-adenine interactions.

The direct evidence for the formation of miniature double helices in A^{s}pU^o comes from the actual experimenal observation of the helix-coil melting curve for A^{s}pU^o.

Helix Coil Transition in A^{s}pU^o

NMR spectroscopy has an advantage over such methods as circular dichroism, ultraviolet absorption and infrared spectroscopy to monitor the helix coil transition in polynucleotides because more than one chemical shift can be followed as a function of temperature. Patel and coworkers[58-60] Cross and Crothers[57] and Arter, *et al.*[61] have used proton chemical shifts to monitor the helix coil transition in tetra-, penta-, and hexanucleotides. From the sigmoidal curve, the melting temperature T$_{1/2}$ has been determined. The transition from coil to helix has been shown[57/58/61] to be accompanied by large chemical shift changes for some of the protons.

We have carried out our melting studies in 2.0 M LiCl (plus 0.01 M sodium cacodylate) the data are plotted in Figures 13 and 14 which show that the H2, H1' and H2' of A^{s}p- part and H1' of -pU^o part show the sigmoidal behavior but H5 and H2' of -pU^o does not. It is important to emphasize that all protons do not have to

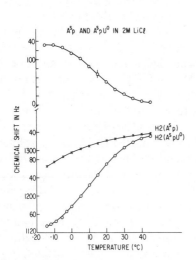

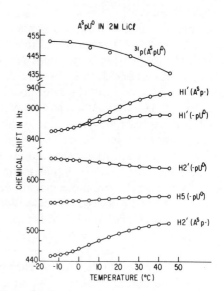

Figure 13. Temperature profiles of the chemical shifts of H2 of A^spU^o and A^sp in 2 M LiCl. The one on the top is the difference curve. Concentration 20 mM.

Figure 14. Temperature profiles of the chemical shifts of ^{31}P and protons other than H2 in A^spU^o in 2 M LiCl, Concentration 20 mM.

display the melting pattern. Chemical shift of a given proton is local geometry and environment dependent and formation of a helix from coil seriously perturbs the chemical shift of certain protons, some protons undergo little perturbations and in other cases a large number of factors such as ring current fields, electric field polarizations, atomic diamagnetic anisotropy, etc. interplay and compensate the shifts. A very important observation is that the ^{31}P chemical shifts (Figure 14) measured at a frequency of 109.3 MHz do not show any sigmoidal behavior and the total change in chemical shifts as a result of helix coil transition is only about 15 hertz. This indicates that during helix- coil transition there is little change in the O-P-O bond angle.[62/63]

One can derive the thermodynamic parameters for the helix-coil transition from the melting curve. In order to obtain good values, it is necessary to correct the melting curve from contributions of aggregation effects. This is particularly so when conditions such as low temperature and high salt concentrations which favor aggregations are employed. We have attempted to correct for this by determining the temperature dependence of the shift of adenine H2 in the mononucleotide A^sp on the assumption that aggregation and column formation in A^sp and A^spU^o may affect the shift values in the same way. This is the best one could do with NMR methodology. The data for A^sp in 2 M LiCl is presented in Figure 13. From the data for A^sp and A^spU^o, the difference curve is plotted and is shown in Figure 13 (top) and the curve is a clean melting curve with a transition temperature of 15°C in 2 M LiCl and the first observed melting curve for a dinucleoside monophosphate. Calculations based on Grella-Crothers[64] expression indicate that the helix coil transition of self-complementry A^spU^o is associated with a reaction enthalpy of 15.7 Kcal/mole. The equilibrium constant calculated by the methods in Patel[58] for the helix coil transition of 20 mM solution in 2 M LiCl at $T_{1/2} = 15°$ is 50 (i.e., $K_{formation}$). The magnitude of reaction enthalpy and the formation constant appears to be reasonable when compared to those reported for tetra-, pentanucleotides, etc.[57-61] In the absence of any such data for dinucleoside monophosphates, it is necessary to state that the reported thermodynamic quantities be treated qualitatively.

Solution of the Controversy and Evidence in Favor
of a Vertical Double Helix for High
Anti Polynucleotides

Evidence presented above unmistakably shows that the high anti self-complementary $A^s pU^o$ forms miniature double helices and in doing so the adenine H2 undergoes overwhelming shielding with little shielding of the H5 of $-pU^o$, i.e., the structure of the miniature double helix is characterized by strong interstrand adenine-adenine interaction with little intramolecular stacking between the adenine and uracil of the same strand, i.e., *significant interstrand base stacking and little intrastrand base stacking* (Figure 12).

Our experimental observation that under conditions in which χ_{CN} is fixed at $\simeq 120°$, $A^s pU^o$ forms miniature double helices with *significant interstrand base stacking* and *little intrastrand base stacking* enables to settle the question about the helix handedness of the duplexes of high anti polynucleotides. **Examination of the Sundaralingam- Yathindra model (Fig. 5ab in Reference 41) clearly reveals that in their model there is strong intrastrand and little interstrand base stacking. This is exactly the reverse of what is observed experimentally in our laboratory, unmistakably leading to the conclusion that the theoretical helix parameters that Sundaralingam and Yathindra[41] have derived from the rigid nucleotide concept is untenable,**and consequently their proposed molecular mechanics of untwisting a helix and switching the sense of the helix[41] are of questionable validity. The observed significant interstrand and little intrastrand base stacking in the miniature duplex of $A^s pU^o$ are also not a feature of A-DNA, B-DNA, C-DNA and D-DNA as is evident from Plates 10-13. In plates 14 and 15 are shown two exploded views of the Olson's right handed vertical double helix for poly(rA)•poly(rU), and one immediately notices that the conspicuous feature of this double helix is the significant interaction between interstrand bases and little intrastrand base-base interactions. **Thus our experimental observations that when χ_{CN} is fixed at $\simeq 120°$the self-complementary $A^s pU^o$ forms miniature double helices with significant interstrand and little intrastrand base-base interactions provide the first direct experimental support for the novel vertical double helix for high anti polynucleotides.**

The arguments that are presented in favor of a vertical double helix are qualitative based on the geometric relationships of the base pairs in a poly(rA)•poly(rU) duplex. However, our system is $A^s pU^o$ and the corresponding polymer is made of the self- complementary poly rArU system. Hence Olson's coordinates for a self-complementary → ApU UpA ← system was derived and these coordinates were

Table IV

(A1pU1)(A2pU2) Calculated and Observed Shifts

for Miniature Double Helix Effect of Adenine (A1)

Moiety	Proton	Calculated Shift in Parts Per Million (ppm)				Observed Shift (ppm)
		Ring Current	Diamagnetic anisotropy	Paramagnetic anisotropy	Total ∂_{cal}	∂_{obs}
A2	H-8	0.125	-0.005	0.050	0.170	--
	H-2	0.292	-0.005	0.099	0.386	0.47
Sugar (A2)	H-1'	0.305	0.003	0.091	0.399	0.14
	H-2'	0.096	0.001	0.049	0.146	0.19
Sugar (U2)	H-1'	-0.085	0.005	-0.017	-0.101	0.12
	H-2'	-0.048	0.001	-0.018	-0.064	-0.06
U2	H-5	-0.079	0.003	-0.047	-0.123	0.02
	H-6	-0.060	0.002	-0.002	-0.081	--
U1	H-5	-0.044	0.002	-0.014	-0.056	0.00
	H-6	-0.032	0.002	-0.004	-0.035	--

utilized to generate the miniature double helix conforming to Olson's torsion angles. This is shown in Figure 15 where the interstrand base stacking can be seen very clearly. For the structure shown in Figure 15 we have made detailed ring current and diamanetic and paramagnetic susceptibility anisotropy calculations. The experimentally observed and theoretically predicted shielding for a miniature double helix conforming to Olson's torsion angles are given in Table IV. Except for the H1'the agreement between the two is remarkable. The lack of agreement of H1' is a direct consequence of the presence of a sulphur or oxygen atom situated very near the H1' in the experimental system of $A^{s}pU^{o}$. From these studies it appears reasonable to conclude for a nucleic acid system if the sugar base torsion angle can adopt a high anti value, it most likely may take up the shape of the vertical double helix proposed by Olson.[45] It should be noted that our experiments disprove the left-handed model of Sundaralingam and Yathindra;[41] but it may be possible to generate left-handed models with significant interstrand and little intrastrand base stacking. This we have not studied.

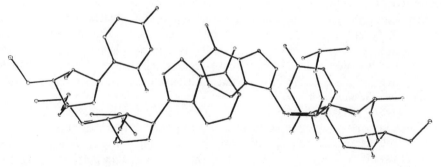

Figure 15. Self complementary miniature double helix of ApU conforming to the torsion angles of vertical double helix.

Vertical Double Helices for Naturally Occurring Polynucleotides?

In her opus on the novel vertical double helix[45] Olson advances it as a potential alternative ordered structure available to naturally occurring nucleic acid systems. If naturally occurring ApU, GpC, etc. have potentialities to become part of self- complementary vertical double helices, they should show proclivities to adopt a high anti sugar base torsion. Detailed NMR studies of all the naturally occurring dinucleoside monophosphates[47-53] indicate that they are flexible in aqueous solution and this flexibility gives rise to conformational pluralism,[49-51] i.e., they exist as a conformational blend in which certain ones predominate over the others. NMR methods have not been able to determine accurately the magnitude of χ_{CN} except that they occupy the anti domain in general. If ApU in aqueous solution has tendencies to adopt high anti χ_{CN}, one indeed would expect ApU to go into high anti double helical arrays like A^{s}pU^o and display shift patterns similar to A^{s}pU^o. In Figure 16, we have plotted the temperature dependence of H2, H8 and H1′ of Ap- and H5, H6 and H1′of -pU fragments of ApU under conditions identical to Figures 13 and 14 for A^{s}pU^o. A comparison of the data for A^{s}pU^o and ApU indicate that they adopt different spatial configurations and ApU has no proclivity whatsoever to adopt a χ_{CN} of $\simeq 120°$ and

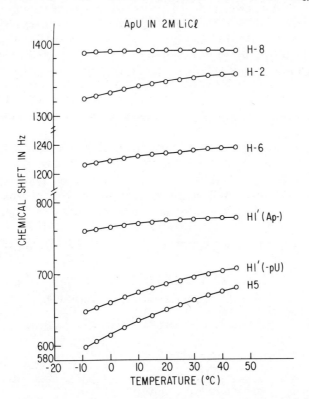

Figure 16. Temperature profiles of the chemical shifts of the various protons in ApU in 2 M LiCl. Concentration 20 mM.

enter into a double helical array similar to A^spU^o. Our argument is that if any real potential existed for ApU to adopt a $\chi_{CN} \simeq 120°$ the imperatives of molecular stereodynamics demand that it adopts such values under conditions which favor double helices formation because such values of χ_{CN} generate a double helix stabilized by adenine-adenine stacking interactions, so stabilizing is the interaction that one is able to observe the helix coil melting curve for the dinucleoside monophosphate. We have done identical experimens with d-pGpC and there was no indication of the formation of miniature double helices of the type in Figure 15. Our conclusion is that self-complementary ApU and d-pGpC in 2 M LiCl have no potentiality to form miniature double helices which can be considered part of an ordered vertical double helix for the corresponding polynucleotide.

This is not surprising because we have shown[47-53] that one of the important contributing conformers in the conformational blend of single stranded naturally occurring dinucleoside monophosphates in aqueous solution is a right handed stacked one (O3'-P and P-O5' torsions $\simeq 290°$ and $\simeq 290°$) with low values for χ_1 and χ_2. If the double helix of such a structure is made a repeating unit of a polynucleotide duplex, this will result in traditional double helical arrays. Our solution results are in eminent agreement with the single crystal X-ray results on dinucleotide monophosphates.[14] Obviously, in the absence of experimental data to the contrary, which are so difficult to obtain, one cannot dismiss the claims in which naturally occurring polynucleotides take up reasonable spatial configurations such as vertical, side by side, kinky, propellor twisted and so on under conditions in which they interact in vivo with other biological structures. And it will be long before one would know the truth.

Acknowledgment

The author thanks   National Cancer Institute DHEW (CA12462) and National Science Foundation (PCM-7822531) for the support of this investigation. The authors are deeply indebted to Irving Geis for the artistic rendition of A-DNA, B-DNA, and tRNA. We thank Wilma Olson for supplying the coordinates and Bernard Pullman and Giessner-Prettre for programs to some of the complex calculations. We are indebted to M. Ikehara for a sample of A^spU^o. The author thanks Alexander Rich for providing a preprint of reference 36 and the figures

References and Footnotes

1. Judson, J. R. *Eighth Day of Creation*, Simon and Schuster, New York (1979).
2. Watson, J.D., and Crick, F.H.C. *Nature 171*, 737 (1953).
3. Rodley, G.A., Scobie, R.S., Bates, R.H.T., and Lewitt, R.M., *Proc. Natl. Acad. Sci. USA 73*, 2959 (1976).
4. Sasisekharan, V. an Pattabiraman, N. *Curr. Sci. 45*, 779 (1976).
5. Sasisekaran, V., Pattabiraman, N. and Gupta, G. *Proc. Natl. Acad. Sci. USA 75*, 4092 (1978) and references therein.
6. Cyriax, B. and Grath, R., *Naturwissenschaften 65*, 106 (1978).
7. Pohl, W. F. and Roberts, G.W. *J. Math. Biol. 6*, 383 (1978).
8. Arnott, S., *Nature 278*, 780 (1979).

9. Crick, F.H.C., Wang, J.C., and Bauer, W.R., *J. Mol. Biol. 129*, 449 (1979).
10. Crick, F.H.C. *Proc. Natl. Acad. Sci., USA 73*, 2639 (1976).
11. White, J.H. *Amer. J. Math. 41*, 693 (1969).
12. Fuller, F.B. *Proc. Natl. Acad. Sci. USA 68*, 814 (1971).
13. The definition is such that for a left-handed double helix, the linking number will be negative.
14. Seeman, N.C., in *Nucleic Acid Geometry and Dynamics*, Ed., Sarma, R.H., Pergamon Press, New York, Oxford ₁ p. 109 (1980).
15. Sobell, H.M., in *Nucleic Acid Geometry and Dynamics*, Ed., Sarma, R.H., Pergamon Press, New York, Oxford p.289 (1980).
16. Rich, A., Quigley, G.J., and Wang, A.H-J. in *Stereodynamics of Molecular Systems*, Ed., Sarma, R.H., Pergamon Press, New York, Oxford p. 315 (1979).
17. Berman, H.M., and Neidle, S., in *Stereodynamics of Molecular Systems*, Ed., Sarma, R.H., Pergamon Press, New York, Oxford p. 367 (1979).
18. Quigley, G. I., Seeman, N. C., Wang, A. H. J., Suddath, F. L. and Rich, A., *Nucleic Acids Research 2*, 2329 (1975).
19. Sussman, J. L., Kim, S. H., *Biochem. Biophys. Res. Commun., 68*, 89 (1976).
20. Ladner, J. E., Jack, A., Robertus, J. D., Brown, R. S., Rhodes, D., Clark, B. F. C., and Klug, A., *Nucleic Acids Research 2*, 1623 (1975).
21. Stout, C. D., Mizuno, H., Rubin, J. Brennen, T., Rao, S. T., and Sundaralingam, M., *Nucleic Acids Research 3*, 1111 (1976).
22. Schevitz, R. W., Podjarny, A. D., Krishnamachari, N., Hughes, J. J., Sigler, P. B. and Sussman, J. L. *Nature 278*, 188 (1979).
23. Franklin, R.E. and Gosling, R.G. *Nature 171*, 740 (1953).
24. Franklin, R.E. and Gosling, R.G. *Acta Cryst. 6*, 637 (1953).
25. Arnott, S., Dover, S. D., and Wonacott, A. J., *Acta Cryst., 25*, 2192 (1969).
26. Arnott, S., Hukins, D. W. L., Dover, S. D., Fuller, W., Hodgson, A. R. *J. Mol. Biol. 81*, 107 (1973).
27. (a) Arnott, S., Chandrasekaran, R., Hukins, D. W. J., Smith, R. S. C., and Watts, L., *J. Mol. Biol., 88*, 523 (1974). (b) Arnott, S., and Selsing, E., *J. Mol. Biol. 98*, 265 (1975).
28. Marvin, D. A., Spencer, M., Wilkins, M. F. H. and Hamilton, L. D. *J. Mol. Biol. 3*, 547 (1961).
29. Levitt, M., *Proc. Natl. Acad. Sci. USA 75*, 640 (1978).
30. Crothers, D.M., Dattagupa, N., and Hogan, M. in *Stereodynamics of Molecular Systems*, Ed., Sarma, R.H., Pergamon Press, New York, Oxford, p. 383 (1979).
31. Klug, A., Jack, A., Viswamitra, M.A., Kennard, O., Shakked, Z. and Steitz, T.A., *J. Mol. Biol. 131*, 669 (1979).
32. Viswamitra, M.A., Kennard, O., Shakked, Z., Jones, P.G., Sheldrick, G.M., Salisbury, S., and Falvello, L. *Nature 273*, 687 (1978).
33. Patel, D.J., in *Nucleic Acid Geometry and Dynamics*, Ed., Sarma, R.H., Pergamon Press, New York, Oxford p. 185 (1980).
34. Sarma, R.H., in *Nucleic Acid Geometry and Dynamics*, Ed., Sarma, R.H., Pergamon Press, New York, Oxford p. 1 (1980).
35. Dhingra, M.M., Sarma, R.H., Uesugi, S., and Ikehara, M., *J. Am. Chem. Soc. 100*, 4669 (1978).
36. Wang, A. H-J., Quigley, G.J., Kolpak, F., Crawford, J.L., van Boom, J.H., van der Marel, G., and Rich, A. *Nature* (in press).
37. Ikehara, M., and Uesugi, S., *J. Am. Chem. Soc., 94*, 9189 (1972).
38. Ikehara, M., and Tezuka, T., *J. Am. Chem. Soc., 95*, 4054 (1973).
39. Fujii, S., and Tomita, K., *Nucleic Acids Research, 3*, 1973 (1976).
40. Yathindra, N., and Sundaralingam, M., *Nucleic Acids Research, 3*, 729 (1976).
41. Sundaralingam, M., and Yathindra, N., *International J. Quant. Chem. QBS 4*, 285 (1977).
42. Evans, F.E. and Sarma, R.H. *Nature 263*, 567 (1976).
43. Neidle, S., Taylor, G., Sanderson, M., Huey- Sheng, S., and Berman, H. M., *Nucleic Acids Research, 5*, 4417 (1978).
44. Olson, W. K., and Dasika, R. D., *J. Am. Chem. Soc., 98*, 5371 (1976).
45. Olson, W. K., *Proc. Natl. Acad. Sci. USA, 74*, 1775 (1977).
46. Dhingra, M. M., Sarma, R. H., Uesugi, S., and Ikehara, M. (submitted).

47. Lee, C. H., Ezra, F. S., Kondo, N. S., Sarma, R. H., and Danyluk, S. S., *Biochemistry 15*, 3627 (1976).
48. Ezra, F. S., Lee, C. H., Kondo, N. S., Danyluk, S. S., and Sarma, R. H., *Biochemistry, 16*, 1977 (1977).
49. Sarma, R. H., and Danyluk, S. S., *International J. Quant. Chem. QBS 4*, 269 (1977).
50. Dhingra, M. M., and Sarma, R. H., "Proceedings, International Symposium on Biomolecular Structure, Conformation, Function and Evolution" R. Srinivasan (Ed.), Pergamon Press, Inc. (in press).
51. Dhingra, M. M., and Sarma, R. H., *International J. Quant. Chem. QBS 6* (in press).
52. Cheng, D. M., and Sarma, R. H., *J. Am. Chem. Soc., 99*, 7333 (1977).
53. Dhingra, M. M., and Sarma, R. H., Giessner- Prettre, C., and Pullman, B., *Biochemistry, 17*, 5815 (1978).
54. Evans, F. E., and Sarma, R. H., *Biopolymers, 13*, 2117 (1974).
55. Krugh, T. R., Laing, J. W., and Young, M. A., *Biochemistry, 15*, 1224 (1976).
56. Young, M. A., and Krugh, T. R., *Biochemistry, 14*, 4841 (1975).
57. Cross, A. D., and Crothers, D. M., *Biochemistry, 10*, 4015 (1971).
58. Patel, D. J., *Biochemistry, 14*, 3984 (1975).
59. Patel, D. J., and Canuel, L., *Proc. Nat. Acad. Sci. USA, 73*, 647 (1976).
60. Patel, D. J., and Tonelli, A. E., *Biochemistry, 14*, 3990 (1975).
61. Arter, D. B., Walker, G. C., Uhelenbeck, O. C., and Schmidt, P. G., *Biochem. Biophys. Res. Commun., 61*, 1089 (1974).
62. Ribas Prado, F., Giessner-Prettre, C., Pullman, B., and Daudey, J-P., *J. Am. Chem. Soc.* (in press).
63. Gorenstein, D. G., "Nuclear Magnetic Resonance Spectroscopy in Molecular Biology" B. Pullman (Ed.) D. Reidel Publishing Company, pp. 1-15 (1978).
64. Grella, J., and Crothers, D. M., *J. Mol. Biol., 78*, 301 (1973).

Crystallographic Investigation of Oligonucleotide Structure

Nadrian C. Seeman
Institute of Biomolecular Stereodynamics,
Center for Biological Macromolecules
Department of Biological Sciences
State University of New York at Albany
Albany, New York 12222

In Science, Address the Few ...
Edward Bulwer-Lytton

Introduction

Nucleic acids are the genetic material of all living organisms. As such, there is intrinsic interest in their structures: Further, since the means by which the information is contained and transmitted by nucleic acid molecules was first deduced from structural principles,[1] the three-dimensional structures of these molecules are of particular interest. Until 1971, the only data that were available on the structures of nucleic acids were low resolution fiber diffraction patterns, and single crystal structures of nucleosides and nucleotides. Since that time, a small number of oligonucleotide structures have been determined by high-resolution single crystal X-ray diffraction analysis, from both the RNA and DNA series. The prominent features of these structures will be summarized in this chapter. We shall also try to emphasize the biological relevance of the quantitative effects which can be seen as trends when the structures are surveyed. An understanding of the material in earlier chapters on the methodology of single crystal crystallography[41] and particularly nucleic acid nomenclature[42] is necessary to appreciate what follows.

It is well known that the genetic information of paramount importance to life is linearly encrypted in the sequence of bases. Thus, the independent unit of genetic information is the mononucleotide. However, because the bases influence each other very strongly, it is not sufficient to study mononucleotides when examining the structural aspects of nucleic acids. In order to consider all relevant parameters, one must include the internucleoside linkage in the study. The minimum portion of a nucleic acid which contains this is the dinucleoside phosphate, which consists of a 5' base, its sugar, the internucleoside linkage, a 3' sugar and its base. Indeed, many of the structural properties of nucleic acids seem to rely less upon the individual bases than upon sequential aspects of the covalent structure. For example, stacking and intercalation properties appear to be dependent on the dinucleoside sequence being examined. The structures of five RNA isolated fragments[2-10] and three isolated DNA fragments[11-13] have been reported. Numerous other fragments have been determined which included intercalative molecules, but these are covered by other chapters within this volume. Only one of the DNA fragments has been reported in complete detail,[11] so the treatment of DNA fragments will be much less comprehensive than that of RNA.

The majority of structures which have been reported contain self-complementary sequences. These sequences were crystallized in the hope that they would form Watson-Crick base paired double helical structures within the crystals. In most of these cases, that has been at least partially true, when crystallization conditions have permitted such pairing. This is in contrast to the situation with monomeric fragments of nucleic acids, particularly those containing A and U or T, which indicates that the backbone constraints tend to promote this form of interaction. In two cases, dpTpT[11] and ApA⁺pA⁺,[9] Watson-Crick base pairing was not possible, and in the case of UpA, the acidic pH of the crystallization conditions also prevented Watson-Crick base pairing.[2-4]

The reader should be aware that structure as described by the crystallographer within a single crystal, is structure within a defined environment, or field. This field constitutes the lattice environment of the molecule under examination. It should be recalled that an observation of the structure within this field is both time averaged (over small oscillations) and space averaged (over the entire crystal).[14] Just because a given material crystallizes with a given conformation, this does not necessarily mean that this is the most favorable conformation under all possible conditions. Rather, it is the favored structure for the formation of the given crystal, under the given

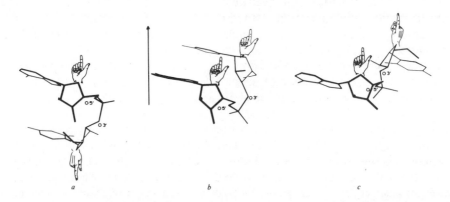

a   *b*   *c*

Figure 1. Views of (a) UpA1, (b) helical RNA with eleven base pairs per turn, and (c) UpA2, positioned with similar orientations on the adenosine portions. In order to clarify the conformational differences between these molecules, a hand has been attached to the 04′ ring oxygen of each ribose. The index finger is indicating the direction of the helix axis for the helix which would include each nucleoside. Note that in (b) the fingers point in the same direction, parallel to the helix axis, which is indicated by the arrow. In (a) it can be seen that the hands, and therefore the riboses of UpA1 are pointing in opposite directions. In (c), the hands point in the same direction, but are rotated differently from those in (b). Thus, the UpA2 molecule shown in (c) is also helical, but it forms a different helix from the one shown in (b). A rotation of 110° of the uridine portion of (c) will align the molecule so that it looks like the RNA-11 fragment shown in (b). Taken from reference 2.

crystallization conditions that prevail at the time that the material leaves solution and becomes part of the crystal. The presence of proteins, variations in the composition of the solution, or even a different crystal lattice may alter the observed conformation of a molecule. For this reason, those interested in biological structure are dependent upon solution probes of structure, such as NMR and various forms of light spectroscopy, in order to complement the results of crystallography. These techniques can yield information about the deviations from, and distortions of, the crystallographically derived structure under solution conditions. Furthermore, understanding the dependence of the structure upon such factors as ionic strength, pH and temperature is crucial to comprehending the complete dynamic structure of nucleic acids. These solution techniques are most powerful, however, when a static crystallographic structure is used as a starting point; there is no substitute for crystallography in determining the three-dimensional coordinates of any molecule at high resolution.

This review will proceed in the following way: We will describe each of the structures briefly: then we will then return to a survey of the common properties of the structures, and some of the generalizations that may be drawn.

The Individual Structures

1. Structures without Watson-Crick base pairing.

a. UpA[2-4]

Although Shefter and his colleagues determined the crystal structure of A(2′,5′)pU[15] in the late 1960's, the first crystal structure of a naturally occurring dinucleoside phosphate, UpA, was not determined until 1971.[12] The structure was of the acid form of UpA, in which the dinucleoside phosphate crystallized as the zwitterionic hemihydrate, with two independent molecules incorporated within the lattice. The negative charge rested, as usual, on the phosphate, while the positive charge was manifest as protonation of N1 of adenine. This protonation precluded the possibility of Watson-Crick base pairing in this structure, so that a mini-double helix was not one of the structures available to acidic UpA. The two molecules were in different conformations, differing both from each other and from the double helical sort of structure, as well, as seen in Figure 1. In that figure (a) shows the conformation of the UpA1 molecule, (b) corresponds to a helical dinucleoside phosphate, and (c) depicts the UpA2 molecule. The hands have been covalently attached to the 04′ ring oxygen atoms of the ribose sugars, and point in the direction that a conventional helix containing that nucleoside would point. The adenosine portions of the molecules are oriented similarly. It can be seen from the figure that any helices attached to the nucleosides in the UpA1 molecule would be pointing in opposite directions, while those similarly attached to UpA2 would be pointing in the same direction. Although useful as a model building aid in building loop structures,[16] the UpA1 conformation was not seen in the structure of Yeast tRNA[phe]. This crystal structure was unique in its low degree of hydration, being only the hemihydrate; this indicates that the field experienced by the molecules was somewhat different from that which exists in solution.

The salient feature of the UpA structure was that the conformations of all four of the independent nucleosides within the crystal were virtually identical. The only major differences between the two molecules were the torsion angles about the phosphoester bonds. All four nucleosides were in the C3'-*endo* conformation, χ_{CN} was in the low-*anti* region, and the interphosphate torsion angles, except for those involving the phosphoester bonds were very similar (see Table I). From examination of this structure, certain investigators were led to claim that the nucleotide, at least in the RNA case, was likely to have a small number of conformational degrees of

Table I
Backbone Torsion Angles For The Fragments
From Oligonucleotide Structures

	$\zeta(5')$ C4'-C3'	α C3'-O3'	β O3'-P	γ P-O5'	δ O5'-C5'	ϵ C5'-C4'	$\zeta(3')$ C4'-C3'	Ref.
1. Double Helical Conformations								
GpC	89	211	292	285	184	50	77	8
GpCl	79	222	294	291	181	47	79	10
GpC2	73	217	291	293	172	57	80	10
GpC3	96	224	290	286	167	63	74	10
GpC4	88	216	293	288	177	57	74	6
ApU1	84	213	293	288	177	57	74	6
ApU2	78	221	284	295	168	58	77	6
ApA⁺	82	223	283	297	160	53	81	9
ApT (1)	90	213	294	293	176	68	134	12
ApT (2)	83	212	284	302	171	64	139	12
2. Single Helical Conformations								
UpA(2)	77	224	164	271	192	54	93	3
TpT	157	252	163	288	187	41	158	11
TpA	134	204	168	286	186	49	83	12
3. Helix Reversing Conformations								
UpA (1)	86	206	81	82	203	55	85	3
A⁺pA⁺	81	207	76	92	186	56	79	9
4. Fiber Studies								
ARNA	95	202	294	294	186	49	95	32
ADNA	83	178	313	285	208	45	83	33
BDNA	156	155	264	314	214	36	156	34
CDNA	141	211	212	315	143	48	141	35
DDNA	156	141	260	298	208	69	156	36

freedom, similar to the situation involving the peptide linkage. This statement was made with both greater[4] and lesser[16] degrees of enthusiasm, with one group even coining the term, "rigid nucleotide,"[4] a concept which is not literally true.[17/18] However, there is a small number of discrete conformations which do dominate nucleotide structure, without forming an exclusive set: most standard nucleotides are found in either the C2'-*endo* or C3'-*endo* conformation, within the polynucleotide context, and dynamic equillibria do exist within solution; this is discussed in detail by Sarma[43] in the next chapter. This proper understanding of the conformational similarity of nucleotide units was of great use in the interpretation of the 3Å electron density map of Yeast tRNA[phe]. The constraints imposed by this model aided the interpretation of the electron density map in several key areas.[19]

The base pairing seen in the UpA structure is shown in Figure 2. The adenines paired with each other in the form proposed for acidic poly-A,[20] as shown in Figure 2a. The reader should note that the riboses on the two pairing adenosines are parallel, and that there is a two-fold axis perpendicular to the plane of the bases. The uracil-uracil pairing shown in Figure 2b had not been seen previously. Note that the riboses of the two uridines are in an anti-parallel configuration; the symmetry operator which relates the two uridine nucleosides has its rotation axis contained approximately within the plane of the bases.

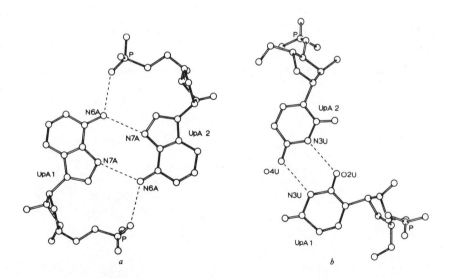

Figure 2. Base pairing seen in UpA. (a) shows the adenines paired with each other, while (b) shows the uracils paired with each other. There was thus no pairing between the adenines and the uracils. The pairing between the adenines is like that seen in the structure[9] of ApA⁺A⁺ and that proposed for the structure of acid poly-A.[20] Note that there is a twofold axis perpendicular to the plane of the paper relating the two adenosine-5' mononucleotides shown. The two uridines shown in (b) are paired in a fashion not previously seen. These nucleosides are related by a symmetry element whose axis lies within the plane of the paper. Taken from reference 2.

b. ApA⁺pA⁺⁹

This acidic trimer was also crystallized in the absence of cations. Because there are only two phosphates in this molecule, there are only two adenosines that must be protonated to balance the negative charge of the phosphates. This protonation is again at the N1 position, as was the case in UpA. The authors have concluded that the positive charges reside on the two adenosines at the 3′ end of the molecule. It was expected that the molecule would form a double helical dimer of the type proposed from fiber diffraction studies for acidic poly-A.[20] The fiber studies indicated a double helical complex of two parallel molecules, which had a two-fold axis relating the two strands, coincident with the helix axis. The A-A pairing which had been seen in the UpA structure was seen locally within the ApApA crystal structure. Adenosine 2

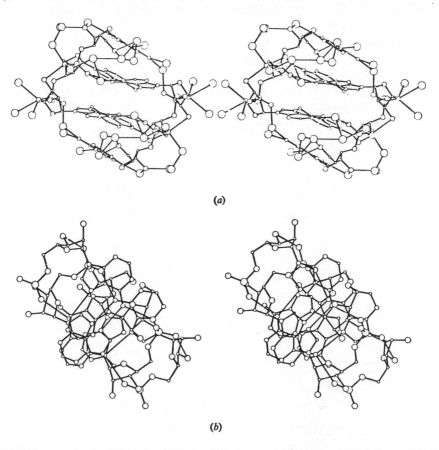

(a)

(b)

Figure 3. Stereoscopic views of the ApA⁺A⁺ dimer. (a) The two-fold axis relating the two trimeric oligonucleotides is perpendicular to the paper. The reader will note the two A-A base pairs which stabilize this complex. Note the reversal of chain direction in the internucleoside linkages between them. This results in the two riboses on each side of each base pair being parallel rather than anti-parallel. (b) The two-fold axis is in the plane of the paper. The largest circles represent water molecules. Taken from reference 9.

of one ApA⁺pA⁺ molecule paired with adenosine 3 of another ApA⁺pA⁺ molecule related to it by crystallographic two-fold symmetry. This is shown in Figure 3, which is a stereo view of the dimeric complex. The adenosine at the 5′ end of the molecule is not involved in base pairing. In order for the riboses of both adenosine 2 and adenosine 3 (in the two-fold related molecule), paired to it, to be parallel, the directions of the chains must be reversed within the internucleotide linkage which connects them. This is shown in Figure 4, where the 3′ terminal adenosine has been highlighted. Thus, as indicated in Figure 3, there is a two-fold axis, between the two sides of the complex, which is perpendicular to planes of the bases, rather than parallel with them as expected. The linkage which accomplishes this reversal of the riboses is qualitatively similar to the one seen in the structure of UpA1 (See Table I). The other linkage is qualitatively helical, while the nucleosides all adopt the C3′-*endo* conformation. The χ_{CN} values are all in the low-*anti* range.

c. The Sodium Salt of d-pTpT[11]

This is the only structure in the deoxy series for which full details have yet been published. This molecule was crystallized as the sodium salt and was the first structure of a dinucleotide to be determined, as well as the first structure of a dimer in the

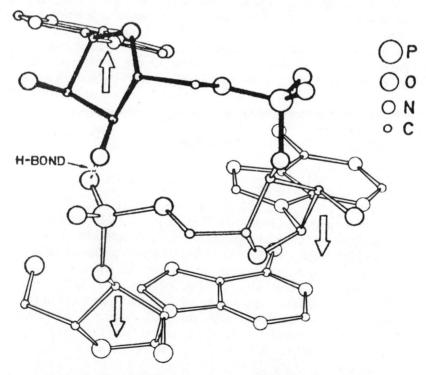

Figure 4. The ApA⁺pA⁺ monomer. The middle and bottom adenosines are related by a helical arrangement, while the 3′ terminal adenosine is pointing in the opposite direction, as indicated by the arrows. A stabilizing hydrogen bond is also indicated. Taken from reference 9.

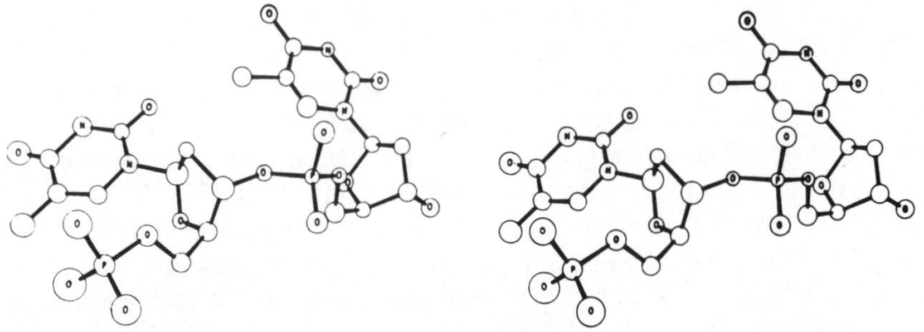

Figure 5. A stereoscopic plot of the d-pTpT molecule. The 5′ terminal nucleotide is on the left, while the 3′ terminal nucleotide is on the right. The relationship between the two portions of the molecule may be noted by inspection of the sugar and base orientations. Taken from reference 11.

deoxy series. The most striking aspect of this structure was the virtual identity of the two nucleotides, dpT1 and dpT2. They are almost exactly superimposable: indeed, the average difference between positions of equivalent atoms is 0.16Å, the largest difference being for the methyl group carbon atoms, 0.32Å. This is illustrated in Figure 5, a stereo pair, showing the conformation of the dinucleotide unit. The similarity is emphasized in Figure 6, which shows the two nucleotides aligned parallel to one another. The backbone torsion angles (see Table I) were very similar to those seen in the UpA2 molecule. However, the structure is quite different from the UpA2 structure, because the nucleosides are in the C2′-*endo* conformation, rather than the C3′-*endo* conformation. This structure was the first example of the C2′-*endo* conformation in the oligonucleotide series. Surprisingly, there was no base pairing at all within this crystal structure. The bases interacted with the sodium cations and with water molecules, but did not pair with themselves.

2. Structures Containing Watson-Crick Base Pairs.

a. The Sodium Salt of ApU[5/6]

The sodium salt of ApU crystallizes as a miniature right-handed anti-parallel RNA double helix within the crystalline lattice. Because it has a self-complementary sequence, two molecules can pair to form the miniature double helix. This particular structure was the first direct visualization of A-U base pairing within a single crystal. It was also the first opportunity to obtain direct quantification of a number of the important parameters associated with the RNA double helix, at the high resolution characteristic of such studies. There was no model building involved in the determination of this crystal structure, and it was minimally inferential in the derivation of the double helical structure. Therefore, it was the first high resolution crystallographic evidence that RNA indeed forms Watson-Crick right-handed antiparallel double helices, as a stable energy minimum. Two views of the structure are shown in Figure 7. The reader will note from Figure 7a that the pairing between the adenines and uracils is of the Watson-Crick type. It may be further noted from that view that there is a rotational relationship between the two ribose molecules on each strand. This is a vital component of the helical relationship which exists between the two nucleosides that compose the short RNA chain. This rotational relationship and the translational relationship are both evident in Figure 7b; this is a view approximately perpendicular to the helix axis, which is vertical in the figure. The reader is looking down the non-crystallographic two-fold axis which relates the two strands of the double helix. The conformational parameters of this structure are not very different from those predicted from fiber diffraction studies of RNA double helices.[21] The nucleoside conformations were also similar to the C3'-*endo*, low-*anti* structures seen elsewhere. The important point about this structure is that it was derived without recourse to the assumptions necessary in fiber work, due to the limited nature of that sort of data. A sodium ion is prominently shown in Figure 7: it

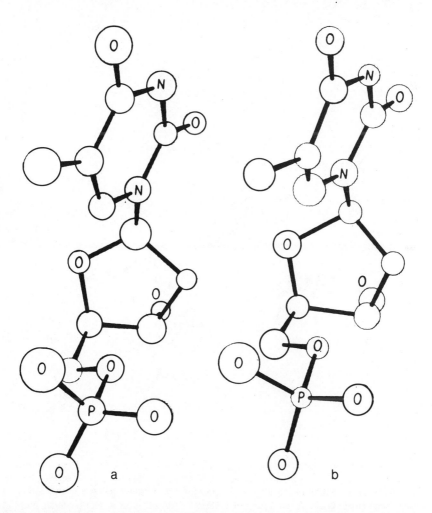

Figure 6. Comparison of the two 5'-mononucleotides which compose d-pTpT. (a) is the 5' terminal nucleotide, while (b) is the 3'-terminal nucleotide. The conformational identity of the two nucleotides is evident from the figure. Taken from reference 11.

is on the dyad axis, in the minor groove, complexed to the two uracil oxygens which do not partake of the Watson-Crick base pairing. Another sodium ion is complexed to the phosphates. A view of the lattice showing the hydration of the structure is

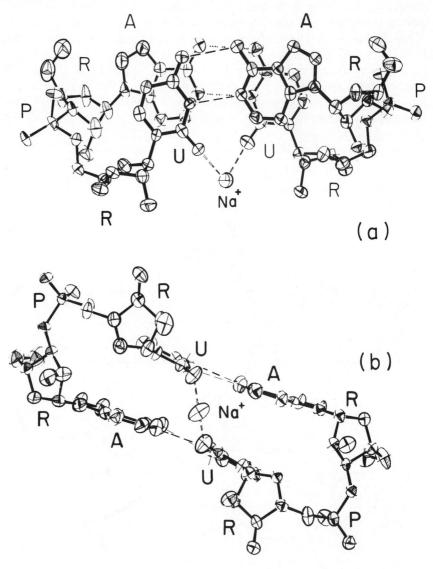

Figure 7. Molecular structure of the ApU double helix. This is an ORTEP[40] plot of the ApU structure. Each atom is represented by an ellipsoid which shows the thermal motion of the atom. Hydrogen bonds and coordination of the sodium ion are shown as broken lines. A is adenine, R is ribose, P is phosphate and U is uracil. (a) A view of the double helical complex approximately perpendicular to the plane of the bases. The major groove of the double helix is at the top of the figure, while the minor groove is at the bottom. (b) A view of the structure down the dyad axis in which the helical axis is vertical and the bases are viewed edge on. Taken from reference 6.

shown in Figure 8. ApU crystallized on the hexahydrate, with 12 independent water molecules in the lattice.

b. The Sodium Salt of GpC[32]

The sodium salt of GpC is very similar to the sodium salt of ApU, as seen in Figure 9. The same views of the GpC molecules are shown in Figure 9 as were shown for ApU in Figure 7. The conformational parameters are very similar to those of ApU, although the helical parameters are slightly different. This self-complementary sequence also crystallizes in a miniature right-handed anti-parallel double helix within the crystal lattice. The packing is quite similar to that seen in the ApU structure, and the only significant differences between the two structures are the sodium ion coordination (only with the phosphates in this case), and the fact that the two GpC strands are related by crystallographic, rather than non-crystallographic symmetry as was the case with ApU. The nucleosides in this structure also assume the

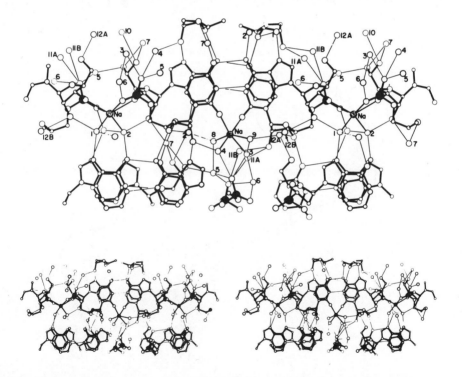

Figure 8. A view of the ApU lattice showing hydration of the double helical complex. A fragment of the lattice is illustrated, viewed approximately perpendicular to the plane of the bases. The water molecules are indicated as open circles. The sodium ions are stippled and the phosphate atoms are solid black. The bonds in the ApU molecule are thicker than those lines indicating hydrogen bonding and co-ordination of sodium. (a) The water molecule sites are numbered 1 to 12. Waters 11 and 12 are partly disordered and the two sites occupied by them are indicated as A and B. (b) A stereoscopic view of the same projection. It should be noted that the water contacts effectively shield one double helical fragment from another in the lattice. Taken from reference 6.

C3′-*endo*, low *anti* conformation. A view of the GpC lattice including the hydration structure is shown in Figure 10.

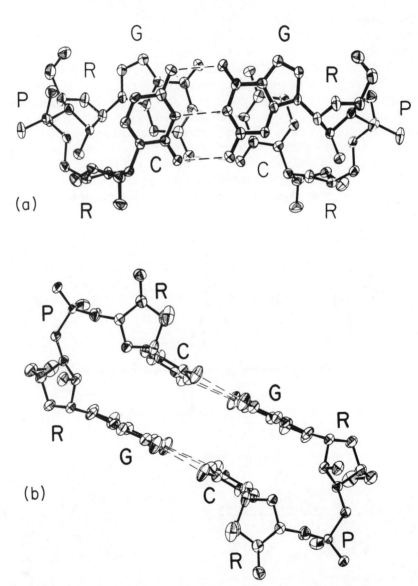

Figure 9. Molecular structure of the sodium GpC double helix. This is an ORTEP[40] plot of the GpC structure. Each atom is represented by an elipsoid which shows the thermal motion of the atom. Hydrogen bonds are shown as broken lines. G is guanine, R is ribose, P is phosphate and C is cytosine. (a) A view of the double helical complex approximately perpendicular to the plane of the bases. The major groove is at the top of the Figure, while the minor groove is at the bottom. (b) A view of the structure down the dyad axis. The helical axis is vertical and the bases are viewed edge on. Taken from reference 8.

c. The Calcium-Salt of GpC[10]

The structure of calcium GpC is qualitatively very similar to the structure of sodium GpC. However, since the charge of calcium is $+2$, rather than $+1$, the lattice has been slightly distorted, and there are 4 independent GpC molecules in the asym-

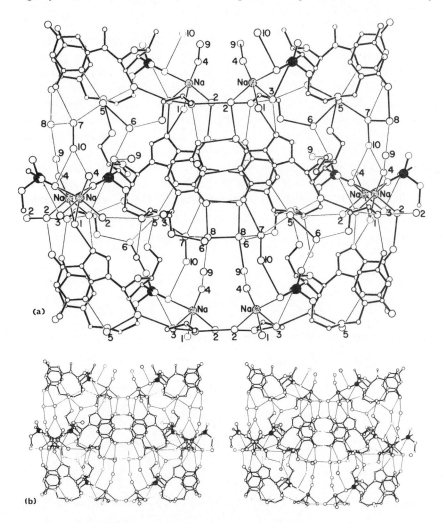

Figure 10. A view of the sodium GpC lattice. The hydration and packing of the lattice are indicated. The view is approximately perpendicular to the plane oof the bases. The water molecules are shown as open circles, while the sodium ions are stippled and the phosphorus atoms are solid black. The covalent bonds of the GpC fragment have been darkened. (a) The water molecule sites are numbered 1 through 10. It should be noted that there is a fairly large solvent channel in the minor groove of the double helix. Note the differences with Figure 8 (sodium ApU) caused by the absence of the sodium ion present in that structure. The double helical fragments are insulated from each other fairly well by the hydration network of water molecules. (b) Stereoscopic view of the hydration, as seen in this projection. Taken from reference 8.

metric unit rather than just one, as in the case of the sodium salt. The helices that can be generated from the calcium GpC molecules are similar to those found in the case of sodium GpC. The ribose conformations were also all C3'-*endo* and low *anti*. Both structures were nonahydrates.

d. The Ammonium Salt of d-pApTpApT[12]

This structure is the first tetranucleotide structure which has been determined, and full crystallographic details are as yet unavailable. The view of the structure shown in Figure 11 indicates the general conformation of the molecule. The base pairing in this structure is shown in Figure 12. As can be seen from that drawing, the firt two bases, A1 and T2 pairs with T4 and A3, respectively, of another tetrameric molecule related to the first by crystallographic two-fold screw symmetry. The screw axis is approximately horizontal in Figure 12. Thus, this tetrameric structure only represents, within this crystal, a Watson-Crick base paired structure which is two residues long. The salient features of this DNA structure include the alternation of sugar conformation: In this miniature double helix, the adenosines have the C3'-*endo* conformation, while the thymidines exhibit the C2'-*endo* conformation. It

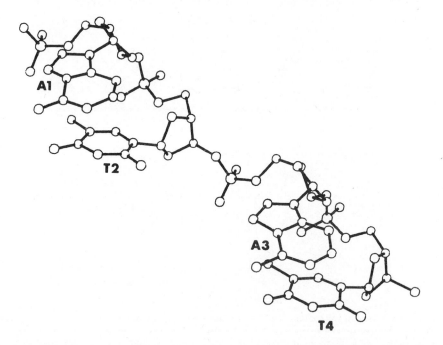

Figure 11. The structure of d-pApTpApT. The molecular structure is shown indicating the conformation which the molecule adopts. Note that the internucleotide linkage between adenosine 1 and thymidine 2 and that between adenosine 3 and thymidine 4 are similar and that the sugar puckers alternate. The linkage between thymidine 2 and adenosine 3 is different from the other two. Taken from reference 12.

should be noted that this structure is the first demonstration of A-T pairing in the Watson-Crick mode to be seen in single crystal analysis. The sugar-base orientation angles follow the alternation in sugar pucker; being low-*anti* values of 5° and -9° for the adenosines and high -*anti* values of 69° and 75° for the thymidines. The conformation of the phosphoester bonds is helical for both the 1-2 and 3-4 linkages. The linkage between the middle two nucleosides is similar to that seen in the cases of UpA2 and d-pTpT.

e. d-CpGpCpGpCpG[13]

This is a very recently solved structure, of which the authors have been kind enough to give us a photograph (see Figure 13). This structure is a hexanucleoside pentaphosphate which was phased by multiple isomorphous replacement techniques. The helix seen is left-handed, and corresponds to a structure containing 12 residues per turn, and has been dubbed Z-DNA by the authors, as there is no fiber precedent for the structure. The repeating part of the helix is actually a dimeric unit, wherein each d-CpGp unit is the unique portion. The alternating cytidines and guanosines are in different conformations, with the guanosines all adopting the unusual *syn* conformation, with the C3'-*endo* sugar conformation, while the cytidines are

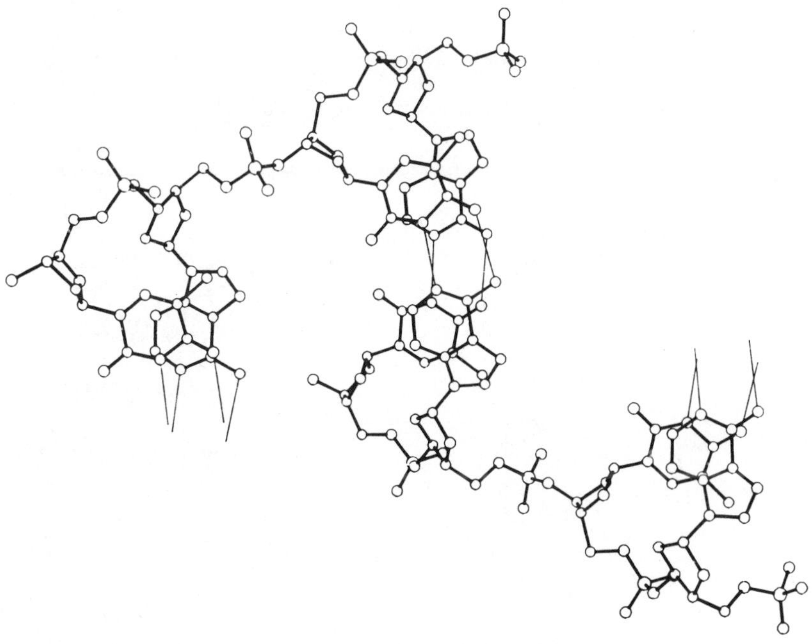

Figure 12. The lattice of d-pApTpApT. The way in which two molecules, which are related by a horizontal two-fold screw axis, form base pairs, is shown. The resultant structure is a miniature DNA double helical fragment of length 2. The hydrogen bonds are indicated by the thin lines, while the molecular structure is darkened. Taken from reference 12.

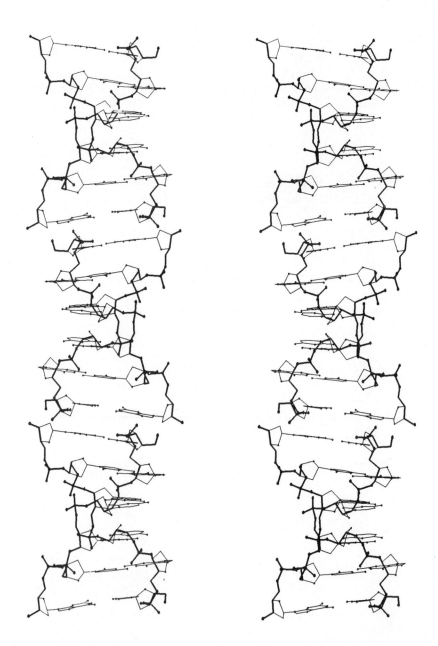

Figure 13. A stereoscopic view of the left handed double helix of d-CpGpCpGpCpG (Z-DNA). The helix shown is constructed from three hexameric units related by crystallographic symmetry. That is the reason for the gaps between hexameric units. Note the alternating internucleotide linkages between d-CpG and d-GpC portions of the structure. The bases are paired in the usual Watson-Crick fashion, but the guanosines adopt the *syn* conformation. Note the ribose-base stacking which stabilizes the structure. Photo generously provided by Dr. A.H.J. Wang and Dr. Alexander Rich.

C2'-*endo* and *anti* in the relationship of the base to the sugar. The 3'-terminal guanosines are C2'-*endo*, however, as an end effect. With the normal Watson-Crick base pairing obtaining throughout the six crystallographically independent base pairs, the backbone torsion angles adopt conformations which result in the left-handed helix. This astounding structure has its helix axis in the minor groove, and the major groove is virtually absent. This is in contrast to A-RNA double helices where the helix axis is in the major groove[22] and B-DNA double helices where the helix axis goes through the bases.[23] The major groove surfaces of the bases protrude directly into the solution and are unprotected by a double layer of phosphates as in the cases of both A-RNA type structures and, to a lesser extent, B-DNA type structures. For detailed summary of the article on Z-DNA by Rich and coworkers[13], see Sarma[44] in this volume.

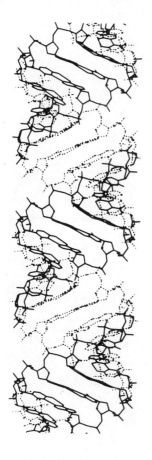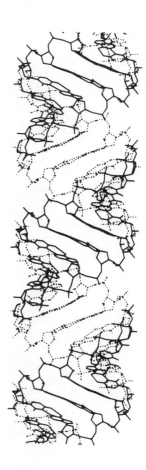

Figure 14. Stereoscopic representation of the RNA double helix derived from the structure of sodium ApU. In this drawing those components of the double helix which are on the side of the helix axis away from the reader have been dotted, while solid lines are used to represent the side of the double helix toward the reader. Taken from reference 22.

General Features of Oligonucleotide Structure

1. Backbone Conformations and Helical Structures.

The internucleotide linkages observed in single crystal structures of oligonucleotides are summarized in Table I. Three qualitative types of linkages have been noted: (1) The double helical type, characterized by beta and gamma angles both in the g^- range; (2) the single helical type, with $\beta\gamma$ as tg^- as seen in UpA2, d-pTpT and the central linkage of d-pApTpApT; (3) the g^+g^+ linkage of UpA1 and the second linkage of ApA$^+$pA$^+$. Otherwise, there is very little qualitative difference in the values of the torsion angles seen in these structures. However, that does not mean that helices generated using these linkages need be similar, since the nucleoside conformations are an important contributor to these extrapolations. In the case of RNA molecules, in the non-intercalative case, the C3'-*endo* conformation with a χ_{CN} near zero ap-

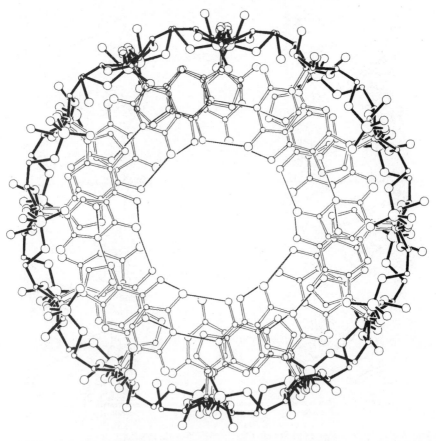

Figure 15. A view of the double helix generated from the ApU structure viewed down the helix axis. It should be noted that there is a hole in the center of the helix of approximately 9Å diameter. In this drawing, the ribose phosphate backbone is drawn with solid bonds, while the bonds in the bases are unshaded. Hydrogen bonds are indicated by single solid lines. Taken from reference 22.

pears to be very heavily preferred for the nucleoside. In the absence of a perturbing agent, such as an intercalative drug, it seems clear that *within the crystal* the C3′-*endo* conformation is likely to be observed for oligoribonucleotides. It is quite clear that in the case of Yeast tRNA[phe], this is not necessarily the case; furthermore, in that molecule, the cannonical 60° torsion angle on epsilon has even been violated.[18] Thus, the reader must be aware that the structure observed at the oligonucleotide level are possibilities which are neither exclusive, nor all-inclusive of the conformations which may be observed in polynucleotides. In the case of Yeast tRNA[phe], some structures were seen which were not observed in oligonucleotides, while some conformations seen at the oligonucleotide level were not observed at all. Thus, in the absence of intercalating agents C2′-*endo* sugars were seen, as were both 180° and 300° conformation angles about epsilon;[18] conversely, the g⁺g⁺ phosphodiester conformation was not seen.[24]

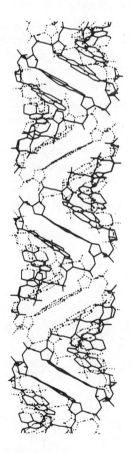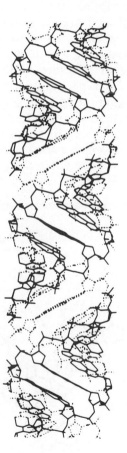

Figure 16. Stereoscopic representation of the RNA double helix generated by the sodium GpC structure. The same conventions are used as in Figure 14. It should be noted that this helix is slightly thinner than that obtained from the ApU structure. This is associated with a somewhat greater tilting of the bases. Taken from reference 22.

Since one has identical backbone subunits in oligomers, one is tempted to generate helices to see what would happen if the structure of the molecule was mathematically polymerized. Least square procedures for this have been developed.[22] The double helical parameters, for the RNA case, fall within the acceptable range determined by fiber work for RNA double helices. The ApU least squares double helix, shown in Figures 14 and 15 is very similar to the RNA double helix, except that it is 11.9-fold, rather than 11-fold. The GpC double helix generated from the coordinates of the sodium salt by least squares procedures was found to be 10.4-fold (Figure 16). The GpC double helices derived from the Calcium salt appear to be about 8-fold, although the least squares mathematical generation procedure was not used.

In the deoxytetramer, d-pApTpApT, the helical repeat is 31°, or 11.6-fold.[12] Helices have also been generated from the UpA2 molecule (Figure 17), the pTpT molecule (Figure 18) and the helical portion of ApA⁺pA⁺ (Figure 19). The helical parameters are all summarized in Table II. The sensitivity of generated helical parameters to small variations within the helical relationship searching "probes" (i.e., the portions assumed to be structurally identical in all subunits[22]) cannot be overestimated. The biological relevance of the helices generated from the single-helical molecules (UpA2, dpTpT and ApA⁺pA⁺) is not at all clear; however, they do give the reader a feeling for the possible structures which nucleic acids can adopt if they are in the appropriate environment to do so. The reader should remember, when considering these structures, that it is a combination of the internal energy minima and the external field which cause nucleic acids to adopt the conformations which they assume. The crystallographic observations are informative about which energy minima are most likely to be assumed; however, these are not the exclusive structures which can be made. Long helical structures of the tg⁻ type indicated by UpA2 and dpTpT are

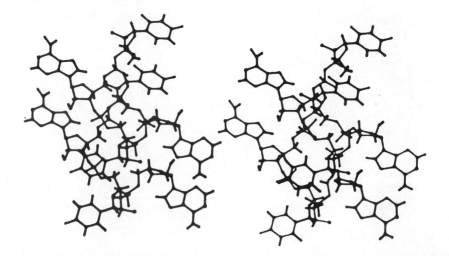

Figure 17. Stereoscopic view of the four-fold single helical structure generated from UpA2. Note that the bases do not stack at all, since they are separated by a distance of 12.4Å. Taken from reference 16.

Table II
Selected Helical Parameters

Material	Residues/Turn	Pitch	Reference
NaApU	11.9	28.1	2
NaGpC	10.4	26.9	2
CaGpC (2)	8	—	10
UpA2	4	12.4	16
ApA⁺	9.0	25.4	9
RNA-A	11	30.9	21
DNA-A	11	32.7	23
DNA-B	10	36.0	23
dpApTpApT	11.6	38.8	12
dpTpT	8	27.0	11

not likely to exist in solution, as there are no stabilizing interactions favoring them. Stabilizing interactions such as intermolecular stacking interactions within the crystal are often an important component of the crystal field, but are not likely to

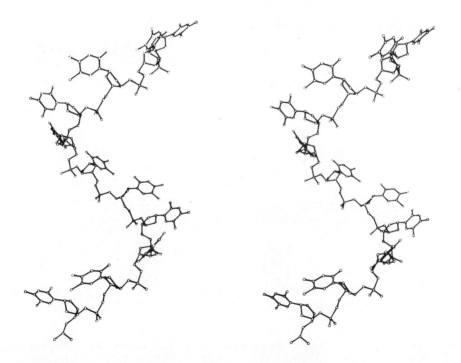

Figure 18. Stereoscopic view of the single helix generated from the structure of dpTpT. The helix axis is vertical. Slightly more than a single turn is shown. Taken from reference 11.

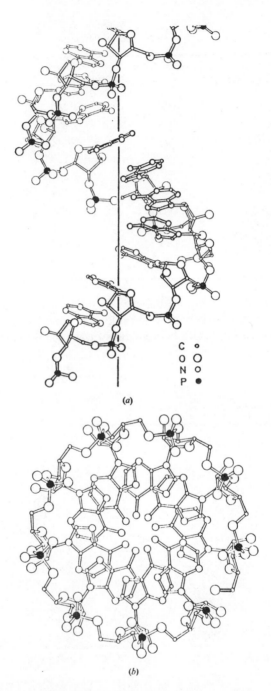

(a)

(b)

Figure 19. The polyA single helix derived from the A1pA2⁺ fragment. (a) A view perpendicular to the helix axis which is indicated by a dark line in the center. (b) A view down the helix axis. Note the lack of a large hole in the center of the helix, in contrast to the situation with ApU and GpC. Taken from reference 9.

exist in solution at very low dilution. On the other hand, such stabilization may well be present in the cell, as shown in the case of d-pTpT, which fits the cleft of Staphlococcal nuclease.[11]

It is clear from these studies that, in the absence of intercalating agents, the DNA molecules have a great deal more variability in structure than do the RNA molecules. The d-CpGpCpGpCpG structure showed all the guanosines to be in the *syn* conformation, a result never seen in the RNA series; d-pApTpApT had its purine nucleosides in the C3'-*endo* conformation, while its pyrimidine nucleosides adopted the C2'-*endo* conformation. Clearly this variability can be associated with biological function. Whether or not this is the case, remains to be seen.

In the case of the RNA molecules, one of the major goals of oligonucleotide single-crystal crystallography has been achieved, in the mini-double helical structures of ApU and GpC: *viz.*, with those structures it was possible to mathematically polymerize the conformation found within a single internucleotide unit and to generate thereby a helix which gave parameters qualitatively similar to those observed for RNA double helical fibers. To date, this goal has not been achieved for the DNA double helix. A base-paired single internucleotide structure has not yet been observed which can be mathematically polymerized to give parameters similar to fibers of A-DNA or B-DNA. It remains to be seen whether, indeed, this will ever be done, or whether it will turn out that the DNA structure itself is not a single repeating backbone structural unit; it may be composed of several backbone units which repeat, or its structure may even be more sequence dependent and perhaps more dynamic than has been commonly imagined.

2. Stacking Interactions.

The importance of the stacking interactions seen in these structures should not be underestimated. Stacking is one of the driving forces for maintaining the stability of polynucleotide chains, both in the form of double helices and in the structure of tRNA molecules. In the crystal structure of Yeast tRNA[phe], only four of the residues are not included in one of the two major stacking domains.[25] Base-base stacking is one of the primary modes of stacking which has been observed in these different structures. The different possibilities for the ApU-form RNA double helix are shown in[26] Figure 20. It should be noted that only the purine-pyrimidine type of stacking, as in the sequence of ApU or GpC has been seen in these structures—the others are extrapolations. However, the sequence dependence of stacking interactions is qualitatively illustrated in the figure: the purine-pyrimidine sequence is dominated by intra-strand stacking, the pyrimidine-purine sequence is dominated by inter-strand stacking, and the third alternative illustrates intermediate amounts of both forms.

Although base-base stacking is a stacking mode of crucial importance, it is not the only one. Some structures, such as that of UpA, are surprisingly bereft of this form of stacking. Such structures are often dominated by ribose-base stacking,[27] which

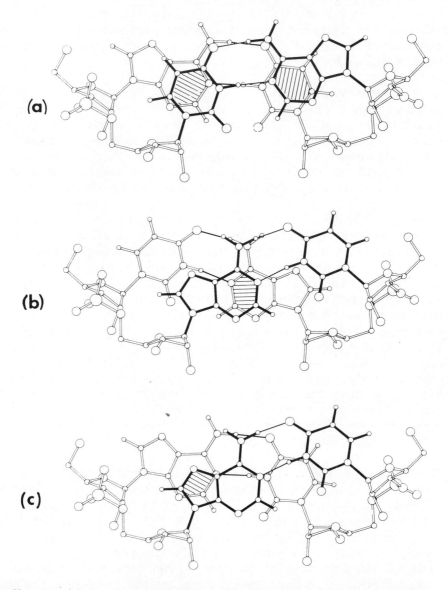

Figure 20. A view of the base stacking in the RNA double helix as a function of base sequence. Three views of double-helical fragments are shown perpendicular to the base plane. The base pair with solid bonds is closer to the reader than the base pair with unshaded bonds. The area of overlap of the purine or pyrimidine rings is indicated by hatching. The ApU double-helical backbone[6] has been used throughout in generating these double-helical fragments. (a) The fragment contains the self-complementary purine-pyrimidine sequence of ApU. The overlap seen would be the same with the self-complementary purine-pyrimidine sequence of GpC. Note that the overlap is intrastrand. (b) The self-complementary pyrimidine-purine sequence, UpA. The largely interstrand overlap seen would be the same with the self-complementary pyrimidine-purine sequence CpG. The sequence illustrated in (c) is the purine-purine sequence paired to the pyrimidine-pyrimidine sequence. The sequence ApA is shown paired to UpU, but the overlap would be similar with GpG paired to CpC. Note that (b) and (c) are extrapolations from (a), and not experimental observations. Taken from reference 26.

was also seen, *inter alia*, in the structures of the mini-double helices (Figure 21) where these interactions helped hold the mini-double helices in the lattice. Indeed, sugar-base stacking is one of the major stabilizing factors in the Z-DNA helix recently observed in the d-CpGpCpGpCpG structure; it may thus turn out to be more important than was originally apparent.

3. Base Pairing

Several types of base pairing have been noted in these structures. The Watson-Crick A-U base pair, whose dimensions are illustrated in Figure 22, has only been seen in this series in the structure of ApU. The Watson-Crick G-C base pair has been seen in both GpC structures and in d-CpGpCpGpCpG as well as in monomer studies. The C1′-C1′ distances are of interest, in that they define the separation of the backbone strands of the double helix. That shown in Figure 23, 10.67Å is typical of the G-C distances seen in monomeric studies as well as two of those seen in the structure of the calcium salt. Two of the distances in that structure, however, are short, comparable to the A-U distance. The reported C1′-C1′ distances for unperturbed Watson-Crick base pairs are summarized in Table III. The A-U distances (about 10.5Å) of ApU and the A-T distances (10.2Å) in the structure of d-pApTpApT are characteristically shorter than the usual G-C distances at 10.7Å. These general trends

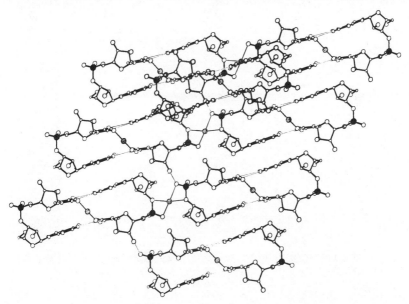

Figure 21. A packing diagram illustrating the stabilization of the ApU lattice. Sodium ions are stippled while phosphorus atoms are solid black. This is a view down the dyad axis. The double helical fragments form lath-like structures which pass from lower left to upper right. At the top of the structure an extra double-helical fragment is placed on a layer one level closer to the reader. Besides the considerable stabilization of the laths by base-base stacking between double helical fragments, the role of ribose-base stacking is evident in the interactions of the uracils with the riboses of adenosines not on the same covalent molecule. Taken from reference 6.

may or may not be biologically significant, as the differences are very small. The base pairs are apparently not exactly the same size, and the G-C pairs are slightly larger on the average than A-T or A-U pairs. The other types of pairing seen in these structures have been shown in Figure 2, with the A-A pairing shown there also noted in the ApA⁺pA⁺ structure.

4. Ionic Interactions.

The ways in which the metallic ions interact with nucleic acids are of great relevance to their biological role as the carriers of genetic information. The nucleic acids are

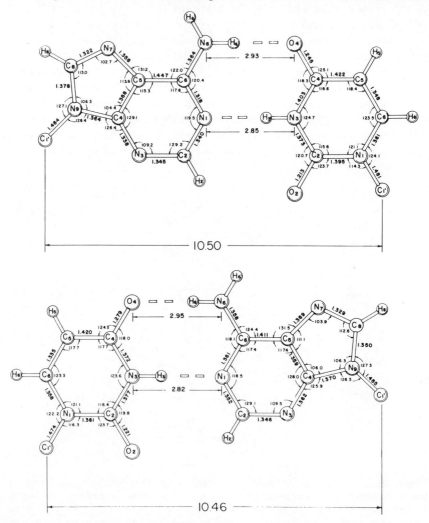

Figure 22. The dimensions of the A-U pair, as seen in the structures of sodium ApU. The C1′-C1′ distances across the helices are indicated in the figure. Note the presence of two hydrogen bonds between the bases. Taken from reference 6.

Table III
C1′-C1′ Distances for Unperturbed
Watson-Crick Base Pairs

Type	Distance	Compound	Reference
A-T	10.2	dpApTpApT	12
A-U	10.46	Na-ApU	6
A-U	10.50	Na-ApU	6
G-C	10.67	Na-GpC	8
G-C	10.59	Ca-GpC	10
G-C	10.42	Ca-GpC	10
G-C	10.53	Ca-GpC	10
G-C	10.66	Ca-GpC	10
G-C	10.68	9-Et-G:1-Me-C	37
G-C	10.67	9-Et-G:1-Me,5-F-C	37
G-C	10.76	dG:5-Br-dC	38
G-C	10.8	9-ET-G:1-Me,5-Br-C	39

negatively charged under physiological conditions and must be neutralized by some species of cation. To date, the best described cases of metallic neutralization are the sodium salts of ApU, GpC and d-pTpT, and the calcium salt of GpC. In the only observed case of divalent cation neutralization, the calcium ions neutralizing GpC

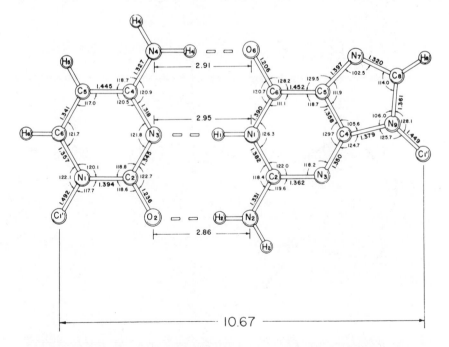

Figure 23. The dimensions of the G-C base pair, as seen in the structure of sodium GpC. The C1′-C1′ distance across the helix is indicated in the figure. Note the presence of three hydrogen bonds between the bases. Taken from reference 6.

are coordinated to two phosphates at one site each. The other ligands are water molecules which are distributed octahedrally, as seen in Figure 24. This is not a surprising interaction since the phosphates are negatively charged and the calciums are positively charged. There is no calcium-base interaction in the GpC structure, although some of the water molecules which coordinate the calcium ions do interact with the bases. The sodium GpC structure is similar to the calcium GpC structure, except that two sodium ions are coordinated to the phosphates in a face-sharing octahedral mode, as shown in Figure 25. Again, the interaction is direct between the positive metallic cation and the negative phosphate anion.

In the case of sodium ApU, again one sodium is coordinated to the phosphates,

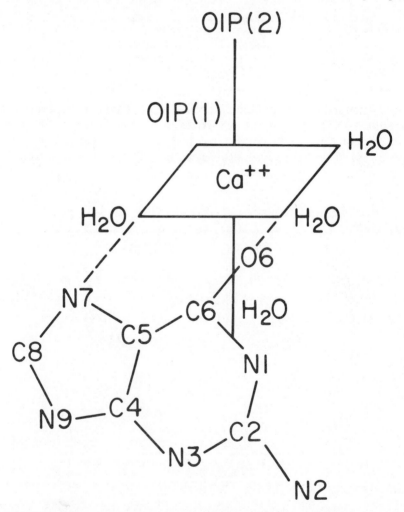

Figure 24. The calcium coordination in calcium GpC. The octahedral coordination of one calcium ion is shown schematically. The coordination of the other calcium ion is identical to this. There is no direct calcium-base interaction. Taken from reference 10.

although this time bidentally, so that 4 of its octahedrally distributed sites are occupied by phosphate oxygen atoms, and the last two are filled by water molecules. This is shown in Figure 26b, and it should be noted that one of the ligands on each side is a phosphoester oxygen atom. The other sodium ion is coordinated directly to the two uracil 02 atoms which are not involved in the Watson-Crick base pairing. This sodium ion, whose coordination is shown in both Figures 26a and 8 is seen in Figure 7 to lie in the minor groove directly on the non-crystallographic dyad axis which relates the two strands. Coordination of this sort is a sequence specific effect, since it was not seen in the GpC structures and it would not be possible for a double helical UpA structure to coordinate the sodium in the same way, since the two uracil oxygens would be about 8Å apart, rather than about 4Å (see Figure 20). Thus, the recognition of bases in a sequence-dependent fashion may be partially dependent upon mediation by metallic ionic species.[28] Although there is no base pairing in the structure of d-pTpT, there is a sodium ion which is coordinated to the two thymine 02 atoms in an analogous fashion, as shown in Figure 27. Note, however, that the details of the geometry are different from that seen with ApU.

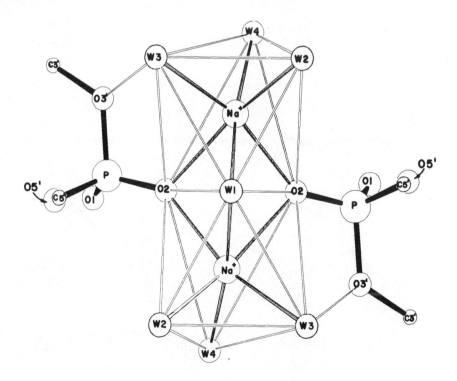

Figure 25. The face-sharing coordination of the sodium ions in soidum GpC. The coordination shown is around the sodium ions which are bonded to the phosphate groups of two different double helical fragments. The coordination around each sodium ion is that of an octahedron and the two octahedra share one face in common. W indicates water molecules. Taken from reference 8.

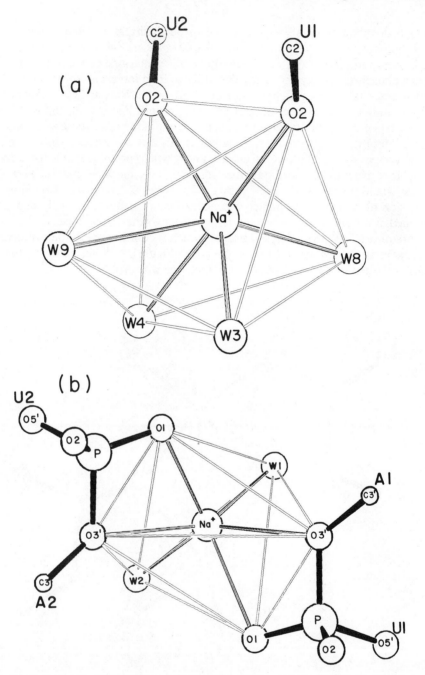

Figure 26. The sodium coordination in sodium ApU. (a) The coordination of the sodium cation complexed to the uracil carbonyl groups in the minor groove of the double helix. The coordination is that of a distorted octahedron. (b) The coordination of the sodium ion bound between two double-helical fragments to two phosphate groups. This coordination is also that of a distorted octahedron. W represents water molecules. Taken from reference 6.

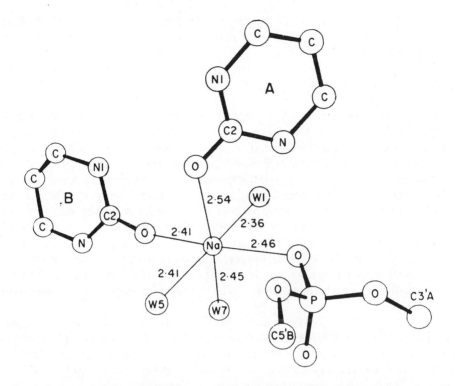

Figure 27. Coordination of the major sodium ion in d-pTpT. The octahedral coordination of this sodium is evident from the figure. The two thymidine 02 ligands are indicated as being from each of the independent thymidines. Note that the geometry is somewhat different from that seen in Figure 26(a), which represents the analogous ion in sodium ApU. Taken from reference 11.

5. Hydration and Its Role in Studying Sequence Specific Recognition.

The hydration structures of nucleic acid molecules are important for us to understand because this is a crude probe of the field of the molecule, and reveals the hydrogen bonding potential of the material under investigation: water molecules can either accept or donate hydrogen bonds; therefore, they can indicate those sites in which other molecules can interact with the major and minor groove portions of the bases, as well as the backbone. In the one double helical case studied in detail, ApU, the hydration structure mimicked the non-crystallographic symmetry of the dimer. Particularly in the cases of double helical molecules, the hydration structures can indicate the alterations of the field of the molecule as sequences vary.

Other aspects of the sequence specific recognition of double helical nucleic acids have been studied, using the detailed geometry of these structures as data.[28] The major conclusions drawn were that the base pairs cannot be uniquely recognized by a single hydrogen bonding probe, and that two sites on the base pair are necessary for recognition. This implies that four types of degeneracy exist amongst the base pairs:

(1) A-T(U)/G-C; (2) A-T(U)/C-G; (3) A-T(U)/T(U)-A; (4) G-C/C-G. The first three types of degeneracy have been noted in the recognition of restriction endonuclease sites.[29]

6. The Past and The Future

These structures taken in sum have contributed a new chapter to our understanding of the structures of nucleic acids. In the case of RNA, the question of whether the double helical structure can be readily parameterized at high resolution has been answered, via the mini-double helices. Furthermore, some of the various conformations available to nucleic acids, both by themselves and, as reported elsewhere in this volume, in the presence of intercalating agents, have been catalogued. This information has been useful in the interpretation of the electron density maps of Yeast tRNA^{phe} at 3Å resolution and many of the structural principles noted from the oligonucleotide studies have been found to be roughly maintained at the polynucleotide level. We would not expect exact maintenance of these features, since the polynucleotide has many more degrees of freedom and thus many more ways of compensating for structures which would be unfavored at the oligonucleotide level.

It should be noted that only purine-pyrimidine sequences have been crystallized in oligomeric RNA double helical structure—nobody has reported high resolution parameters for double helices containing purine-purine sequences or pyrimidine-pyrimidine sequences, or (in the absence of intercalators) even pyrimidine-purine sequences. The details of these structures remain to be worked out.

There are still many questions which are unanswered at the oligonucleotide level, involving RNA. In particular, those questions involving codon-anticodon interactions must be explored at high resolution. The fidelity of wobble pairing[30] or two-out-of-three interactions[31] remain to be explored. Also, the details of the translation of the genetic code in tRNA-synthetase interactions remain to be modelled. There are also many RNA model structures relevant to the processing of mRNA in eukaryotic systems.

In the area of DNA, the story is only just beginning. The bizarre structures observed by Viswamitra[12] and his colleagues and by Wang and his colleagues[13] indicate that there is much more to be learned about the "ordinary" structure of DNA; is its backbone truly double helical? Is it a dynamic or stable structure? What other variations on the double helical theme are available to this molecule? What is the sequence dependence of those structures? These are only some of the obvious questions for the future of this exploding field.

Acknowledgements

I would like to thank Drs. Andrew H. J. Wang and Alexander Rich for a pre-publication picture of dCpGpCpGpCpG and Dr. M. A. Viswamitra for a pre-publication picture of dpApTpApT.

This work was aided by a Basil O'Connor Starter Research Grant from The National Foundation-March of Dimes, by University Award Numer 7421 from the Research Foundation of the State University of New York and by Grant GM26467 from the National Institutes of Health.

References and Footnotes

1. Watson, J.D., and Crick, F.H.C. *Nature 171*, 737-738 (1953).
2. Seeman, N.C., Sussman, J.L., Berman, H.M., and Kim, S.H. *Nature New Biology 233*, 90-92 (1971).
3. Sussman, J.L., Seeman, N.C., Kim, S.H., and Berman, H.M. *J. Mol. Biol. 66*, 403-422 (1972).
4. Rubin, J., Brennan, T. and Sundaralingam, M. *Biochemistry 11*, 3112-3128 (1972).
5. Rosenberg, J. M., Seeman, N.C., Kim, J.J.P., Suddath, F.L., Nicholas, H.B., and Rich, A. *Nature, 243*, 150-154 (1973).
6. Seeman, N.C., Rosenberg, J.M., Suddath, F.L., Kim, J.J.P., and Rich, A. *J. Mol. Biol., 104*, 109-144 (1976).
7. Day, R.O., Seeman, N.C., Rosenberg, J.J., and Rich, A. *Proc. Nat. Acad. Sci. U.S.A., 70*, 849-853 (1973).
8. Rosenberg, J.M., Seeman, N.C., Day, R.O., and Rich, A. *J. Mol. Biol., 104*, 145-167 (1976).
9. Suck, D., Manor, P.C., and Saenger, W. *Acta Cryst., B32*, 1727-1737 (1976).
10. Hingerty, B., Subramanian, E., Stellman, S.D., Sato, T., Broyde, S.B., and Langridge, R. *Acta Cryst., B32*, 2998-3013 (1976).
11. Camerman, N., Fawcett, J.K., and Camerman, A. *J. Mol. Biol., 107*, 601-621 (1976).
12. Viswamitra, M.A., Kennard, O., Jones, P.G., Sheldrick, G.M., Salisbury, S., Falvello, L., and Shakked, Z. *Nature, 273*, 687-688 (1978).
13. Wang, A.H.-J., Quigley, G.J., Kolpak, F., Crawford, J.L., van Boom, J.H., van der Marel, G., and Rich, A., *Nature* (in press)
14. Seeman, N.C., in *Stereodynamics of Molecular Systems*, ed. by R.H. Sarma, Pergamon Press, New York, pp. 75-109 (1979).
15. Shefter, E., Barlow, M., Sparks, R.A., and Trueblood, K.N. *Acta Cryst., B25*, 895 (1969).
16. Kim, S.H., Berman, H.M., Seeman, N.C., and Newton, M.D. *Acta Cryst., B29*, 703-709 (1973).
17. Evans, F.E., and Sarma, R.M. *Nature, 263*, 567-572 (1976).
18. Jack, A., Klug, A., and Ladner, J.E. *Nature, 261*, 250-251 (1976).
19. Kim, S.H., Suddath, F.L., Quigley, G.J., McPherson, A., Sussman, J.L., Wang, A.H.J., Seeman, N.C., and Rich, A. *Science, 185*, 435-440 (1974).
20. Rich, A., Davies, D.R., Crick, F.H.C., and Watson, J.D. *J. Mol. Biol. 3*, 71-86 (1961).
21. Arnott, S., Hukins, D.W.L., and Dover, S.D. *Biochem. Biophys. Res. Comm., 48*, 1392-1399 (1972).
22. Rosenberg, J.M., Seeman, N.C., Day, R.O., and Rich, A. *Biochem. Biophys. Res. Comm., 69*, 979-987 (1976).
23. Arnott, S., and Hukins, D.W.L. *Biochem. Biophys. Res. Comm., 47*, 1504-1509 (1972).
24. Quigley, G.J., Seeman, N.C., Wang, A.H.J., Suddath, F.L., and Rich, A. *Nucleic Acids Research, 2*, 2329-2342 (1975).
25. Kim, S.H., Sussman, J.L., Suddath, F.L., Quigley, G.J., McPherson, A., Wang, A.H.J., Seeman, N.C., and Rich, A. *Proc. Nat. Acad. Sci., 71*, 4970-4974 (1974).
26. Rich, A., Seeman, N.C., and Rosenberg, J.M. in *Nucleic Acid-Protein Recognition*, ed. by H.G. Vogel, Academic Press, New York, pp. 361-374 (1977).
27. Bugg, C.E., Thomas, J.M., Sundaralingam, M., and Rao, S.T. *Biopolymers, 10*, 175-219 (1971).
28. Seeman, N.C., Rosenberg, J.M., and Rich, A. *Proc. Nat. Acad. Sci., 73*, 804-808 (1976).
29. Smith, H.O. *Science, 205*, 455-462 (1979).
30. Crick, F.H.C. *J. Mol. Biol., 19*, 548-555 (1966).
31. Lagerkvist, U. *Proc. Nat. Acad. Sci., 75*, 1759-1762 (1978).
32. Arnott, S., Hukins, D.W.L., Dover, S.D., Fuller, W., and Hodgson, A.R. *J. Mol. Biol., 81*, 107-122 (1973).
33. Arnott, S., Dover, S.D., and Wonacott, A. *Acta Cryst., B25*, 2192-2206 (1969).
34. Arnott, S., and Hukins, D.W.L. *J. Mol. Biol., 81*, 93-105 (1973).

35. Marvin, D.A., Spencer, M., Wilkins, M.F.H., and Hamilton, L.D. *J. Mol. Biol., 3*, 547-565 (1961).
36. Arnott, S., Chandrasekaran, R., Hukins, D.W.L., Smith, R.S.C., and Watts, L. *J. Mol. Biol., 88*, 523-533 (1974).
37. O'Brien, E.J. *Acta Cryst., 23*, 92-106 (1967).
38. Haschemeyer, A.E.V., and Sobell, H.M. *Acta Cryst., 19*, 125-130 (1965).
39. Sobell, H.M., Tomita, Y., and Rich, A. *Proc. Nat. Acad. Sci., 49*, 885-891 (1963).
40. Johnson, C.K. ORTEP, ORNL-3794, Oak Ridge National Laboratory, Oak, Ridge, Tennessee (1965).
41. Seeman, N.C., in *Nucleic Acid Geometry and Dynamics*, ed. Sarma, R.H., Pergamon Press, New York, Oxford, pp. 47-82 (1980).
42. Sarma, R.H., in *Nucleic Acid Geometry and Dynamics*, ed., Sarma, R.H., Pergamon Press, New York, Oxford, pp. 1-46 (1980).
43. Sarma, R.H., in *Nucleic Acid Geometry and Dynamics*, ed., Sarma, R.H., Pergamon Press, New York, Oxford, pp. 143-184 (1980).
44. Sarma, R.H., in *Nucleic Acid Geometry and Dynamics,* ed., Sarma, R.H., Pergamon Press, New York, Oxford, pp. 83-108 (1980).

Spatial Configuration of Oligonucleotides in Solution as Reflected in Nuclear Magnetic Resonance Studies

Ramaswamy H. Sarma
Institute of Biomolecular Stereodynamics
State University of New York at Albany
Albany, New York 12222

Introduction

Present day spectroscopic methods are usually unable to provide the intimate details of the molecular structures of very large biological macromolecules. In order to solve this problem, the chemist usually studies the conformational properties of relatively small segments. Such a study, in addition to providing the details of their molecular conformation, enables us to understand the stereochemical principles which govern the structure of such segments. This information can in turn be used to project the spatial configuration of the corresponding large biological molecular systems. In the case of nucleic acids, the small segments are oligonucleotides.

Crystal structures of oligonucleotides that have been determined to date have been discussed by Seeman[1] in an earlier chapter. Here we take up their structure in aqueous solution as determined by NMR spectroscopy. Early NMR work on oligonucleotide conformations in solution was limited to investigations of base stacking and sugar ring pucker from base proton chemical shifts and anomeric proton spin coupling constants.[2-5] In these earlier days the extreme complexity of the spectra as well as poor signal to noise ratios prevented a complete analysis. However, the development of high frequency Fourier transform NMR systems, highly sophisticated interactive computer simulation capabilities, of selective deuteration techniques as well as hetero and homo nuclear decoupling in Fourier transform mode have enabled to solve this problem to some extent. As a result of this, the high resolution NMR spectra of many short single stranded oligonucleotides (i.e., up to trimer level) have been completely analyzed[6-12] as has been discussed earlier.[13] Although increasing the number of nucleotidyl units beyond the trimer level makes the unambiguous assignment and complete analysis difficult, they do bring forth the prospects of studying the base-pairing interactions as well as the dynamics of helix-coil transition.

This study will illustrate how NMR in its existing state of the art is being utilized to probe the conformational properties of single stranded and base-paired short oligonucleotides in aqueous solution, but before these a few words about the mononucleotides. An understanding of the material in an earlier chapter[13] on the

143

methodology of NMR and nucleic acid nomenclature is necessary to appreciate what follows.

The 3′ and 5′ Mononucleotides

In Figure 1 is shown the experimentally observed and computer simulated ^{1}H NMR spectra of a mononucleotide. The simulation provides accurate information about

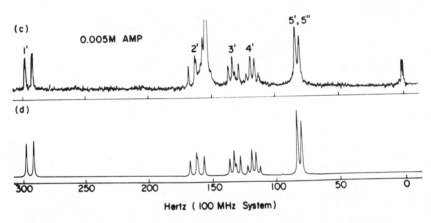

Figure 1. The sugar proton region of the 100 MHz ^{1}H NMR spectra of 5′AMP under conditions in which ^{31}P was decoupled. The top one is the observed one and the bottom one computer simulated. The shifts are in hertz upfield from internal tetramethylammonium chloride.

the coupling constants and chemical shifts which can be translated into conformational parameters using methodology described elsewhere.[13] The data summarized in Table I have several general features. Thus all 5′ mononucleotides irrespective of the nature of the base or the sugar ring display a great deal of tendency to orient the C4′-C5′ and C5′-O5′ bonds in the gg ($\epsilon = 60°$) and g′g′ ($\delta = 180°$) conformation. (See the earlier section, i.e., reference 13 for nomenclature). The sugar ring of the deoxyribo- systems show outspoken preference for the ^{2}E pucker in solution as in B-DNA. In the ribo series the ring pucker is sensitive to the constitution of the base, i.e., the purines tend to shift to the left and the pyrimidine to the right the equilibrium ^{2}E $\rightleftharpoons$ ^{3}E. The deoxyribo- and ribo-systems also show important differences about the C3′-O3′ torsion. In the ribo series one has an equilibrium,

$$^3E\alpha^- \rightleftharpoons {}^2E\alpha^+ \qquad \text{(Figure 19 in Reference 13)}$$

but in the deoxyribo series, C3′-O3′ essentially occupies domains centered around 200°. In the above equilibrium α^- and α^+ are domains centered around 205° and 275° respectively. NMR spectroscopy does not enable a precise determination of the sugar base torsion angle, χ_{CN}, in mononucleotides. However, the several qualitative methods described elsewhere clearly show that for common 5′ and 3′ monomers[13], it lies in the anti domain and that within the anti domain ^{3}E sugar pucker is associated

Table I
Conformational Parameters for Ribo and Deoxyribo Mononucleotides
in Aqueous Solution at 20°C pH = 5.0 From References 6-13

Nucleoside	Sugar Ring $\%\,^3E$	Backbone $\%gg(\epsilon = 60°)$	$\%g'g'(\delta \simeq 180°)$	α
5'AMP	40	77	72	—
5'GMP	37	71	71	—
5'UMP	46	89	75	—
5'CMP	55	77	75	—
3'AMP	31	82	—	$240 \pm 36°$
3'GMP	33	72	—	$240 \pm 37°$
3'UMP	56	68	—	$240 \pm 35°$
3'CMP	60	71	—	$240 \pm 35°$
d-5'AMP	28	63	70	—
d-5'GMP	32	63	67	—
d-5'TMP	32	59	74	—
d-5'CMP	33	70	75	—
d-3'AMP	23	71	—	202°
d-3'GMP	26	57	—	202°
d-3'TMP	33	57	—	199°
d-3'CMP	35	58	—	202°

with a value of χ_{CN} smaller than what is associated with a ^2E pucker.

It has been discussed earlier[13] that an interdependence exists between rotational preferences about C4'-C5' and C5'-O5' bonds (Figure 17 reference 13). Study of mononucleotides which contain modified bases has shown that there is an interrelation between glycosidic torsion, sugar pucker and backbone conformation. Some of the modified mononucleotides studied are: 6-aza-5'UMP, 8-aza-5'AMP, 8-aza-5'GMP, 8-Br-5'AMP, 8-CH₃S-5'AMP, 7-methyl 5'GMP, etc. (Figure 2). These studies[14-16] indicated that introduction of bulky substituents or a high electron density probe at C8 of purine or C6 of pyrimidine considerably perturbs the conformation. In the case of 8-Br- and 8-CH₃S-5'AMP, the preferred conformations were syn-^2E-g/t (g/t = gt and tg) (Figure 3). In Figure 4 are shown the 100 MHz ^{1}HNMR spectra of 5'GMP and 8-aza-5'GMP. Comparison of the 4' and 5' regions of the spectra clearly reveals that 8-aza substitution has clearly affected the backbone conformation. Particularly noteworthy is the fine structure in the 4' region of 5'GMP and the lack of it in 8-aza-5'GMP indicating the influence of four bond zig-zag coupling between H4' and P. This is illustrated strikingly in the ^{31}P NMR spectra of 8-aza-5'GMP and 5'GMP. (Figure 5) The data on the aza analogs indicated that they show preference to exist in (syn $\rightleftharpoons$ anti)-($^2E \rightleftharpoons {}^3E$) - g/t-g'g' conformations. The aza substitution causes increase in ^3E sugar populations and in the populations of g't' conformations, in addition to orienting the C4'-C5' predominantly in g/t conformations. In the 8-aza analogs repulsive electrostatic interactions would prevail between -N= at the 8 position and the negatively charged

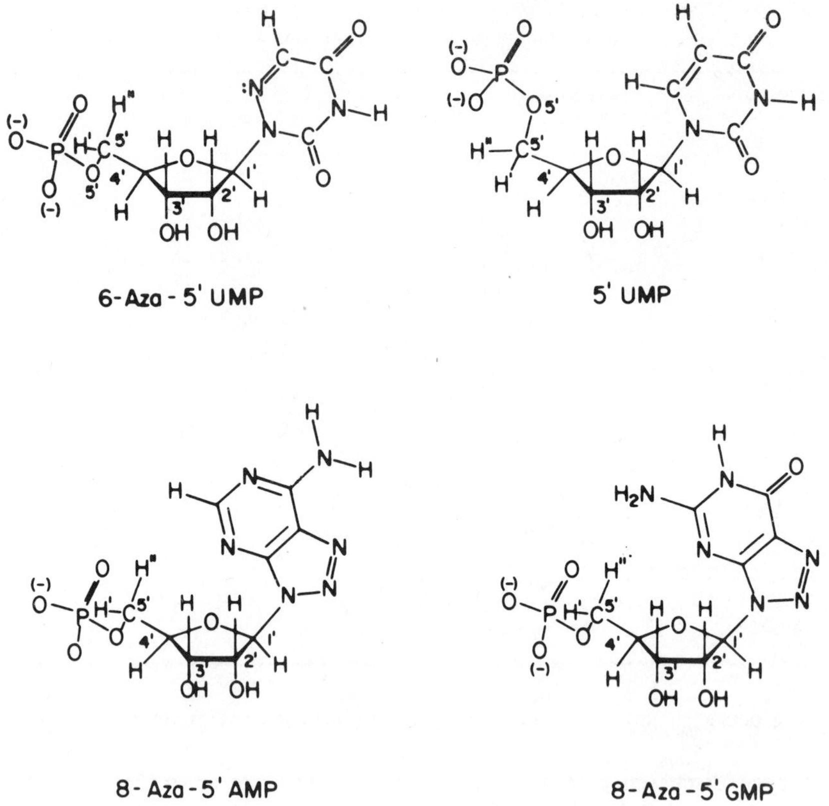

Figure 2. Structures of some modified nucleotides.

phosphate group, if the molecule existed in the anti-gg orientation. Such electrostatic repulsions can be relieved by rotating the C4'-C5' bond from gg to g/t
orientations as well as by torsional variation about the glycosidic linkage from anti
to syn conformation. Rotation from anti to syn orientation, for steric and electrostatic reasons, necessitates a simultaneous rotation about the C4'-C5' bond from
gg to g/t conformation. The observation that the 8-aza substitution in 5'-β-purine
nucleotides causes a depopulation of gg and anti conformers with corresponding increase in the population of g/t and syn orientation seems to support the above
thesis. The finding that rotational variation about a bond such as C4'-C5' is accompanied by torsional variation about C5'-O5' and the glycosidic bonds, as well as
changes in the endocyclic torsion angles of the ribose moiety gives important insight
into the engineering of nucleic acid components; even though they in general prefer
certain conformations, there is enough flexibility present in their molecular
framework, that the entire system can undergo conformational adjustment in
response to a perturbation. Some of the conformational adjustments that are possible for 8-aza-5'AMP are shown in Figure 6.

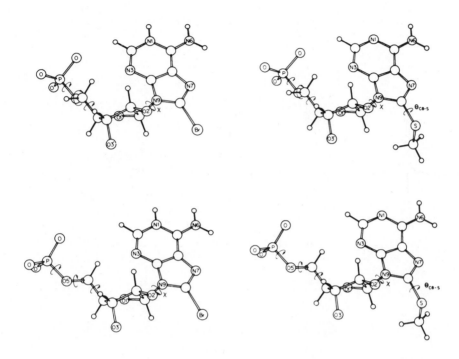

Figure 3. The experimentally observed spatial configurations of 8-Br-5'AMP (left) and 8-CH₃S-5'AMP (right). In the top pair χ is syn (220°) and C4'-C5' is gt (ε = 180°); in the bottom pair χ is syn (220°) and C4'-C5' is tg (ε = 300°).

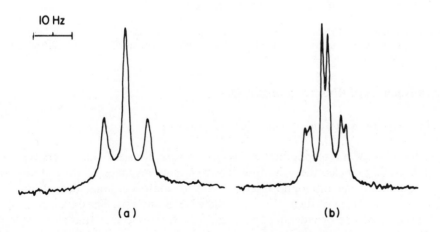

Figure 5. ³¹P NMR spectra of 8-aza-5'AMP (a) and 5'AMP (b). Note the fine structure in the spectrum of 5'AMP due to the four bond H4'-P coupling.

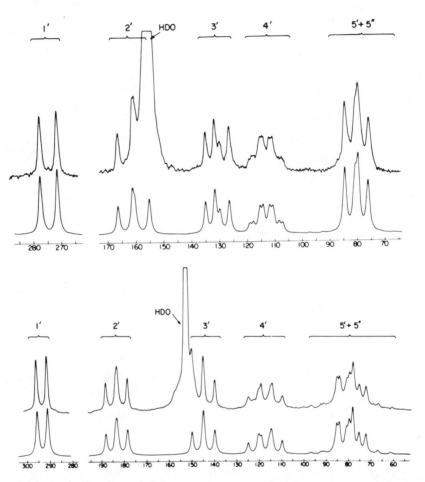

Figure 4. The 100 MHz ¹H NMR spectra of 5′GMP (top pair) and 8-aza-5′AMP (bottom pair). In each pair the bottom spectra are computer simulated. The chemical shifts are in hertz upfield from internal tetramethylammonium chloride.

Stereodynamics of RNA Oligonucleotides

1. The Common Ribodinucleoside Monophosphates.

The formation of a phosphodiester linkage between 3′ and 5′ monomers results in the shortest oligonucleotide—the dinucleoside monophosphate. Of the sixteen possible ribodinucleoside monophosphates, detailed conformational deductions have been made for fifteen of them[7-10] by complete NMR analysis. The data is missing on GpG because of line broadening complications. A sample spectrum along with the simulation for a ribo dimer is shown in Figure 7a. Complete conformational data for the dinucleoside monophosphates and the corresponding monomers are given in Tables II and III.

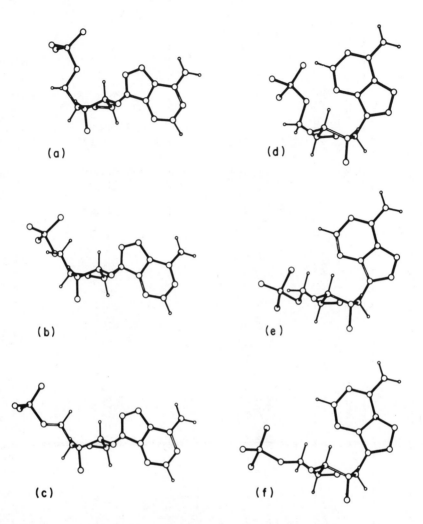

Figure 6. The conformational possibilities for 8-aza-5′AMP (a) anti-gg, electrostatic repulsion between -N= at 8 position and negative charges on phosphate group may cause it to adopt (b) and (c), i.e., anti-g/t. In (d), syn-gg, steric and electrostatic repulsion between phosphate oxygens and the adenine may cause it to adopt (e) and (f), i.e., syn-g/t. Note that the ribose is kept in ²E in (a), (b) and (c) and ³E in (d), (e), and (f).

The data unmistakably indicate that in all dimers irrespective of the nature of the base and sequence, the C4′-C5′ and C4′-O5′ frame displays overwhelming preference for the gg and g′g′ conformation. The ribose rings exist as a ²E ⇌ ³E equilibrium with bias for the ³E pucker in most cases. Orientation about the C3′-O3′ bond is coupled to the ribose conformational equilibrium and the system exists with a bias for the ³Eα⁻ coupled conformation (Figure 19, reference 13). In most cases elevation of temperature[10] led to the following three experimental observations: (i) reduction in the shielding between the adjacent bases indicating destacking (*vide*

Table II

Population Distribution of Conformers in Dinucleoside Monophosphates and Their Components.

Nucleotide	T (°C)	% Stacked[e]	Dimer					Monomer[b]				
			Ribose Ring[a]		Backbone[c]			Ribose Ring[a]		Backbone[c]		
			% ³E	K_{eq}[d]	%gg	%g/g'	ΘPH[g]	% ³E	K_{eq}[d]	%gg	%g/g'	ΘPH[g]
ApA Ap-	20	38 ± 2	58	1.4	79	90	±37	31	0.5	82	72	±36
-pA			61	1.6	74			40	0.7	77		
ApA Ap-	72	19 ± 4	46	0.9	73	73	±33					
-pA			48	0.9	64							
ApG Ap-	20	25 ± 2	49	1.0	81	82	±35	31	0.5	82	71	±36
-pG			51	1.0	75			37	0.6	71		
ApG Ap	80	16 ± 3	40	0.7	72	72	±36					
-pG			48	0.9	66							
GpA Gp-	20	30 ± 2	55	1.2	72	84	±37	33	0.5	72	72	±37
-pA			56	1.3	85			40	0.7	77		
UpU Up-	20	8 ± 5	56	1.3	75	77	±35	56	1.3	68	75	±35
-pU			53	1.1	83			46	0.8	89		
UpU Up-	89	9 ± 2	52	1.1	71	73	±35					
-pU			52	1.1	77							
CpC Cp-	20	35 ± 1	74	2.9	81	84	±35	60	1.5	71	75	±35
-pC			69	2.2	92			55	1.2	77		
UpC Up-	20	18 ± 2	62	1.6	73	83	±35	56	1.3	68	75	±35
-pC			63	1.7	87			55	1.2	77		
UpC Up-	80	0–4[f]	55	1.2	67	70	±35					
-pC			65	1.9	69							
CpU Cp-	20	33 ± 2	66	1.9	78	85	±33	60	1.5	71	75	±35
-pU			64	1.8	90			46	0.8	89		

[a] Computed by using $J_{1'2'} + J_{3'4'} = 9.5$ Hz for the dimers and $J_{1'2'} + J_{3'4'} = 9.3$ for the monomers. [b] Monomer data for solutions at pD = 5.4. [c] Rotamer equations used: gg = $(13.7 - \Sigma)/9.7$; g'g' = $(25 - \Sigma')/20.8$. [d] K_{eq}, 2E ⇌ 3E; estimated errors in K_{eq} are ±0.2. [e] % stacked = $(J_{3'4'}(\text{dimer}) - J_{3'4'}(\text{monomer}))/(9.5 - J_{3'4'}(\text{monomer}))$. [f] Computed from the magnitude of $J_{1'2'}$ and $J_{3'4'}$. [g] Up-ribose moiety. [g] In degrees.

Table III

Population Distribution of Conformers in Dinucleoside Monophosphates and Their Components.

Nucleotide	Temp (°C)	% Stacked[e]	Dimer Ribose Ring[a] % ^3E	Dimer Ribose Ring[a] K_{eq}[d]	Dimer Backbone[c] % gg	Dimer Backbone[c] % g'g'	Dimer Backbone[c] θPH	Monomer[b] Ribose Ring[a] % ^3E	Monomer[b] Ribose Ring[a] K_{eq}[d]	Monomer[b] Backbone[c] % gg	Monomer[b] Backbone[c] % g'g'	Monomer[b] Backbone[c] θPH
ApU Ap--pU	20	34 ± 3	57 / 59	1.32 / 1.44	89 / 95	85	±34	31 / 46	0.45 / 0.85	82 / 89	75	±36
ApC Ap--pC	20	38 ± 2	64 / 75	1.78 / 3.00	85 / 93	85	±33	31 / 55	0.45 / 1.22	82 / 77	75	±36
ApC Ap--pC	80		44 / 61	0.79 / 1.56	69 / 69	70	±36					
GpU Gp--pU	20	27 ± 5	49 / 62	0.92 / 1.63	80 / 92	82	±34	33 / 46	0.49 / 0.85	72 / 89	75	±37
GpC Gp--pC	20	45 ± 4	71 / 79	2.45 / 3.76	76 / 96	89	±33	33 / 55	0.49 / 1.22	72 / 77	75	±37
UpA Up--pA	20	15 ± 3	53 / 51	1.13 / 1.04	74 / 79	80	±35	56 / 40	1.27 / 0.67	68 / 77	72	±35
CpA Cp--pA	20	24 ± 2	71 / 57	2.45 / 1.33	79 / 90	84	±35	60 / 40	1.50 / 0.67	71 / 77	72	±35
UpG Up--pG	20	10 ± 5	54 / 50	1.17 / 1.00	75 / 80	81	±35	56 / 37	1.27 / 0.59	68 / 71	71	±35
CpG Cp--pG	20	25 ± 3	71 / 57	2.45 / 1.33	79 / 85	85	±35	60 / 37	1.50 / 0.59	71 / 81	71	±35
CpG Cp--pG	80		53 / 48	1.13 / 0.92	63 / 67	72	±35					

[a] Computed by using $J_{1'2'} + J_{3'4'} = 9.5$ Hz for the dimers and $J_{1'2'} + J_{3'4'} = 9.3$ for monomers. [b] Monomer data for solutions at pD = 5.4. [c] Rotamer equations used, gg = $(13.7 - \Sigma)/9.7$; g'g' = $(25 - \Sigma')/20.8$. [d] K_{eq}, $^2E \rightleftharpoons {}^3E$; estimated errors in K_{eq} are ±0.2. [e] % stacked = $(J_{1'2'}(\text{monomer}) - J_{1'2'}(\text{dimer}))/J_{1'2'}(\text{monomer})$ or $(J_{3'4'}(\text{dimer}) - J_{3'4'}(\text{monomer}))/(9.5 - J_{3'4'}(\text{monomer}))$. The values are the average % stacked for the purine and pyrimidine fragments.

infra); (ii) increase in J1'2' and decrease in J3'4' with the sum J1'2' + J3'4' and J2'3' remaining constant; (iii) observation of a fine four bond coupling between H2' and the phosphorus atom. These observations indicate that there is a conformational interconnectedness between C3'-O3' torsion, sugar pucker and base-base interaction and led to the postulation of **stereochemical domino effects**.[17-18] We envision a coupled series of conformational events at the onset of stacking made feasible by the swivel nature of the O3'-P-O5' bridge, i.e., torsion about β and γ enables the bases to stack, this causes a reduction in χ_{CN} with an accompanying increase in ^3E populations, the latter in turn shifting the C3'-O3' to α^- domains.

The individual nucleotidyl units in the dimers differ in several key ways from the corresponding monomer conformations (Tables II and III). Specifically, the ribose ring becomes increasingly ^3E, C4'-C5' and C5'-O5' increasingly gg-g'g', i.e., the molecule attempts to achieve conformational purity and identity with specific conformations upon oligomerization.

Above we have described the conformational properties of the individual segments of a dinucleoside monophosphate. However, to arrive at their intramolecular order and molecular topology it is also necessary to determine the sugar base torsions χ_1

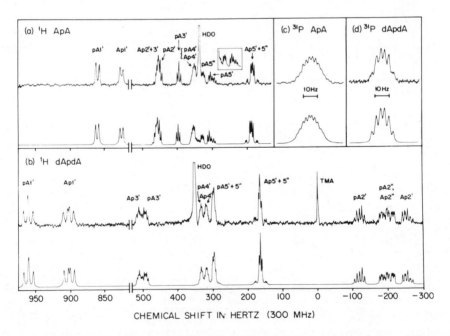

Figure 7. (a-b) 300 MHz ^{1}H NMR spectrum of the ribose region (a) ApA and (b) d-ApdA at 72° respectively with chemical shifts expressed in Hz downfield from internal tetramethylammonium chloride in D$_2$O. The simulations are presented beneath each NMR spectrum. (a) also contains an insert of the pA 5' region at 68°. (c-d) The 40.48 MHz ^{31}P FT NMR spectrum of (c) ApA at 40° and (d) d-ApdA at 27° along with simulations. Concentrations are 0.05M, pD 7.4.

and χ_2 as well as the phosphodiester torsions β and γ. All the above four angles are closely interrelated because changes in β and γ (under conditions in which α, δ and ϵ and sugar pucker occupy their preferred torsions) can bring the bases closer together to stack which in turn will changes χ_1 and χ_2.

No satisfactory methods exist to determine χ_1, χ_2, β and γ from coupling constants. An elaborate method based on the calculation of contribution to shielding from ring current, diamagnetic anisotropy and paramagnetic anisotropy from x, y, z coordinates has been developed by Sarma and coworkers and this was discussed extensively earlier.[13] This approach in addition to providing hard data for χ_1, χ_2, β and γ, unmistakably has shown that in aqueous solution dinucleoside monophosphates are **conformationally pluralistic**. This pluralism results because of the presence of a flexible conformational framework. While flexibility is allowed and alternate conformations are accessible, these molecules nevertheless attempt to achieve conformational identity by showing preferences—sometimes overwhelming preferences—for certain orientations. The composition of the blend and the preferred intramolecular order are largely determined by the constitution and sequence which control the torsional and flexural movements about the various bonds in the molecular system.

For example, UpA principally exists as a conformational blend of stacked, skewed, and extended arrays involving g⁻g⁻ ($\beta/\gamma \simeq 290°/290°$), g⁺t ($\beta/\gamma \simeq 80°/180°$), g⁺g⁺ ($\beta/\gamma \simeq 80°/80°$), tg⁺ ($\beta/\gamma \simeq 180°/80°$) and tg⁻ ($\beta/\gamma \simeq 180°/290°$) conformers in which the g⁻g⁻ and g⁺t predominates.[19] On the other hand, UpU exists primarily in an extended array. We have described elsewhere[9,10,17-19] the conformational properties of common ribodinucleoside monophosphates in detail.

2. The Modified Ribodinucleoside Monophosphates

A comparison of the conformational properties of common dinucleoside monophosphates with those containing modified bases and sugar rings has provided some important information about the effect of these modifications on the composition of the conformational blend. It was shown earlier[13] that substitution of a common base by ϵA usually increases the population of the skewed array g⁺t (Figures 23, 26, reference 13). Even modification of the base by methylation can have significant influence on the conformer distribution. NMR studies of m¹Apm¹A, m¹ApU and Upm¹A indicate[20] that methylation causes a shift in the $^2E \rightleftharpoons {}^3E$ equilibrium to the left and increases the population of disordered unstacked structures, with little effect on α, δ, and ϵ torsions. The 7-methylation of guanosine in 5'-GMP shifts the $^2E \rightleftharpoons {}^3E$ equilibrium so much to the right that 7-methyl-5'GMP preferentially exists in ^3E and 5'GMP in ^2E conformation. Comparison of the data for CpC and CmpC as well as ApA and AmpA[22,23] indicates that methylation at the 2' of the ribose can influence the conformational properties of these molecules. Thus 2'-0-methylation of the Cp- of CpC causes: (i) a reduction in the magnitude of χ_1, (ii) an increase in the population of ^3E pucker at Cp- and (iii) perturbations of β and γ. In the case of ApA, the significant effect of 2'-0 methylation of Ap- is destacking, shift of $^2E \rightleftharpoons {}^3E$ to the left for Ap- and change of α. In Figure 8 is shown the preferred conformation of AmpAp. Comparison of this with the preferred spatial configuration of ApA (Figure 24, reference 13) clearly shows that the major effect of 2′-0-methylation in ApA is chain extension, creating a cavity between the two adenine nucleotidyl units. The implications of these findings with respect to biologically functional polynucleotides will be discussed later.

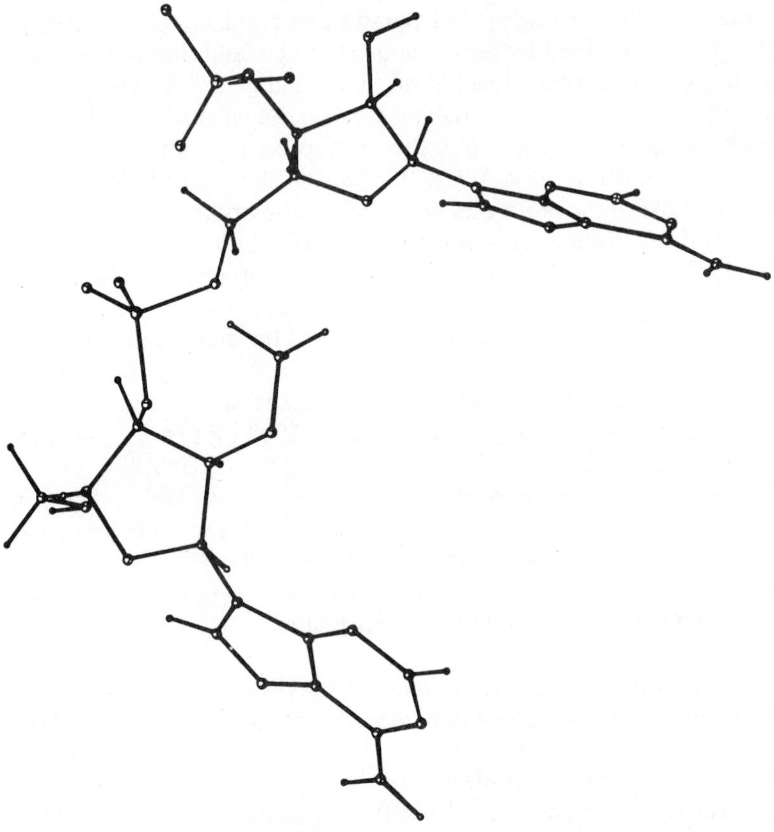

Figure 8. The preferred spatial configuration of AmpAp. The ribose of Amp- is ^2E, that of -pAp- = ^3E; $\chi_1 = 40°$, $\beta = 270°$, $\gamma = 220°$, $\delta = 195°$, $\epsilon = 60°$ and $\chi_2 = 100°$.

3. The Ribotrinucleoside Diphosphates —
The Looped Out and Bulged Configurations

The conformational deductions of ribodinucleoside monophosphates have shown that the spatial configuration properties of RNA oligomers are strongly influenced by the chain length and sequence and the monomeric units undergo interdependent stereodynamical changes as they become integrated into the framework of ribodimers. A study of the trinucleoside monophosphates will indicate how these conformational changes are transmitted and propagated beyond the dimer level.

Theoretically there are 64 possible ribotrinucleoside diphosphates. A complete analysis of an ^{1}H NMR spectrum of a ribotrimer is a formidable task. However, with the help of selective deuteration, homo- and hetero nuclear decouplings and high field NMR studies so far complete analysis of the 270 MHz ^{1}H NMR spectra of three ribotrimers have been completed.[11/24]

The ^{1}H NMR spectra of ApApA along with its complete simulation and assignments are illustrated earlier.[13] Analysis of the NMR data from the [24] trimers ApApA, CpCpA and ApCpC indicates that the dominant spatial array in the conformational blend is what is expected from the trend one has seen in progressing from monomer → dimer → trimer, i.e., in each residue the sugar pucker is ^{3}E, $\alpha \simeq 205°$, $\beta\gamma = g^-g^-$, $\delta \simeq 180°$, $\epsilon \simeq 60°$ and sugar base torsion anti. Such a spatial configuration[24] for CpCpA is illustrated in Figure 9e.

The data[24] reveal that as *one progresses from monomer to dimer and dimer to trimer the major conformational purification takes place at the first step, i.e., from dimer to trimer the magnitude of changes are small.* The conformational freedom and flexibility in the trimer enables it to assume a variety of spatial configurations as the less dominant ones. For example, in reference 24, in the cases of CpCpA and ApCpC, evidence is presented for the presence of unusual looped out spatial configurations in which the central cytidine unit -pCp- is bulged out enabling stacking interactions between the terminal adenine and cytosine bases. Some of the details of these will be discussed in connection with deoxyribotrinucleoside diphosphates (*vida infra*).

Further a comparison of the spatial configuration data for ApCpC and CpCpA provides insights regarding the effect of sequence on local conformational flexibility. For example, in CpCpA the population of $\epsilon_2 = 60°$ conformers is noticeably less compared to CpA; movement of adenine from the 5' end (i.e., ApCpC) to the 3' end (i.e., CpCpA) shifts the ^{2}E $\rightleftharpoons$ ^{3}E equilibrium toward ^{2}E so much so that the sugar ring of -pA of CpCpA displays equal proclivity for ^{2}E and ^{3}E conformations. This information along with the observation (*vide supra*) that in aqueous solution ^{2}E is associated with $\alpha \simeq 275°$ and ^{3}E with $\alpha \simeq 205°$, and in the anti domain χ_{CN} is larger for the ^{2}E systems compared to the ^{3}E, clearly indicates that in addition to the spatial array shown in Figure 9e, conformers in which the geometric details of the adenosine moiety display considerable variation from other residues make noticeable contribution to the conformational blend of CpCpA, the acceptor end of tRNA.

The Conformational Destiny of the Ribomononucleotide as it Becomes Part of the Ribopolynucleotide. The Untenability of the Rigid Nucleotide Concept in Solution

From the extensive examination of solid-state data Sundaralingam[25/26] has concluded that nucleotides are considerably more rigid than nucleosides and that nucleotides essentially maintain their isolated conformations when they become part of a polynucleotide. This is the concept of a rigid nucleotide. The concept also states that the nucleic acids may achieve conformational changes by torsional variations about

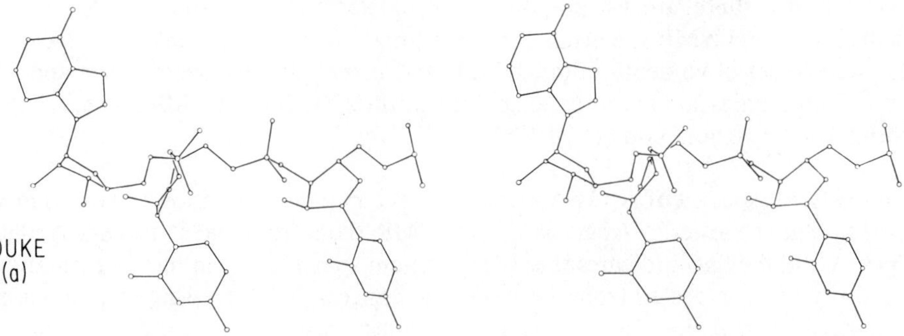

DUKE
(a)

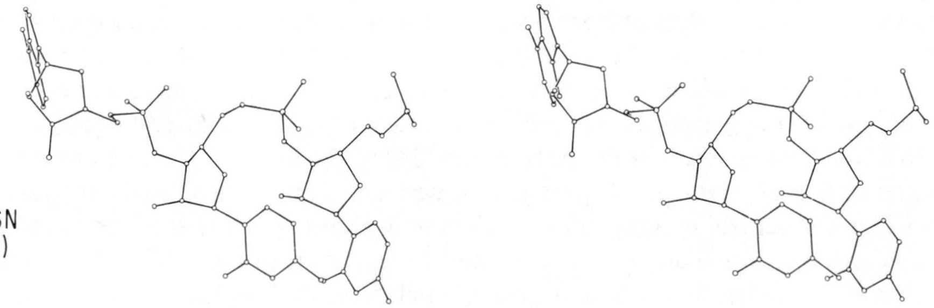

MSN
(b)

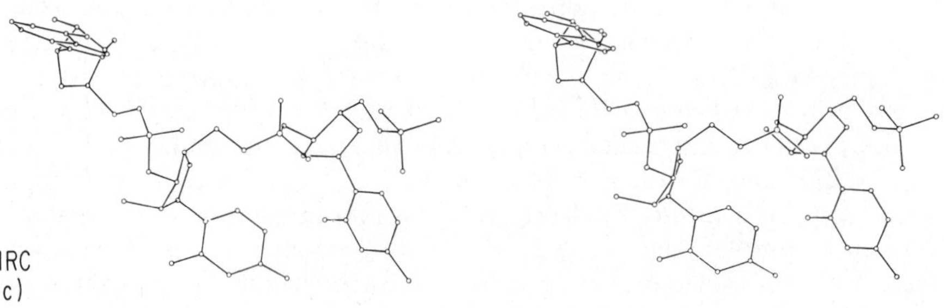

MRC
(c)

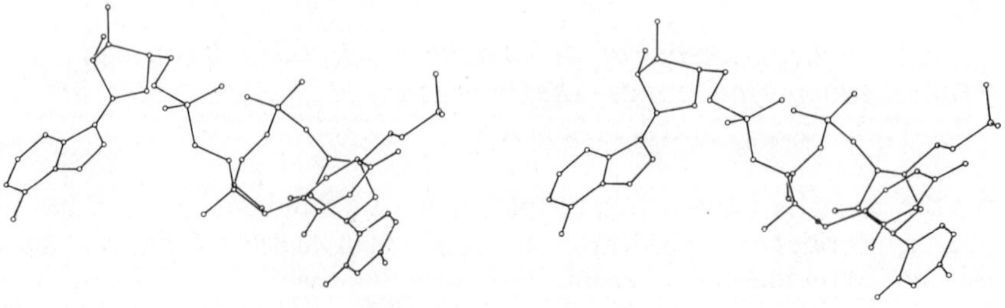

MIT
(d)

Figure continued, next page

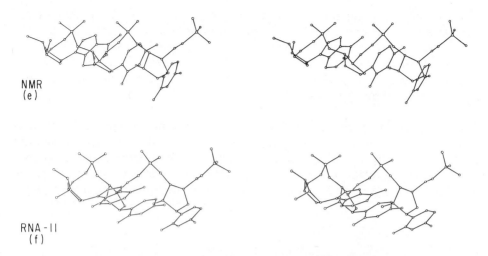

Figure 9. The spatial configurations of CpCpA terminus of tRNA. The perspectives are in stereo. They are drawn from the published coordinates from the four laboratories identified in the diagram. Also given is the preferred configuration of CpCpA determined from NMR studies (e). The model from the NMR data is unrefined and will undergo some changes on refinement. A possible model based on RNA 11 coordinates is shown in (f). There is remarkable similarity between (e) and (f).

O3′-P and P-O5′ bonds.

Aqueous solution data on purine ribonucleosides, and the corresponding 3′ and 5′ mononucleotides as well as 3′, 5′ nucleoside diphosphates clearly reveal that the backbone conformations of ribo nucleosides are as flexible as that of the corresponding nucleotides.[11] Examination of the ¹H NMR parameters in adenosine, 3′AMP, 5′AMP, 3′,5′ADP, ApA, ApApA, and poly A under various conditions reveals that **the conformational destiny of mononucleotides as they become part of polynucleotides is ordained by their isolated conformations only when significant amounts of the polymer are in the destacked state. The conformational conservation breaks down as soon as the mononucleotides become integrated into the backbone framework of a biologically functional base stacked polynucleotide.**This is dramatically illustrated in Figure 10 where the sugar conformation data are presented for adenosine, 3′AMP, 5′AMP, 3′5′ADP, ApA, ApApA and poly A under conditions of high base stacking (low temp) and low base stacking (high temp). Base stacking interactions have a profound impact on the conformational properties of ribo nucleic acids; in aqueous solution their major effect is to diminish conformational freedom. Comparison of the computed percentage populations of ³E conformers for stacked poly A, ApApA, and ApA with that of the component monomers (Figure 10) reveals that the ribose ring in the mostly stacked oligomers and polymer displays a conformational preference dramatically different from that of the monomeric components. The ³E populations in the monomers are about 40%, while the corresponding populations for ApA, ApApA, and poly A in conditions which favor stacking lie in the range of 60-80%. The higher end of the range is preferred by the central -pAp- unit of

ApApA and the nucleotidyl units in poly A. It should be noted that *as the monomeric components become integrated into the framework of a biologically functional stacked polymer, not only is there a significant increase in the populations of ³E conformers, but also there is a shift in the kind of pucker they prefer, that is, the monomers prefer ²E sugar pucker (≃ 60%), but the oligomers and polymer prefer ³E sugar pucker.*

From the foregoing discussion, it is clear that the conformational properties of common ribonucleic acid components, oligo- and polynucleotides in aqueous solution are not governed by a common principle such as the rigid nucleotide concept. Indeed recent crystallographic studies by Neidle, Berman and their coworkers[27/28] of the complex between ApA and proflavine indicate that the structure has little in com-

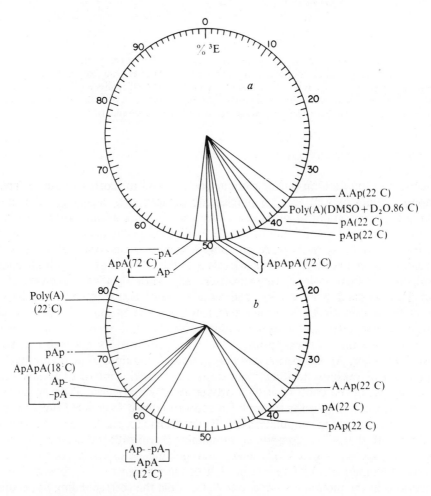

Figure 10. Computed percentage populations of the ³E ribose conformers in adenine nucleotide systems under destacked (a) and stacked (b) conditions. From reference 11.

mon with other dinucleoside phosphates and violates one of the tenets of the rigid nucleotide concept that the phosphodiester torsions are the key flexible domains in nucleic acids.

Relevance to Ribonucleic Acid Structure

The aqueous solution study of the stereodynamical properties of the oligoribonucleotides have revealed beyond any doubt that constitution, sequence and chain length have a profound effect on the spatial configurations they adopt in solution. Is this of any value to obtain information about biologically functional polynucleotides? At the outset, it should be stated that generalizations from solution conformational data of oligomers to derive the *intimate details and nuances* of polynucleotide behavior is not valid. However, the studies unequivocally provide the stereodynamical principles which go into the engineering and crafting of a living polymer. There is no reason to believe that the conformational interconnectedness and the stereochemical domino effects which we have discussed for the oligomers cannot operate at the polymer level. Our observation that major conformational changes happen at the first step in oligomerization, i.e., from monomer to dimer and that at the level of a trimer there is considerable conformational flexibility suggests that one cannot expect in aqueous solution polynucleotides with single rigid spatial configuration and that they must possess a great deal of flexibility. Recent examination of the anatomy of tRNA *viz-a-viz* polynucleotide flexibility by Rich, *et al.*[29] support this thesis, where they find the molecule adopting unusual conformations associated with chain extension, changes in direction of the polynucleotide chain or with the accommodation of other bases which intercalate into the chains. The observation that in aqueous solution the oligonucleotide CpCpA is structured in such a way that it is capable of adopting a variety of spatial configurations with considerable flexibility for the terminal adenine nucleotide unit suggests that it may be so for the acceptor end of tRNA. Examination of the four reported crystal structures of t-RNA—two from orthorhombic[30/31] and two from monoclinic[32/33] indicate considerable variation in the structure of residues 74, 75 and 76 (Figure 9) which could be a manifestation of the intrinsic flexibility of the CpCpA oligomer. Such conformational variability may be implicated in recognition of amino-acyl tRNA synthetase.

The observation of the effect of methylation of the base or the sugar on the conformational properties of dimers suggest that one of the roles of methylation may be to induce local conformational changes in the polymer. Is it not striking that residue 58 in tRNA[Phe] is N[1] methylated adenine nucleotide and it is the linking nucleotidyl residue between TψC arm (stem) and TψC loop regions where the ordered stacked structure of the stem region alters into a more open loop conformation and that N1 methylated adenylate dimers themselves have a disordered spatial configuration.

Study of the conformational effects of 2'-O-methylation of ribose and 7-methylation of guanosine have been of great value to arrive at the spatial configuration of the bizarre 5' terminus of mammalian messenger. In Figure 11 is shown the

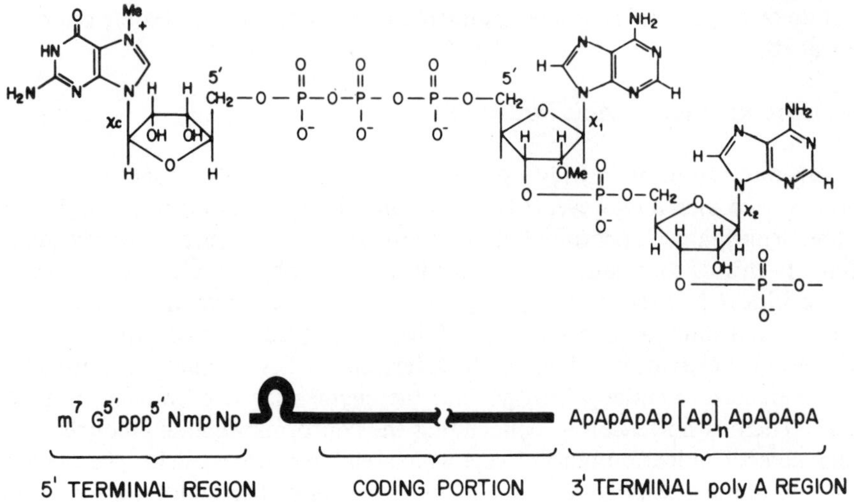

Figure 11. The sequence of mRNA at the 5′ region. The above sequence can be abbreviated as m⁷G⁵ppp⁵AmpA. Bottom part shows the schematic representation of mRNA connecting the terminus with the coding portion via a looped region.

sequence of mRNA at the 5′-region, and in Figure 12 is depicted a proposed model for the terminus in which the 7-methylated guanosine segment is shown to inter-calate in between the two adenine nucleotidyl units.[21] As was discussed before, the presence of the 2′-O-methyl group in the AmpA segment causes chain extension and creates a cavity into which the 7-methyl guanosine group can "slip" in and inter-calate. It should be noted that in the intercalated trimer (Figure 12) the 2′-O-methyl, the 7-methyl, H2′ and H3′ of Amp- and m⁷G⁵p- parts are in close vicinity creating hydrophobic interactions and a water free channel. Details are in reference 21.

Stereodynamics of DNA Oligonucleotides

From the extensive NMR studies of ribooligonucleotides reported above, we have obtained information about their conformational features, properties and dynamics. One is curious to know whether the simple substitution of 2′OH by hydrogen in the sugar ring will have far-reaching repercussions with respect to the three dimensional spatial configuration and whether such conformational effects may enable us to understand the profound differences that exist in the physico-chemical and physiological properties between ribo- and deoxyribonucleic acid structures. This is attempted below.

1. The Deoxyribodinucleoside Monophosphates

Of the sixteen possible deoxyribodinucleoside monophosphates, detailed conforma-tional deductions have been made for fifteen of them by complete analysis of their

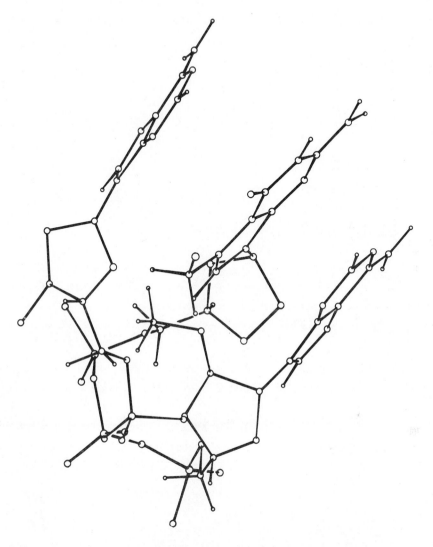

Figure 12. A proposed model for the 5'-terminus m^{7}G$^{5'}$ppp$^{5'}$AmpA of mRNA. For details see reference 21.

NMR spectra.[6] A sample spectrum along with the simulation for a dexoyribo dimer is shown in Figure 7 bottom. The conformational parameters for the dimers and monomers are given in Table IV. The data indicate that the deoxy dimers exist as a conformational blend in which certain spatial arrays predominate. The pentose ring exists as a ^{2}E $\rightleftharpoons$ ^{3}E equilibrium with significant preference for the ^{2}E pucker in all cases irrespective of the nature of the base and sequence. The C4'-C5' and C5'-O5' bonds form a conformational network in which $\delta \simeq 180°$ and $\epsilon \simeq 60°$ are preferred. The C3'-O3' bond occupies a domain in which $\alpha \simeq 200°$. Except for d-TpT, d-TpC, and d-CpT the dimers exist significantly in stacked arrays in which $\beta\gamma$ is in the g$^-$g$^-$

Table IV

Population Distribution of Conformers in Deoxyribodinucleoside Monophosphates and Their Components

Nucleotide		Temp, °C	Dimer					Monomer[d]				
			Sugar ring[a]		Backbone[b]			Sugar ring[a]		Backbone[b]		
			% ³E	K_{eq}[c]	% gg	% g'g'	θPH, deg	% ³E	K_{eq}[c]	% gg	% g'g'	θPH, deg
d-ApA	dAp-	27	22	0.3	73	85	±45	23	0.3	71	70	±38
	-pdA		37	0.6	87			28	0.4	63		
d-ApA	dAp-	72	20	0.3	59	76	±43					
	-pdA		37	0.6	76							
d-GpG	dGp-	80	28	0.4	57	86	±45	23	0.3	71	67	±38
	-pdG		39	0.6	90			32	0.5	63		
d-ApG	dAp-	27	19	0.2	65	86	±45					
	-pdG		31	0.4	94							
d-ApG	dAp-	72	26	0.4	57	76	±44	26	0.4	57	70	±38
	-pdG		37	0.6	76			28	0.4	63		
d-GpA	dGp-	20	33	0.5	69	86	±40					
	-pdA		37	0.6	84							
d-GpA	dGp-	80	28	0.4	58	74	±42					
	-pdA		32	0.5	53							
d-CpC	dCp-	20	33	0.5	67	76	±40	35	0.5	58	75	±38
	-pdC		32	0.5	75			33	0.5	70		
d-CpC	dCp-	80	33	0.5	67	69	±40					
	-pdC		38	0.6	61							
d-TpT	dTp-	18	30	0.4	63	82	±41	33	0.5	57	74	±41
	-pdT		37	0.6	74			32	0.5	59		
d-TpT	dTp-	65	33	0.5	49	77	±39					
	-pdT		34	0.5	64							
d-CpT	dCp	20	33	0.5	63	74	±37	35	0.5	58	74	±38
	-pdT		32	0.5	71			32	0.5	59		
d-CpT	dCp-	80	31	0.4	68	76	±37					
	-pdT		37	0.6	73							
d-TpC	dTp-	20	27	0.4	56	77	±38	33	0.5	57	75	±41
	-pdC		32	0.5	79			33	0.5	70		
d-TpC	dTp-	80	32	0.5	56	71	±38					
	-pdC		34	0.5	69							

Table IV *(continued)*

Nucleotide	Temp, °C	%^3E	K_{eq}[c]	%gg	%$g'g'$	θPH, deg	%^3E	K_{eq}[c]	%gg	%$g'g'$	θPH, deg
d-ApC											
dAp-	27	26	0.4	71			23	0.3	71		
-pdC		37	0.6	63	82	±41	33	0.5	70	75	±38
dAp-	80	31	0.5	61							
-pdC		37	0.6	53	72	±41					
d-ApT											
dAp-	27	32	0.5	71			23	0.3	71		
-pdT		41	0.7	86	89	±44	32	0.5	59	74	±38
dAp-	80	26	0.4	65							
-pdT		49	1.0	42	72	±39					
d-GpC											
dGp-	20	31	0.5	65			26	0.4	57		
-pdC		37	0.6	73	82	±40	33	0.5	70	75	±38
dGp-	80	31	0.5	63							
-pdC		37	0.6	63	73	±41					
d-GpT											
dGp-	20	30	0.4	55			26	0.4	57		
-pdT		44	0.8	81	82	±39	32	0.5	59	74	±38
dGp-	80	33	0.5	55							
-pdT		31	0.5	81	82	±41					
d-CpA											
dCp-	20	28	0.4	58			35	0.5	58		
-pdA		36	0.6	90	86	±40	28	0.4	63	70	±38
dCp-	80	28	0.4	58							
-pdA		36	0.6	63	75	±39					
d-CpG											
dCp-	20	33	0.5	55			35	0.5	58		
-pdG		37	0.6	75	86	±40	32	0.5	63	67	±38
dCp-	80	25	0.3	65							
-pdG		37	0.6	56	75	±44					
d-TpA											
dTp-	20	24	0.3	55			33	0.5	57		
-pdA		36	0.6	90	86	±37	28	0.4	63	70	±41
dTp-	80	33	0.5	63							
-pdA		36	0.6	61	80	±40					
d-TpG											
dTp-	20	24	0.3	55			33	0.5	57		
-pdG		35	0.5	75	86	±37	32	0.5	63	67	±41
dTp-	80	24	0.3	55							
-pdG		29	0.4	79	88	±35					

[a] Computed by using $J_{1'2'} + J_{3'4'} = 10.8$ Hz for the dimers and $J_{1'2'} + J_{3'4'}$ for the monomers. [b] Rotamer equations used: $gg = (13.7 - \Sigma')/9.7$; $g'g' = (25 - \Sigma')/20.8$. [c] K_{eq} $^2E \rightleftharpoons {}^3E$. [d] Monomer data for solutions at pD 5.4.

domain. Elevation of temperature and consequent destacking have only minor effects on pentose conformation and that about α. However, it leads consistently to changes of β from g⁻ to t domains with little effect on γ. The dimers d-TpT and d-

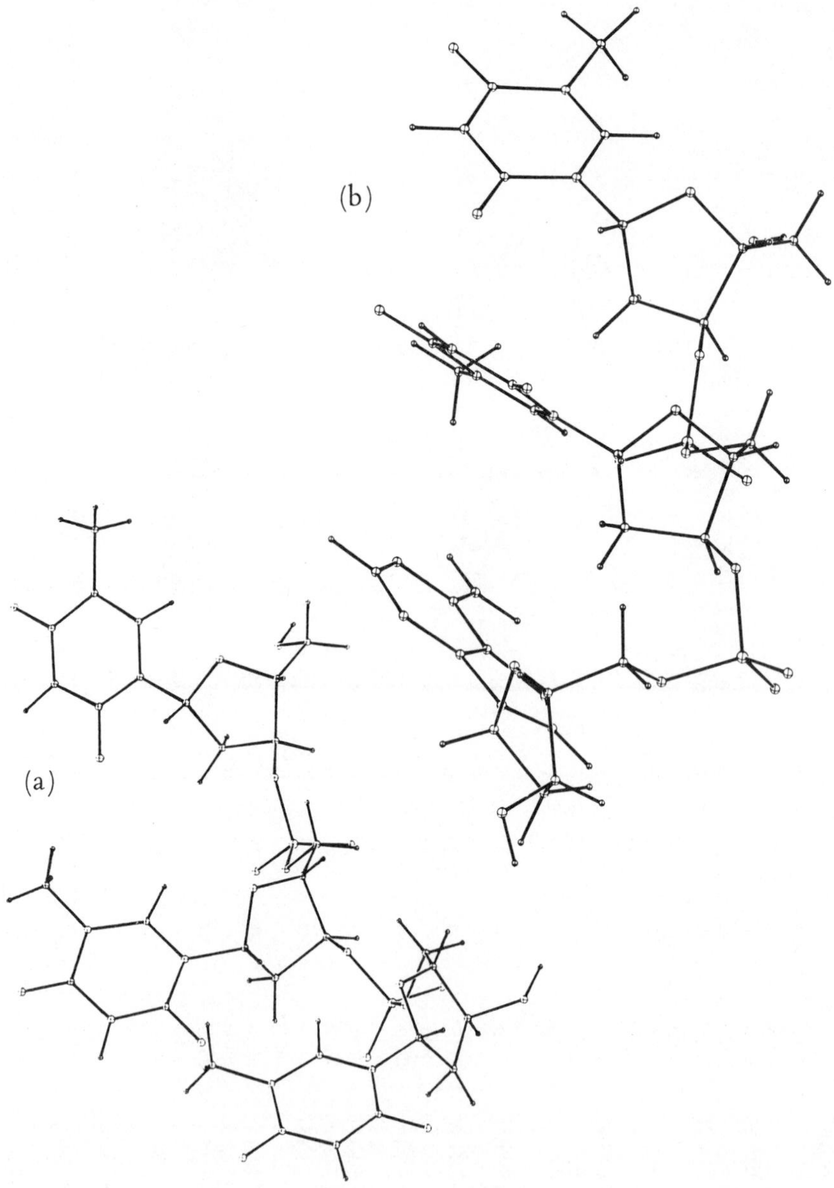

Figure 13. (a) The preferred spatial configuration of d-TpTpT in aqueous solution. In each of the individual nucleotidyl units χ_{CN} = anti and sugar pucker is ^2E; in addition ϵ', ϵ_1, ϵ_2 = 60°, δ_1, δ_2 = 180°, α_1, α_2 = 199°, $\beta_1\gamma_1$ and $\beta_2\gamma_2$ = tg⁻ (b) The preferred spatial configuration of d-TpTpA in aqueous solution. Except for $\beta_2\gamma_2$ = g⁻g⁻, all angles are the same as in Figure 13(a).

TpC display tg⁻ conformation for the phosphodiester bonds. Oligomerization of monomer to dimer results in significant increase in the population of gg and g'g' conformers about the internucleotide C4'-C5' and C5'-O5' bonds with no noticeably important changes in other conformational parameters.

2. *The Deoxyribotrinucleoside Phosphates.*
The Bulged and Looped Out Configurations

Complete analysis of the ¹H NMR spectra of two deoxyribotrinucleoside diphosphates (d-TpTpT and d-TpTpC) at 20° and 80°C have been reported so far.[12] The observed NMR parameters indicate that the conformational properties of the trimers are very similar to those of constituent dimers, i.e., in the deoxy systems the dimers conserve their intrinsic conformational features when they are incorporated into the oligomers. The preferred spatial configuration of d-TpTpT is shown in Figure 13a.

The conclusion that the trimers d-TpTpT and d-TpTpC essentially maintain the isolated conformations of the corresponding dimers agrees with similar conclusions by Kan *et al.*[34] about d-ApTpT and d-TpTpA from an incomplete analysis of their NMR spectra. However, Kan, *et al.* conclude that the dimer segments of d-ApTpT and d-TpTpA significantly populate in the naturally preferred right handed arrangement ($\beta_1/\gamma_1 \simeq 290°/290°$ and $\beta_2/\gamma_2 \simeq 290°/290°$). Recent extensive ¹H NMR studies[6] of d-TpT, d-ApT and d-TpA have shown that d-TpT prefers the tg⁻ domain ($\beta/\gamma \simeq 180°/290°$) and d-ApT and d-TpA the g⁻g⁻ ($\beta/\gamma \simeq 290°/290°$) domain. Based on this new information, one should conclude that the phosphodiester bonds of the d-TpT segment of d-TpTpA and d-ApTpT preferentially exist in tg⁻ orientation and that of the d-TpA and d-ApT segments prefer the traditional g⁻g⁻ domain. (Figure 13b) In the conformation shown in Figure 13b, the magnitude of cylindrical coordinates are such that H6, H1' and the methyl protons of dTp-residue will experience no shielding from the adenine moiety. This means that the actually observed distant shielding of the terminal dTp-residue of d-TpTpA by the terminal adenine[34] does not originate from the preferred conformation of the trimer (Figure 13b). Kan *et al.*[34] have suggested that the distant shielding may originate from right-handed stacks. In such stacks, $\beta_1\gamma_1$ and $\beta_2\gamma_2$ occupy g⁻g⁻ domains, and bases are stacked, and it is clear from the cylindrical coordinates[12] z, ϱ_5 and ϱ_6, the bases of the end nucleotidyl units are too far apart to have any mutual shielding effects. Hence the observed distant shieldings cannot originate from conformations in which both the phosphodiester bonds occupy g⁻g⁻ domains, even though such conformers may make minor contribution to the conformational blend. We have shown earlier[13] that the presence of bulged and looped out structures in which the central nucleotidyl unit bulges out enabling the end ones to interact, (Figure 29, reference 13) can be used to rationalize the distant shielding data. Ts'o and coworkers[35-37] have made attempts to study oligomers longer than trimers and have been successful in assigning the base protons by means of the incremental procedure.

Comparison Between RNA and DNA Single Stranded Oligomers

and Relevance to Biological Structures

In the *deoxy* ribo-3' and -5' mononucleotides, as well as in the dimers, irrespective of the nature of the base and sequence, the pentose ring shows a clear preference to exist in ^2E conformation. This is the case in each and every naturally occurring constituent of deoxyribonucleic acids. Furthermore, the conformational distribution of the pentose does not show any meaningful sensitivity to such strong intramolecular perturbations such as stacking/destacking interactions. From these, one is reasonably sure to conclude that the conformation of the pentose moieties in single stranded hetero- and homodeoxyribopolynucleotides in aqueous solution resembles that in their constituents and that the pentose moieties in them will show a great deal of preference to the ^2E mode of pucker. We realize that in crystals both ^2E and ^3E DNAs have been reported. We do not have any evidence that there is any chance that a given sequence of single stranded DNA alone in aqueous solution could populate predominantly in the ^3E form.

This is in sharp contrast to the behavior of ribonucleic acid structures where in aqueous solution the kind of sugar pucker a system prefers is determined by the nature of the base and stacking interactions. For example, the purine ribomononucleotides prefer ^2E pucker, the pyrimidines ^3E pucker. Stacking interactions cause a shift of $^2E \rightarrow {}^3E$ pucker and under conditions in which significant amounts of the polymer exist destacked the preferred pucker for the sugar is ^2E.

In the ribo series it has been shown that torsional variation about C3'-O3' is coupled to the ribose conformation and governed by the equilibrium.

$$^3E\alpha^- \rightleftharpoons {}^2E\alpha^+ \qquad \text{(Figure 19, reference 13)}$$

In the deoxyribo series the favored conformational coupling is between ^2E and α^-. The majority of ribo- and deoxyribodinucleoside monophosphates prefer to exist in stacked arrays in which $\beta\gamma$ lie in the g^-g^- domain. Because of the difference in the sugar pucker, it is obligatory that there should be local differences between the two classes within the g^-g^- conformation space.

An important observation is that temperature induced unstacking causes changes in β from g^- to t with little effect on γ. This suggests that the main torsional event at the onset of unwinding of double helical DNA and RNA is that about O3'-P. In both ribo and deoxyribo series changes in $\beta\gamma$ are accompanied by destacking and χ_1 and χ_2 changes. Such $\chi_1\chi_2$ changes alter the mode of sugar pucker and torsion about C3'-O3' in the ribose series, but no such changes are noticeable for the deoxy systems (compare Figures 14 and 15). This suggests that polyribonucleotides are engineered in such a way that their conformations are very sensitive to minor perturbations and that it is possible that they are capable of fulfilling their multifunctional biological roles as tRNA, mRNA, and rRNA because of this built-in potentiality in their constitution for conformational versatility and pluralism. Even though the solution data project that deoxyribopolynucleotides may be conforma-

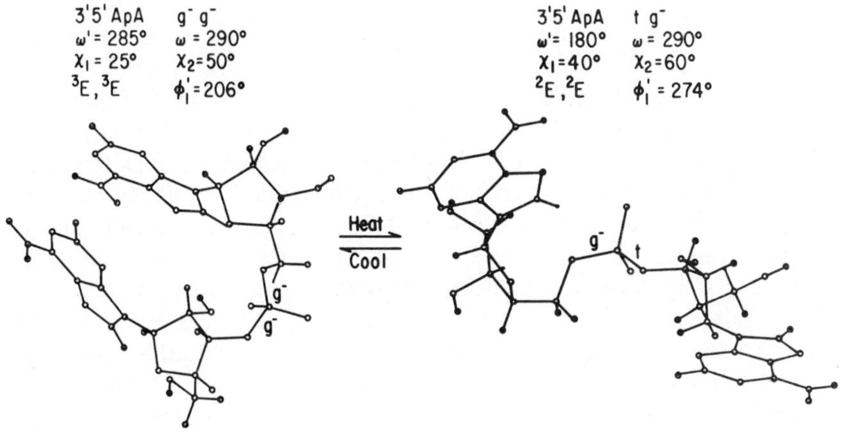

Figure 14. A schematic representation of the conformational events which accompany elevation of temperature for ApA.

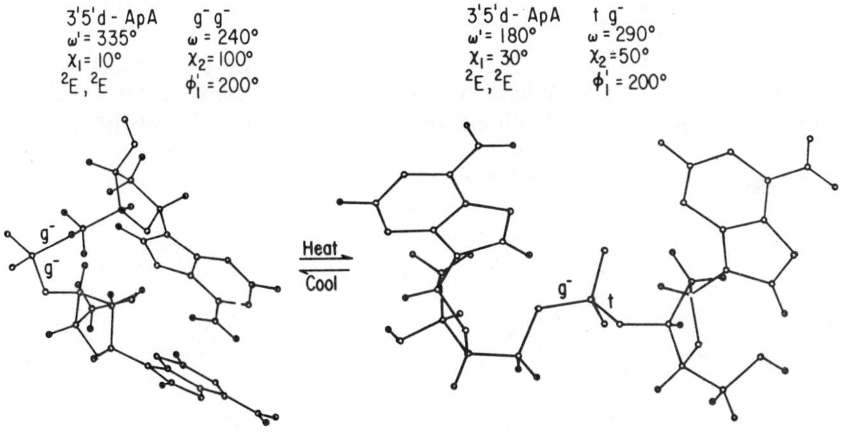

Figure 15. A schematic representation of the conformational events which accompany elevation of temperature for d-ApA.

tionally less versatile than the ribopolymers, crystal data indicate that the deoxys exhibit greater flexibility in the sugar pucker as in the A and B forms of DNA. They are permitted both ^2E and ^3E pucker. The former is sterically disfavored in the RNAs. Consequently in crystalline tRNA, for example, only a few of the approximately 80 residues have sugar pucker ^2E and the reason for this is elaborated by Rich, *et al.*.[29] The reader is directed to the chapters by Seeman[1] and Patel[71] in this volume for a discussion of the conformational versatility of DNA.

Evolution of the 3'5' Internucleotide Linkage

There have been several attempts to answer the questions: Why do nucleic acids have 3'5' and not 2'5' bonds? Does the 3'5' internucleotide bond ordain the nucleic acids with some special spatial configurations? Thermal polymerization of nucleotides in solution, as carried out by Sulston, *et al.*[38] indicated the formation of large fractions of the 2'5' and 3'5' isomers. In view of the exclusive presence of the latter in nucleic acids these workers recognized that total specificity of the 3'5' linkage would be difficult to achieve under prebiotic conditions. To circumvent this problem, it was suggested that a preferential formation of the 3'5' isomer might have resulted from a solid surface of nonbiological origin functioning as a contemporary enzyme. Alternatively, an initial mixture formed indiscriminately of 2'5' and 3'5' isomers might ultimately through an evolutionary process in the presence of a subsequently formed enzyme, result in the predominance of the 3'5' isomer. Recent differential scanning calorimetric studies on 2', 3' and 5' phosphate derivatives of nucleic acid systems[39] indicated that the intrinsic thermal properties of these isomers are significantly different and that the 2' isomers condense to form 2'5' nucleotidyl linkages at a rate considerably slower compared with their 3' and 5' counterparts. Usher[40] and Usher and McHale[41/42] considered the same question some years ago and predicted that the 2'5' bond in oligonucleotides would be more susceptible to non-enzymatic chain breakage than would the 3'5' bond. Thus during the early chemical evolution of nucleic acids, they suggested, there existed a strong selective pressure for the 3'5' bond, and against the 2'5' bond. Very recently we[43] attempted to answer the problem from spatial configuration considerations. We determined the intimate details of the geometry of seven synthetic 2'5' dinucleoside monophosphates and the correspon-

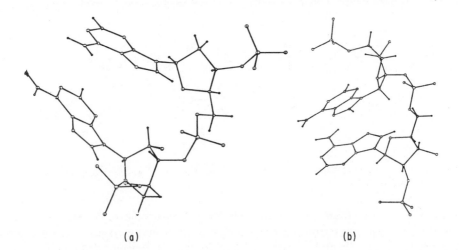

(a)　　　　　　　　　　　　　　　　　(b)

Figure 16. (a) The stacked spatial configuration of 3'5'pApAp in aqueous solution. The ribose of both units are ³E; the various torsion angles are: $\chi_1 = 25°$, $\alpha = 206°$, $\beta = 285°$, $\gamma = 290°$, $\delta = 180°$, $\epsilon = 60°$ and $\chi_2 = 50°$. (b) The stacked spatial configuration of 2'5'pApAp in aqueous solution; the ribose of both units are ³E; the torsion angles are: $\chi_1 = 310°$, $\chi_2 = 60°$, $\alpha = 209°$, $\beta = 295°$, $\gamma = 300°$, $\delta = 180°$ and $\epsilon = 60°$.

ding 3'5' systems, by extensive high frequency NMR studies. The preferred spatial configuration of 3'5'pApAp and 2'5'pApAp are shown in Figure 16a,b. These geometries were then utilized to derive the spatial configuration of 2'5' and 3'5' linked polynucleotides which were considered to be made of repeating units of dimer structures. The study revealed that the intrinsic geometries of the 2'5' systems cannot become part of stable helical configurations with base-base stacking interactions whereas the geometries of the 3'5' systems can. We[43] argued that polynucleotides must be present either completely or significantly in stable helical configurations in order to participate in the fundamental processes of information storage, transfer and retrieval. The exclusive presence of 3'5' phosphodiester bonds in nucleic acids is due to the built in potentiality in their constitution to generate biologically functional stable helices with base-base stacking interactions; 2'5' linkages are not encountered because of their inability to produce comparable helical configurations.

However, in molecular biology nucleic acids function mostly in double helical configurations and the arguments of Dhingra and Sarma[43] are valid only for single stranded structures because they did not consider the perturbations that could be introduced by the presence of a complementary strand, i.e., interactions with a com-

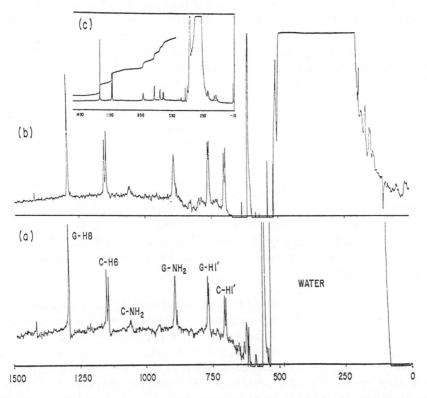

Figure 17. 270 MHz ^{1}H NMR spectra of 2'5' CpG in water at 2°C at neutral pH (a) 5mM, (b) 15mM, and (c) 40mM at 30°C (inset).

plementary strand may introduce sufficient perturbations to the intrinsic geometries of 2'5' systems so much so they may form double helices. In fact, it has been shown that 2'5' linked oligomers can complex with 3'5' linked complementary polymers forming triple helix[44] of the type 1A.2U, although the melting temperatures of these complexes do tend to be lower than in the case of an all 3'5' linked complexes and it is not known whether the complex formation involves Watson-Crick hydrogen bonding.

The tendencies of nucleic acids to form double helices can be probed directly by monitoring the chemical shifts of the amino and iminohydrogens of the bases which are involved in Watson-Crick base pairing. Since these protons exchange with the conventional solvent D_2O, their chemical shifts have to be determined in H_2O solutions and at temperatures at which the exchange rate is sufficiently low to make feasible the observation of their chemical shifts. 270 MHz ^1H NMR spectra of self complementary 2'5'CpG and an equimolar mixture of 2'5'CpC and 2'5'GpG and the corresponding mononucleotides were taken under water suppression conditions using the methods of Bleich and Glasel.[45] In Figure 17 are shown the observed spectra of 2'5'CpG in H_2O at 2°C, 5mM and 15mM. It is obvious that the spectra are noisy and there are phase problems near the huge suppressed water resonance. This is simply because these are extremely difficult experiments to perform at the employed concentrations and temperature in water. As can be seen from the inset in Figure 17 (c) increase in concentration and elevation of temperature results in noise free clean spectra. However, the spectra at lower concentrations at 2°C (Figure 17 ab) are sufficiently clear to obtain accurate NMR parameters.

Krugh, *et al.*[46] demonstrated that 3'5' CpG forms miniature duplexes

$$2(CpG) \rightleftharpoons \begin{matrix} \longrightarrow \\ CpG \\ GpC \\ \longleftarrow \end{matrix}$$

by following the shift trends of guanosine and cytosine NH_2 in water. Increasing the concentration from 0 to 20 mM at 4°C caused the guanosine NH_2 to move by 0.15 ppm (i.e., $\simeq$ 40 hertz in our 270 MHz system) and cytosine NH_2 by 0.5 ppm (i.e., 135 hertz in our 270 MHz system) to lower fields indicating that at 4°C increasing concentration causes an increase in the population of the self-complementary duplexes of 3'5' CpG.

In Figure 18 are presented the variation of the chemical shifts of the relevant protons of 2'5'CpG as a function of concentration at 2°C. Lowest possible temperature was employed because miniature double helices are expected to be much stabler at low temperatures. A striking feature of the plot is the invariance of the chemical shift of guanosine NH_2 and cytosine NH_2 of 2'5'CpG in the concentration range of 5-20 mM, a range in which the same protons of 3'5'CpG displayed pronounced downfield shifts of 40 and 135 hertz.[46] At higher concentrations when vertical stacking and guanine aggregation become important, the base protons, H1' and guanine NH_2

begin to show shifts (Figure 18). The constancy of the guanine NH_2 and cytosine NH_2 chemical shifts of 2'5'CpG in the relevant concentration range (5-20 mM) and low temperature (2°C) clearly indicates that no Watson-Crick base paired self complementary miniature double helices are formed for 2'5'CpG. To determine whether this observation has any general validity, we performed similar experiments using complementary mixtures of 2'5'CpC and 2'5'GpG. In this experiment as is generally

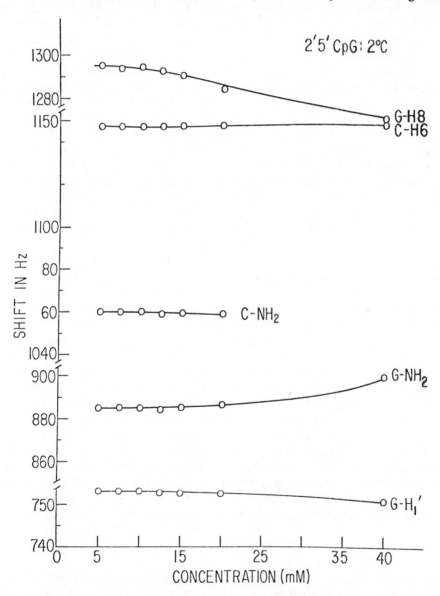

Figure 18. Concentration dependence of the chemical shifts of several protons of 2'5'CpG at 2°C in water.

true of GpG systems, we have been unable to locate the guanosine NH$_2$ resonances. However, the chemical shift of the observable cytosine NH$_2$ in 2'5'CpC + 2'5'GpG equimolar mixture was found to be invariant with concentrations in the 5-20 mM range at 2°C. This observation also suggests the inability of 2'5' systems to engage in Watson-Crick hydrogen bonding network to form miniature double helices.

The data discussed indicate that 2'5' CpG as well as equimolar mixtures of 2'5'CpC and 2'5'GpG do not form Watson-Crick hydrogen bonded miniature double helices under conditions in which their 3'5' analogs have been shown to form such complexes. The reported equilibrium constant for the formation of self-complementary double helices[46] by 3'5'CpG is 14.0 ± 2.5 M^{-1}. No such data for the 2'5'CpG system can be obtained because we do not detect miniature duplex formation. In this connection, it is important to note that in the reported crystal structure of 2'5'ApU no miniature double helices have been observed.[47] We have argued elsewhere[43] that nucleic acids have 3'5' and not 2'5' phosphodiester bonds because the intrinsic molecular stereodynamics of 2'5' internucleotide systems are such that they cannot support stable helical structures. The present study reveals that the availability of a complementary strand does not introduce enough perturbation to the intrinsic stereochemistry of 2'5' systems to induce the formation of double helices. In most instances in biology, polynucleotides express themselves in double helical configurations and this probably provided the 3'5' linkages an evolutionary advantage over their 2'5' analogs.

The Double Helical Oligonucleotides

While the application of NMR methods has yielded a wealth of information about the geometry and dynamics of single stranded oligonucleotides, their application to base paired oligonucleotides has not yielded comparable information. This is primarily because of the difficulty to analyze complex spectra from the duplexes from large oligomers. So far there have been only a single complete analysis of the spectra of a duplex and this is for a miniature self complementary duplex of an analog of ApU in which the adenine and uracil were chemically linked to the sugar moieties[48/49] at a χ_{CN} of approximately 120°. This study provided the first experimental support for vertically stabilized double helices in which the base planes are parallel to the helical axis and the backbone is right handed for high antipolynucleotides as has been projected by Olson from theoretical calculations.[50] Most of the NMR studies of the duplexes have been concerned with effect of composition, sequence and chain length and helix coil transition. From these some qualitative information about their spatial configuration has emerged.

1. Base Paired Oligomers of the RNA Family

Krugh, et al.[46/51] have monitored the shift of the amino protons in a number of self complementary ribodinucleoside monophosphates CpG, GpC, UpA and the complementary mixture GpU + ApC as a function of concentration at low temperatures (Figure 19). They observed a large downfield shift for the guanine amino and one of

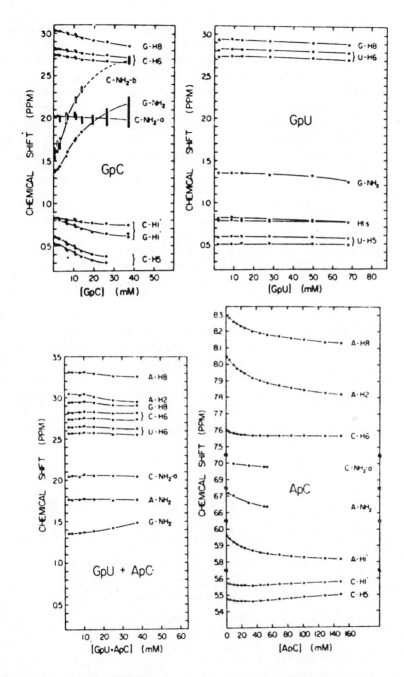

Figure 19. The concentration dependence of the chemical shifts of: GpC in H_2O solution (4°C); GpU in H_2O (0°C); GpU + ApC (1:1) in H_2O (1°C); and ApC in D_2O (4°C, solid circles). The ApC in D_2O chemical shifts are given relative to an external hexamethyldisiloxane capillary reference. Several ApC spectra were also recorded in H_2O solution (4°C, open circles) in order to measure the concentration dependence of the amino resonances. Reproduced with permission from Tom Krugh.

the cytosine amino protons as a function of increasing concentration in CpG and GpC at 1 °C and this led them to conclude that self complementary ribo dimers form Watson-Crick base paired duplexes. In the complementary mixture GpU + ApC, the effect of hydrogen bonding on the shift of guanine amino protons were observed, but the adenine amino protons were not affected. The data led them to conclude that the (GpU)•(CpA) miniature double helix is much less stable than either of the self complementary double helices (GpC)•(GpC) or (CpG)•(CpG). This is an indication of the effect of the number of GC pairs on stability. The self complementary UpA was found not to associate again suggesting the stability differences between GC and AU pairs.

The base protons were found to shift upfield with increasing concentration in the range of 1 to 100 mM[46/51] (Figure 19). They observed that the magnitude of the changes in H5 of the cytidine residue are significantly larger than those predicted for various double helical conformations of nucleic acids, possibly due to inter-molecular aggregation. Such complications prevent the extraction of detailed geometric information concerning the conformation of these hydrogen bonded complexes.

Arter, *et al.*[52] observed only two resonances at 13.18 and 1.245 ppm[53] due to GC pairs in the self complementary r-CpCpGpG suggesting the formation of a symmetrical double helix. Direct evidence for the formation of the duplex was provided by the observation of a melting curve (T_m = 51 ± 2°C) for the base protons H8 and H2. In order to obtain some information about the geometry of the complex, they calculated the shielding[52] in a variety of helical configurations and compared them with the observed shifts. The authors report that the observed shifts are consistent with those calculated for RNA-11.

2. Base Paired Oligomers of the DNA Family

During the last few years, there have been a large number of studies of the double helical properties of many deoxyoligonucleotides. In most of the studies, the shift trends of the exchangeable imino and amino protons and the nonexchangeable base protons have been monitored.

Young and Krugh[51] demonstrated that the self complementary deoxydimers d-pGpC, d-pCpG, d-GpC and d-CpG and the complementary mixtures d-pGpG + d-pCpC and d-pGpT + d-pApC form Watson Crick base paired miniature duplexes at low temperature by following the downfield shifts of the guanine and cytosine amino resonances as a function of concentration in the range of 1-100 mM. Their results showed the following order in stability (d-pGpG)•(d-pCpC) ≥ (d-pGpC)•(d-pGpC) > (d-pCpG)•(d-pCpG) > (d-pGpT)•(d-pGpT). This reflects the effect of nucleotide sequence and composition on helix stability.

Patel has studied the self-associative properties of the sequence isomers d-CpGpCpG,[54] d-CpCpGpG,[55] d-GpGpCpC[55] and d-GpCpGpC[56] using ¹H and

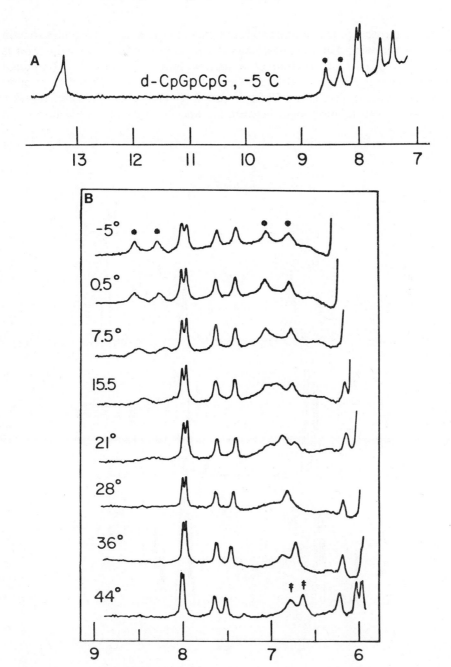

Figure 20. The 360 MHz continuous wave proton NMR spectra at 20 mM (in strands) d-CpGpCpG in 0.1M phosphate, H₂O, neutral pH.[40/50] (a) The spectral region 7 to 14 ppm at -5°C. The guanosine H-1 Watson-Crick imino proton resonates between 13.0 to 13.5 ppm. (b) The spectral region 6 to 9 ppm as a function of temperature (-5° to 44°C). The cytidine 4-amino protons are designated by * while the guanosine 2-amino protons are designated by ≠.

^{31}PNMR spectroscopy. Due to the two fold symmetry these tetranucleotides contain equivalent terminal and internal base pairs. These sequences form stable duplexes as demonstrated by the observation of a narrow (internal) and broad (terminal) guanosine 1-imino Watson-Crick exchangeable resonances at 13.15 and 13.25 ppm, respectively, in the spectrum of d-CpGpCpG at -5°C (Figure 20a). The exchangeable amino protons can be monitored between 6.5 and 8.5 ppm and are well resolved

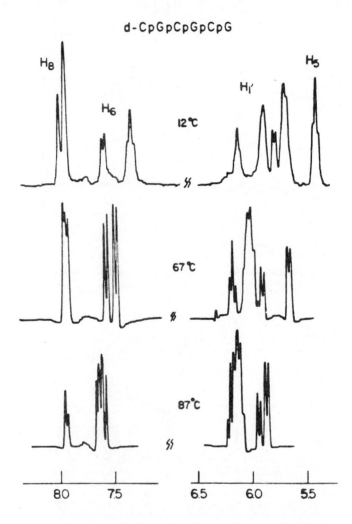

d-CpGpCpGpCpG

Figure 21. The 270 MHz Fourier transform (recorded under conditions of HDO suppression and resolution enhancement) proton NMR spectra of 10.5 mM (in strands) d-CpGpCpGpCpG in 0.1M cacodylate, 1 mM EDTA, [2]H₂O, pH 6.25 between 5.5 and 8.5 ppm at 12°, 67°, and 87°C.

from the nonexchangeable protons in this region (Figure 20b). Slow rotation about the C-N bond of cytidine results in the observation of separate resonances for the Watson-Crick hydrogen-bonded (8.0 to 8.25 ppm) and exposed (6.75 to 7.05 ppm)

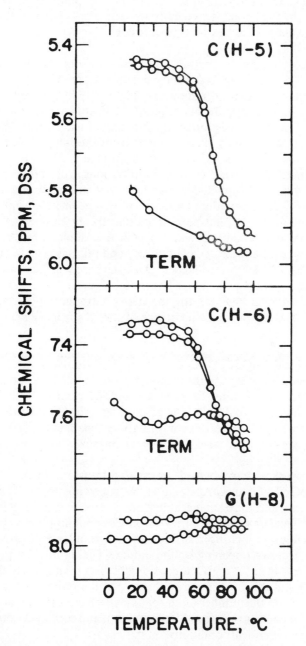

Figure 22. A plot of the base proton chemical shifts for d-CpGpCpGpCpG between 0° and 100°C under conditions in Figure 21.

cytidine 4-amino protons of d-CpGpCpG at -5°C (Figure 20b). These protons broaden on raising the temperature to 21°C with the onset of intermediate rotation rates relative to the 1.2 ppm chemical shift separation. Average resonances are observed for the guanosine 2-amino protons (6.8 to 7.1 ppm) indicative of fast rotation rates about the C-N bond with the broad resonances observed at 21°C narrowing on raising the temperature to 44°C (Figure 20b).

The nonexchangeable proton spectra (5.0 to 8.5 ppm) for the self-complementary hexanucleotide d-CpGpCpGpCpG which contains terminal, internal and central base pairs exhibits well resolved resonances (Figure 21) in the duplex (12°C) state, the strand (87°C) state and at the midpoint of the melting transition (67.5°C).[54] Individual chemical shifts can be plotted as a function of temperature (Figure 22) with the terminal base pairs readily differentiated from the interior base pairs for the cytidine H-5 and H-6 resonances. The upfield shifts at the base protons on duplex formation originate in the ring current contributions from adjacent overlapping base pairs[57] and it has been concluded that these tetranucleotide and hexanucleotide self-complementary duplexes adopt base pair overlap geometries consistent with the B-DNA conformation.[54-56] It has been noted that the guanosine H-8 shifts upfield by 0.6 to 0.8 ppm on formation of the r-CpCpGpG duplex[52] but by less than 0.1 ppm on formation of the d-CpCpGpG duplex[55] and this has been interpreted on the basis that the RNA and DNA tetramers of the same base sequence must be adopting different base pair overlap geometries. The coupling sum $J_{1'2'} + J_{1'2''}$ was observed to be in the range 14.5 to 17.0 Hz for the deoxy tetranucleotide duplexes at low temperature[54-56] demonstrating the existence of the ^2E sugar pucker as found in B-DNA.

The hexanucleotide d-CpGpCpGpCpG contains five internucleotide phosphates and all five resonances are resolved in the proton noise decoupled phosphorous spectrum in the duplex and strand states between 4.0 and 4.3 ppm upfield from standard trimethylphosphate (Figure 23a).[54] The resonances shift downfield on raising the temperature (Figure 23b) and reflect conformational changes in the sugar-phosphate backbone on conversion from stacked duplex to unstacked strands.

Crothers and coworkers have studied the melting transition of the complementary duplex (d-TpTpGpTpT)•(d-ApApCpApA) by following the shift trends of the methyl resonances[58] and the exchangeable imino protons[59] as well as the absorbance melting curve.[59] They observe that the Watson-Crick imino protons broaden out at temperatures below the melting transition of the complex indicative of exchange with solvent by transient opening of the duplex. The T_m for the pentanucleotide system of Crothers, *et al.*[58/59] (9°C) is considerably less than that for the tetramer d-CpCpGpG (42°) and this in turn may indicate that the stability of the duplex state depends very much on the number of GC pairs. However, it should be noted that T_m is very sensitive to concentration and ionic strength and a comparison of T_m should be made only under identical conditions.

The double helix unwinding mechanism can be probed by studying the transition

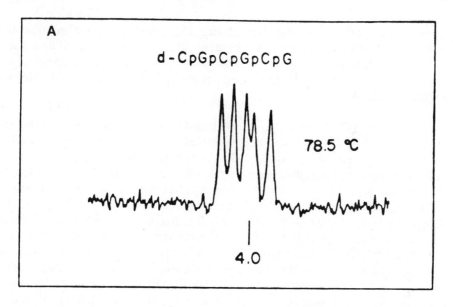

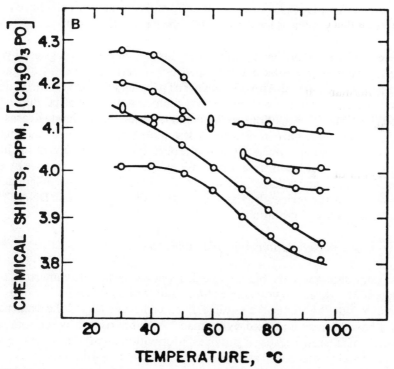

Figure 23. (a) The 145.7 MHz Fourier transform ^{31}P NMR spectrum (proton noise decoupled at 360 MHz) of 10.5 mM (in strands) d-CpGpCpGpCpG in 0.1M cacodylate, 10 mM EDTA, ^2H_2O, pH 6.5 at 78.5°C. The chemical shifts are upfield from internal standard trimethylphosphate. (b) The temperature dependence of the five internucleotide phosphates are plotted between 30° and 95°C.

temperatures at individual base pairs during the melting of an oligonucleotide duplex. For example, in a duplex of the hexamer, there are three kinds of base-pairs, i.e., the terminal, internal and central. The temperature and pH dependence of the resonances from these three kinds of base pairs may reflect the sequence in which the base-pairs open.

Patel and coworkers[60-63] found that the Watson-Crick imino resonances of the self-complementary d-ApTpGpCpApT duplex shifted to higher field and broadened in a sequential order from the ends of the helix on increasing the temperature or raising the pH. The sequential broadening process is consistent with fraying predominantly at the terminal A•T base pairs and to some extent at the internal A•T base pairs.[62] The fraying process represents a rapid opening and closing of the Watson-Crick hydrogen bonds and occurs on a time scale much shorter than the melting transition of the duplex monitored by the central G•C base pairs. The formation of the hexanucleotide duplex was characterized in terms of nucleation at the central G•C base pairs and propagation at the A•T base pairs from additional studies of the exchange process catalyzed by phosphate solution.[63] NMR studies of the nonexchangeable protons at individual base pairs confirmed the sequential opening of the d-ApTpGpCpApT duplex from it ends.[60/61] It is of interest to note that in the duplex of the hexamer d-CpGpCpGpCpG the internal and central base pairs have identical T_m suggesting that fraying is localized only at the terminal base pair.

Kallenbach and coworkers have studied the Watson-Crick and Hoogsteen hydrogen bonding interactions in double and triple helices in solution.[64] Their investigation on the self-complementary d-ApApApGpCpTpTpT duplex[64] and contributions of Patel and Canuel on the self-complementary d-GpGpApApTpTpCpC duplex[65] have extended the NMR investigations to the octanucleotide duplex level. Kearns, Wells and their collaborators have investigated **telestability** (transmission of stability along the duplex) in block copolymers[66] and also probed the conformation at the junction of A and B nucleic acid conformations.[67] We suggest reading the reviews by Kallenbach and Berman[68] and Kearns[69] on structural studies of the above oligonucleotide systems as well as the reviews by Patel on NMR studies of synthetic DNA's[70/71] and and synthetic RNA's[71] in solution.

Relationship Between Solution and Solid State Data

Single crystal data are available on several mononucleotides (summarized by Sundaralingam[25]) three ribodinucleoside monophosphates,[72-77] one deoxyribodinucleotide,[78] a trimer,[79] a tetramer,[80] and a hexamer[81] and the details of their geometry have been summarized by Seeman.[1] A comparison between solution and solid data is of interest because it provides information about the effect of forces in solution and solid state as well as aggregation on molecular structure.

In the solid state the ribofuranose ring of 3' and 5'-purine and pyrimidine mononucleotides displays either a ${}^3E$ or a ${}^2E$ pucker with no regularity. The softness of the pucker can be seen from the fact that molecule 1 of 5'ATP is ${}^3E$ whereas molecule 2 is ^{2}E.[21/82] In aqueous solution they exist as an equilibrium blend of ^{2}E $\rightleftharpoons$ ^{3}E with the purines showing a slight preference for the ^{2}E and the pyrimidines for ^{3}E. For common mononucleotides in the solid state, the magnitude of χ_{CN} lies in the anti domain; in solution one observes an equilibrium between anti and syn orientations with preference for anti orientation. The C3′-O3′ torsion of ribo-3′-mononucleotides lie in a broad range from 237-269° in the solid state. In solution the C3′-O3′ torsion is coupled to sugar ring conformation by the equilibrium ^{2}Eα^+ $\rightleftharpoons$ ^{3}Eα^- where α^+ and α^- are centered around 275 and 205°.

In addition, the overlap features of adjacent molecules in crystals and aqueous solution show conspicuous differences. Evans and Sarma[83] have shown that the specific intermolecular base stacking geometries in 5′AMP solutions do not agree with those reported in 5′AMP crystals.[84/85] It was also shown that[83] the base stacking orientations of neutral adenosine in the solid state is considerably different from the average orientations in aqueous solution. One reason for this is that in aqueous solution bases aggregate due to base-stacking interactions and not due to hydrogen-bonding. In the adenosine crystal[86] there is a myriad of interstack, intermolecular hydrogen bonds; for example, each nucleoside is hydrogen bonded to six neighboring molecules. This strong interaction, which is peculiar to the solid state, is expected to stabilize otherwise unnatural intermolecular orientations. Furthermore, it is expected that such hydrogen bonding may in some cases alter the preferred intramolecular conformation as well. For example, in the case of adenosine there is a hydrogen-bond stabilizing the $\epsilon \approx 180°$ orientation about the C4′-C5′ bond, whereas in aqueous solution it is clear that the $\epsilon \approx 60°$ orientation is strongly preferred. The differences in the immediate environments most likely account for the observed drastic discrepancy between the solid-state and solution conformation of 6-thiopurine riboside and 4-thiouridine.[87-89]

Despite these differences between solid-state and solution conformations of mononucleotides, there is an area of excellent agreement. In both solution and solid-state studies one invariably finds that the C4′-C5′ and C5′-O5′ bonds form a stable conformational network in which the magnitudes of δ and ϵ are respectively, about $\approx 60°$ and $\approx 180°$. This suggests the possibility that in solution oligonucleotides may achieve conformational versatility by torsional variation about χ_{CN}, C3′-O3′, O3′-P, O5′-P bonds as well as by variation in ribose pucker.

We have noted earlier that the first step in oligomerization, i.e., monomer to dimer causes major changes in conformation and the dimers hence attempt to become conformationally pure. Because of this one may expect a great deal of common stereochemical features between the solution and solid state conformations for the higher oligomers. In fact, the preferred spatial configuration for GpC, ApU and d-pTpT in solution is the same as that reported for the corresponding molecules in the solid state. Further the solid state data on ApA$^+$pA$^+$ and UpA where g$^+$g$^+$ and tg$^-$ phosphodiester torsions have been detected can be considered to be in agreement with the solution findings in addition to the g$^-$g$^-$ helical arrays, of arrays in which the phosphodiester assumes different arrangements such as g$^+$g$^+$, tg$^-$, g$^+$t. However, the solution data on d-ApT and d-TpA cannot correctly predict the observed solid state structure for d-pApTpApT which is characterized by alternating sugar puckers as well as g^-g^-, tg^-, and g^-g^- phosphodiester torsions. No solution data to date of this tetramer is available.

Epilog

In these pages, we have attempted to bring together the available stereochemical information about oligonucleotides in aqueous solution as determined by NMR spectroscopy. Because of the space limitations a detailed documentation was not possible and hence attempt was made to drive home concepts and principles. We urge the students and interested scientists consult the original literature for details.

Acknowledgement

This research was supported by Grant CA12462 from National Cancer Institute of NIH and Grant PCM-7822531 from National Science Foundation. This research was also supported by Grant 1-PO7-PR-PR00798 from the Division of Research Resources. The author thanks Thomas R. Krugh for providing Figure 19 and Dinshaw J. Patel for providing Figures 20, 21, 22 and 23. We are grateful to Dinshaw J. Patel for being kind enough to carry out extensive revision of the section on base-paired oligonucleotides of the DNA family.

References and Footnotes

1. Seeman, N.C., in *Nucleic Acid Geometry and Dynamics*, Ed., Sarma, R.H., Pergamon Press, New York, Oxford, Frankfurt, Paris, p.   (1980).
2. Ts'o, P.O.P., Kondo, N.S., Schweizer, M.P., and Hollis, D.P., *Biochemistry, 8*, 997 (1969).
3. Hruska, F.E., and Danyluk, S.S., *J. Amer. Chem. Soc., 90*, 3266 (1968).
4. Chan, S.I., and Nelson, J.H., *J. Amer. Chem. Soc. 91*, 168 (1969).
5. Bangerter, B.W., and Chan, S.I. (1969), *J. Amer. Chem. Soc., 91*, 3910.
6. Cheng, D.M., and Sarma, R.H., *J. Am. Chem. Soc., 99*, 7333 (1977).
7. Evans, F.E., Lee, C.H., and Sarma, R.H., *Biochem. Biophys. Res. Commun., 63*, 106 (1975).
8. Lee, C.H., Evans, F.E., and Sarma, R.H., *FEBS Lett., 51*, 73 (1975).
9. Lee, C.H., Ezra, F.S., Kondo, N.S., Sarma, R.H., and Danyluk, S.S., *Biochemistry, 15*, 3627 (1976).
10. Ezra, F.S., Lee, C.H., Kondo, N.S., Danyluk, S.S., and Sarma, R.H., *Biochemistry, 16*, 1977 (1977).
11. Evans, F.E., and Sarma, R.H., *Nature, 263*, 567 (1976).
12. Cheng, D.M., Dhingra, M.M., and Sarma, R.H., *Nucleic Acids Res., 5*, 4399 (1978).
13. Sarma, R.H., in *Nucleic Acid Geometry and Dynamics*, Ed., Sarma, R.H., Pergamon Press, New York, Oxford, Frankfurt, Paris, p. 1 (1980).
14. Lee, C.H., Evans, F.E., and Sarma, R.H., *J. Biol. Chem. 250*, 1290 (1975).
15. Sarma, R.H., Lee, C.H., Evans, F.E., Yathindra, N., and Sundaralingam, M., *J. Am. Chem. Soc., 96*, 7337 (1974).
16. Wood, D. J., Hruska, F.E., Mynott, R.J., and Sarma, R.H., *Can. J. Chem., 51*, 2571 (1973).
17. Sarma, R.H., and Danyluk, S.S., *Internatl. J. Quant. Chem. QBS 4*, 269 (1977).
18. Dhingra, M.M., and Sarma, R.H., *Internatl. J. Quant. Chem. QBS 6* (in press).
19. Dhingra, M.M., Sarma, R.H., Geissner-Prettre, C. and Pullman, B., *Biochemistry, 17*, 5815 (1978).
20. Danyluk, S.S., Ainsworth, C.F., and MacCoss, M., *Nuclear Magnetic Resonance in Molecular Biology*. Ed. Pullman, B. Reidel Publishing, Holland, p. 111 (1978).

21. Kim, C.H., and Sarma, R.H., *J. Am. Chem. Soc., 100*, 1571 (1978).
22. Singh, H., Herbut, M.H., Lee, C.H., and Sarma, R.H., *Biopolymers, 15*, 2167 (1976).
23. Cheng, D.M., and Sarma, R.H., *Biopolymers, 16*, 1687 (1977).
24. Cheng, D.M., Danyluk, S.S., Dhingra, M.M., Ezra, F.S., MacCoss, M., Mitra, C.K., and Sarma, R.H. (submitted).
25. Sundaralingam, M., in *Jerusalem Symposia on Quantum Chemistry and Biochemistry, 5*, 417 (1973).
26. Sundaralingam, M., in *Structure and Conformation of Nucleic Acid and Protein-Nucleic Acid Interactions*, Ed. Sundaralingam, M. and Rao, S.T., University Park Press, Baltimore, p.487 (1975).
27. Neidle, S., Taylor, G., Sanderson, M., Shieh, H.S., and Berman, H.M., *Nucleic Acids Res. 5*, 4417 (1978).
28. Berman, H.M. and Neidle, S., in *Stereodynamics of Molecular Systems*, Ed. Sarma, R.H. Pergamon Press, Oxford, New York, Frankfurt, Paris, p. 367 (1979).
29. Rich,A., Quigley, G.J., and Wang, A. H.-J., in *Stereodynamics of Molecular Systems*, Ed. Sarma, R.H., Pergamon Press, Oxford, New York, Frankfurt, Paris, p. 315 (1979).
30. Quigley, G.J., Seeman, N.C., Wang, A. H.-J., Suddath, F.L., and Rich, A., *Nucleic Acids Research, 2*, 2329 (1975).
31. Sussman, J.L., and Kim, S.H., *Biochem. Biophys. Res. Comm. 68*, 89 (1976).
32. Ladner, J.E., Jack, A., Robertus, J.D., Brown, R.S., Rhodes, D., Clark, B.J.C., and Klug, A., *Nucleic Acids Research , 2*, 1623 (1975).
33. Stout, D.C., Mizuno, H., Rubin, J., Brennen, T., Rao, S.T., and Sundaralingam, M., *Nucleic Acids Research, 3*, 1111 (1976).
34. Kan, L.S., Barrett, J.C., and Ts'o, P.O.P., *Biopolymers, 12*, 2409 (1973).
35. Kan, L.S., Borer, P.N., and Ts'o, P.O.P., *Biochemistry, 14*, 4865 (1975).
36. Borer, P.N., Kan, L.S., and Ts'o, P.O.P., *Biochemistry, 14*, 4847 (1975).
37. Kan, L.S., Ts'o, P.O.P., Vander Haar, F., Speinzl, M., and Cramer, F., *Biochemistry, 14*, 3278 (1975).
38. Sulston, J., Lohrmann, R., Orgel, L.E., Schneider-Bernhoefer, H., Weiman, B.J., and Miles, H.T., *J. Mol. Biol., 40*, 227 (1969).
39. Bryan, A.M., and Olafsson, P.G., *Analytical Chemistry, 3*, 685 (1974).
40. Usher, D.A., *Nature New Biology, 235*, 207 (1972).
41. Usher, D.A., *Science, 196*, 211 (1977).
42. Usher, D.A., and McHale, A.H., *Proc. Natl. Acad. Sci. USA, 73*, 1149 (1976).
43. Dhingra, M.M., and Sarma, R.H., *Nature, 272*, 798 (1978).
44. Michelson, A.M. and Monny, C., *Biochem. Biophys. Acta, 149*, 107 (1967).
45. Bleich, H.E., and Glasel, J.A., *J. Mag. Res., 18*, 401 (1975).
46. Krugh, T.R., Laing, J.W., and Young, M.A., *Biochemistry, 15*, 1224 (1976).
47. Shefter, E., Barlow, M., Sparks, R.A., and Trueblood, K.N., *Acta Crystallogra. B25*, 895 (1969).
48. Sarma, R.H., Dhingra, M.M., and Feldmann, R.J., in *Stereodynamics of Molecular Systems*, Ed. Sarma, R.H., Pergamon Press, New York, Oxford, Frankfurt, Paris, p. 251 (1979).
49. Sarma, R.H., in *Nucleic Acid Geometry and Dynamics*, Ed. Sarma, R.H., Pergamon Press, New York, Oxford, Frankfurt, Paris, p. 83 (1980).
50. Olson, W.K., *Proc. Natl. Acad. Sci., USA, 74*, 1775 (1977).
51. Young, M.A., and Krugh, T.R., *Biochemistry, 14*, 4841 (1975).
52. Arter, D.B., Walker, G.C., Uhlenbeck, O.C., and Schmidt, P.G., *Biochem. Biophys. Res. Comm. 61*, 1089 (1974).
53. All shifts are with respect to DSS in the section on base paired oligonucleotides.
54. Patel, D.J., *Biopolymers, 15*, 533 (1976).
55. Patel, D.J., *Biopolymers, 16*, 1635 (1977).
56. Patel, D.J., *Biopolymers, 18,* 553 (1979).
57. Giessner-Prettre, C., and Pullman, B., *Biochem. Biophys. Res. Comm., 70*, 578 (1976).
58. Cross, A.D. and Crothers, D.M., *Biochemistry, 10*, 6015 (1974).
59. Crothers, D.M., Hilbers, C.W., and Shulman, R.G., *Proc. Natl. Acad. Sci. USA, 70*, 2899 (1973).
60. Patel, D.J., and Tonelli, E.A., *Biochemistry, 14*, 3991 (1975).
61. Patel, D.J., *Biochemistry, 14*, 3985 (1975).
62. Patel, D.J., and Hilbers, C.W., *Biochemistry, 14*, 2651 (1975).

63. Hilbers, C.W., and Patel, D.J., *Biochemistry, 14*, 2656 (1975).
64. Kallenbach, N.R., Daniel, W.E., Jr., and Kaminker, M.A., *Biochemistry, 15*, 1218 (1976).
65. Patel, D.J., and Canuel, L.L., *Eur. J. Biochem., 96*, 267 (1979).
66. Early, T.A., Kearns, D.R., Burd, J.F., Larson, J.E., and Wells, R.D., *Biochemistry, 16*, 541 (1977).
67. Selsing, E., Wells, R.D., Early, T.A., and Kearns, D.R., *Nature, 275*, 249 (1978).
68. Kallenbach, N.R., and Berman, H.M., *Q. Revs. Biophys., 6*, 477 (1977).
69. Kearns, D.R., *Ann. Rev. Biophys. Bioeng., 6*, 477 (1977).
70. Patel, D.J., *Acc. Chem. Research, 12*, 118 (1979).
71. Patel, D.J., in *Nucleic Acid Geometry and Dynamics*, Sarma, R.H. (Ed.), Pergamon Press, Oxford, New York, Frankfurt, Paris, p. 185 (1980).
72. Seeman, N.C., Day, R.O., and Rich, A., *Nature, 253*, 324 (1975).
73. Seeman, N.C., Rosenberg, J.M., Suddath, F.L., Kim, J.P., and Rich, A., *J. Mol. Biol., 104*, 109 (1976).
74. Seeman, N.C., Sussman, J.L., Berman, H.M., and Kim, S.H., *Nature New Biology, 233*, 90 (1971).
75. Rosenberg, J.M., Seeman, N.C., Day, R.O., and Rich, A., *J. Mol. Biol., 104*, 145 (1976).
76. Hingerty, B., Subramanian, E., Stellman, S.D., Broyde, S.B., Sato, T., and Langridge, R., *Biopolymers, 14*, 227 (1975).
77. Sussman, J.L., Seeman, N.C., Kim, S.H., and Berman, H.M., *J. Mol. Biol., 66*, 403 (1972).
78. Camerman, N., Fawcett, J.K., and Camerman, A., *J. Mol. Biol., 107*, 601 (1976).
79. Suck, D., Manor, P.C., German, G., Schwalbe, C.H., Weiman, G., and Saenger, W., *Nature New Biology, 246*, 161 (1973).
80. Viswamitra, M.W., Kennard, O., Jones, P.G., Sheldrick, G.M., Salisbury, S., Falvello, L., and Shakked, Z., *Nature, 273*, 687 (1978).
81. Wang, A., Quigley, G.J., Kolpak, R., Van Boom, J.H., and Rich, A., Summer Meeting of Crystallography, Boston, August, 1979.
82. Kennard, O., Isaacs, N.W., Motherwell, W.D.S., Coppola, J.C., Wampler, D.L., Larson, A.C., and Watson, D.G., *Proc. R. Soc. London, Ser. A.*, 325, 401 (1971).
83. Evans, F.E., and Sarma, R.H., *Biopolymers, 13*, 2117 (1974).
84. Kraut, J., and Jensen, L.H., *Acta Cryst, 16*, 79 (1963).
85. Bugg, C.E., in *Purines: Theory and Experiment*, Pullman, B. (Ed.), Academic Press (1971).
86. Lai, T.F., and Marsh, R.E., *Acta Cryst B28*, 1982 (1972).
87. Evans, F.E., and Sarma, R.H., *J. Am. Chem. Soc., 97*, 3215 (1975).
88. Shefter, E., *J. Pharm. Sci. 57*, 1157 (1968).
89. Saenger, W., and Scheit, K.H., *J. Mol. Biol., 50*, 153 (1970).

NMR Studies of Structure and Dynamics of Synthetic DNA and RNA Duplexes and Their Antibiotic Intercalation Complexes in Solution

Dinshaw Patel
Bell Laboratories
600 Mountain Avenue
Murray Hill, New Jersey

Introduction

The observation of narrow resolvable proton resonances from the base and sugar rings and the phosphorus resonances from the backbone have permitted the successful application of nuclear magnetic resonance spectroscopy (NMR) to elucidate the structure and dynamics of stable oligonucleotide duplexes. Extensive research has been undertaken on deoxy- and ribo-oligonucleotides extending from the tetramer to the octamer level[1-6] as well as on block copolymers[7/8] with the sequences summarized in Table I. Our own contribution has focussed on the self-complementary sequences dA-dT-dG-dC-dA-dT with dG•dC base pairs in the interior of the duplex[9] and dG-dG-dA-dA-dT-dT-dC-dC containing a central cluster of dA•dT base pairs.[10]

The present understanding of nucleic acid conformations at the polynucleotide duplex level is based primarily on fiber X-ray diffraction studies of natural and synthetic DNA's as a function of humidity and counter-ion.[11] The early studies using circular dichroism spectroscopy monitored DNA conformational transitions in films[12/13] and in solution[14/15] but lacked the probes necessary to differentiate between contributions due to hydrogen-bonding, base stacking and the torsion angle changes in the sugar-phosphate backbone. This contribution reviews recent efforts to investigate the structure and dynamics of synthetic DNA's and RNA's of defined sequence[16-21] and their intercalation antibiotic complexes by high resolution NMR spectroscopy in solution.

Synthetic RNA and DNA Duplexes

Thermodynamic and kinetic studies demonstrate that the alternating purine-pyrimidine polynucleotide poly(dA-dT) folds into smaller duplexes which melt independently of each other in solution.[22/23] The NMR spectrum should be considerably simplified since all base pairs are structurally equivalent due to the symmetry of the alternating purine-pyrimidine sequence. Further, the NMR resonances may exhibit moderate line widths in the duplex state due to segmental mobility resulting from the rapid migration of the branched duplexes along the polynucleotide backbone. Finally, since the smaller duplexes melt independently of

Table I
Summary of Oligonucleotide Sequences
Studies by NMR Spectroscopy

Sequence	Reference
DNA Fragments	
(dT-dT-dG-dT-dT)•(dA-dA-dC-dA-dA)	1
dA-dT-dG-dC-dA-dT	9
dC-dG-dC-dG	2
dC-dC-dG-dG	103
dG-dG-dC-dC	103
dG-dC-dG-dC	68
dA-dA-dA-dG-dC-dT-dT-dT	3
dG-dG-dA-dA-dT-dT-dC-dC	10
$(dC_{15}\text{-}dA_{15})\bullet(dT_{15}\text{-}dG_{15})$	7
RNA Fragments	
C-C-G-G	4
A-A-G-C-U-U	6
$(A_{20})\bullet(U_{20})$	5
DNA•RNA Hybrids	
$(dG_n)\bullet(C_{11}\text{-}dC_{16})$	8

each other,[22] the duplex dissociation rate constants are on the NMR time scale, so that the resonances may shift as average peaks through the melting transition. These features should result in well resolved base and sugar resonances for high molecular weight alternating purine-pyrimidine polynucleotides in the duplex state and this has been confirmed experimentally in our studies of the duplex to strand transition of poly(dA-dT) (mol. wt. $\sim 10^6$, ~ 2000 base pairs) as a function of temperature.[16] The details of melting transition of this synthetic DNA and its complexes with the intercalating mutagen proflavine, the steroid diamine dipyrandium (which partially inserts between tilted base pairs) and the groove binding agent netropsin (which perturbs base pair regions distant from its binding site) have been reviewed recently.[24]

It would be useful to directly compare the NMR parameters for RNA and DNA duplexes with the same sequence. This has not yet been undertaken at the oligonucleotide level since the deoxy- and ribo-oligomers studied to date exhibit different sequences (Table I). We compare below the spectral parameters of the synthetic RNA poly(I-C) and the synthetic DNA poly(dI-dC) under similar solution conditions.[18]

Hydrogen-Bonding

The base paired duplex state in nucleic acids can be readily characterized by monitoring the exchangeable imino protons in H_2O solution. Earlier studies have demonstrated that the purine H-1 imino proton and the pyrimidine H-3 imino proton of nonterminal base pairs in nucleic acid duplexes are in slow exchange with solvent H_2O and resonate between 12 and 15 ppm.[25-28]

The hydrogen-bonded inosine H-1 proton is observed at ~15.1 ppm for poly(I-C) and poly(dI-dC) in the duplex state with no additional resonances observed between 8 and 17 ppm (Table II). These chemical shifts may be compared with a value of 12.9

ppm for the guanosine H-1 proton in alternating guanosine-cytidine polynucleotides and a value of ~13.0 ppm for the uracil H-3 proton in alternating adenosine-uridine polynucleotides (Table II). These data demonstrate that the specific replacement of guanosine by inosine in double-stranded DNA should result in the introduction of a well-resolved imino resonance of a dI•dC base pair resonating ~2 ppm downfield from the overlapping resonance positions of imino protons in dA•dU and dG•dC base pairs.

The cytidine amino H-4 protons are observed at ~7.6 ppm and ~6.8 ppm for poly(I-C) and at ~7.4 ppm and ~6.4 ppm for poly(dI-dC) in the duplex state. The resonance at lower field (7.4 to 7.6 ppm) is assigned to the hydrogen-bonded cytidine H-4 amino proton while the resonance at higher field (6.4 to 6.8 ppm) is assigned to the exposed cytidine H-4 amino proton for the synthetic RNA and DNA duplexes.

Table II
Chemical Shifts of the Imino Watson-Crick
Hydrogen Bonds in the Synthetic DNA and RNA Duplexes

	Chemical Shifts, ppm
Synthetic DNA	
poly(dI-dC)[1]	15.15 ppm
poly(dA-dU)[1]	13.0 ppm
poly(dG-dC)[2]	12.9 ppm
Synthetic RNA	
poly(I-C)[1]	15.1 ppm
poly(A-U)[1]	13.05 ppm

[1]Buffer: 0.1M phosphate. Temperature 25°C.
[2]Buffer: 10 mM NaCl, 1 mM phosphate. Temperature 25°C.

This suggests that amino group rotation about the C-N bond at position 4 of cytidine in the inosine-cytidine base pair is slow relative to the ~1 ppm chemical shift difference between hydrogen-bonded and exposed positions.

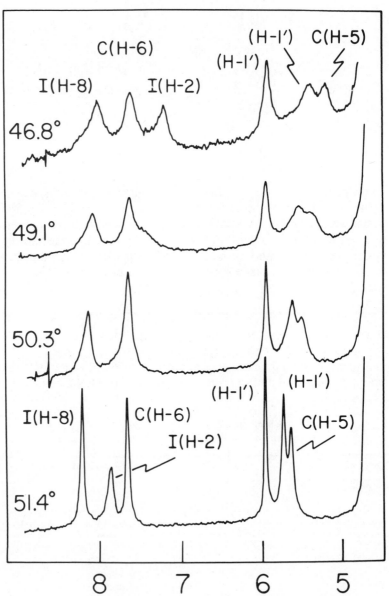

Figure 1. The temperature dependence of the 360 MHz Fourier transform proton NMR spectra (4.5 to 9.0 ppm) of 18.6 mM poly(I-C) in 0.1 M phosphate, 1 mM EDTA, ^2H_2O, pH 7.0 between 46° to 52°C.

Duplex to Strand Transition

The proton NMR spectra (4 to 9 ppm) of poly(I-C) in 0.1 M phosphate, ^2H_2O solution through the melting transition are presented in Figure 1. The base resonances can be readily assigned from their chemical shift positions at high temperature though the sugar H-1′ cannot be definitively differentiated between the inosine and cytidine residues. The base and sugar resonances of poly(I-C) shift as average peaks between the duplex and strand states and can be monitored through the melting transition in 0.1 M phosphate solution (Figure 1). By contrast, the base and sugar resonances of poly(dI-dC) were in intermediate to slow exchange between duplex and strand states and hence could not be monitored through the melting transition in 0.1 M phosphate solution. The resonance assignments for the base resonances of poly(dI-dC) in the duplex state were deduced by comparison with related assignments for poly(dA-dT) in the duplex state.

The chemical shifts of the base and sugar H-1′ protons of poly(I-C) and poly(dI-dC) in 0.1 M phosphate between 0° and 100°C are plotted in Figure 2. The resonances shift between 5° and 45°C (premelting transition), 45° and 55°C (melting transition) and 55° to 95°C (postmelting transition). The chemical shifts versus temperature plots were reversible in successive heating and cooling cycles for the alternating inosine-cytidine RNA and DNA synthetic polynucleotides in contrast to the hysterisis effects observed with natural DNA in solution. The chemical shift parameters are summarized in Table III.

Base Pair Overlaps

Table III

Chemical Shift Parameters Associated with the Melting Transition
of Poly(I-C) and Poly(dI-dC) in 0.1 M Phosphate Solution[1]

	Poly(I-C) $t_{1/2} = 50.1°C$		Poly(dI-dC) $t_{1/2} = 53.5°C$	
	δ_d,ppm	$\Delta\delta$,ppm	δ_d,ppm	$\Delta\delta$,ppm
I(H-8)	7.977	0.321	8.137	0.107
I(H-2)	7.183	0.913	7.313	0.716
u(H-1′)[2]	5.344	0.501	5.542	0.533
C(H-6)	7.575	0.124	7.129	0.402
C(H-5)	5.142	0.656	5.226	0.559
d(H-1′)[2]	5.891	0.098	6.038	0.208

[1]Chemical shift in the duplex state, δ_d, is defined as the extrapolation of the temperature dependent premelting shift to its value at the transition midpoint. The duplex to strand transition chemical shift change $\Delta\delta$, is defined as the chemical shift difference following extrapolation of the temperature dependent premelting and postmelting shifts to their values at the transition midpoints. These $\Delta\delta$ values only approximate the total shift change on proceeding from a stacked duplex to unstacked strands.

[2]u(H-1′) and d(H-1′) represent the upfield and downfield sugar protons, respectively.

We have also investigated the alternating adenosine-uridine synthetic RNA's and DNA's by NMR spectroscopy[19] in order to compare the experimental parameters with those for the corresponding inosine-cytidine analogs.[18]

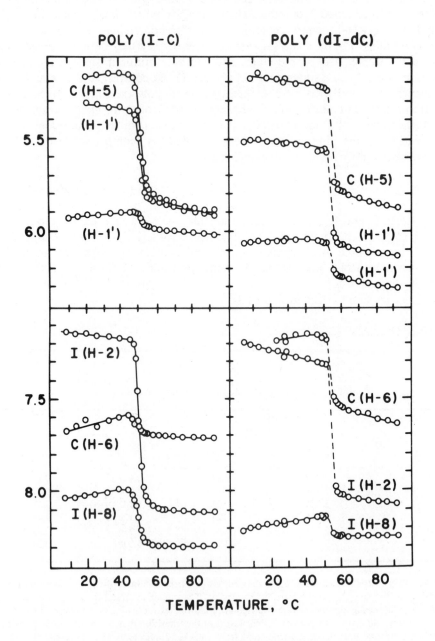

Figure 2. The temperature dependence (5° to 95°C) of the base and sugar H-1′ proton chemical shifts of poly(I-C) and poly(dI-dC) in 0.1 M phosphate, 1 mM EDTA, ^2H_2O solution.

The chemical shift parameters associated with the melting transition of poly(A-U) and poly(dA-dU) are summarized in Table IV. The similar chemical shift changes associated with the melting transition of poly(I-C) and poly(A-U) in 0.1 M phosphate solution (Tables III and IV) suggest that the structure of the alternating purine-pyrimidine synthetic RNA's is independent of base composition. A similar conclusion is reached for the synthetic DNA's based on a comparison of the chemical shift changes associated with the melting transition of poly(dI-dC) and poly(dA-dU) in 0.1 M phosphate solution (Tables III and IV).

The purine H-1 and pyrimidine H-3 Watson-Crick exchangeable protons located in the center of the duplex and the purine H-2 and pyrimidine H-5 nonexchangeable protons located at the base pair edges in the minor and major groove, respectively exhibit similar chemical shifts for the synthetic RNA and DNA in the duplex state (Tables II - IV). The base pair overlaps for the B-DNA and A-RNA conformations deduced from fiber X-ray diffraction data are different[11] and the calculated ring current contributions[29/30] based on these overlaps suggest large differences at the pyrimidine H-3 and H-5 positions (Table V). This discrepancy between the experimental NMR data and the calculated upfield ring current shifts is not understood at this time.

Table IV
Chemical Shift Parameters Associated With the Melting
Transition of Poly(A-U) and Poly(dA-dU) in 0.1 M Phosphate
Solution[1]

	Poly(A-U) $t_{1/2} = 66.5°C$		Poly(dA-dU) $t_{1/2} = 59.0°C$	
	δ_d,ppm	$\Delta\delta$,ppm	δ_d,ppm	$\Delta\delta$,ppm
A(H-8)	7.983	0.323	8.122	0.170
A(H-2)	6.973	1.126	7.166	0.920
u(H-1′)	5.335	0.431	5.615	0.341
U(H-6)	7.462	0.200	7.166	0.344
U(H-5)	4.935	0.790	4.964	0.678
d(H-1′)	5.895	0.097	6.050	0.241

[1]δ_d and $\Delta\delta$ are as defined in caption to Table III.

Table V
Computed Upfield Ring Current Shifts,
ppm, on Alternating Adenosine-Uridine
Polynucleotide Duplex Formation[1]

	A-RNA	B-DNA
A(H-8)	0.12	0.04
A(H-2)	0.80	1.06
U(H-6)	0.45	0.06
U(H-5)	1.42	0.62
U(H-3)	1.00	1.88

[1]The computations are based on nearest neighbor, next-nearest neighbor and cross strand ring current contributions tabulated by Arter and Schmidt.[30]

The purine H-8 and pyrimidine H-6 protons are directed towards the phosphate backbone and hence their chemical shifts are sensitive to changes in the glycosidic torsion angles in addition to changes in the base pair overlaps. Experimentally, the pyrimidine H-6 proton resonates ∿0.4 ppm to higher field and the purine H-8 proton resonates ∿0.15 ppm to lower field in the duplex state of the synthetic DNA's compared to the synthetic RNA's (Tables III and IV). Since the remaining base protons exhibit similar shifts in the duplex state it appears likely that the differences observed at the purine H-8 and pyrimidine H-6 duplex chemical shifts between the synthetic RNA's and DNA's reflect differences in the glycosidic torsion angles in these two systems. It should be noted that the glycosidic torsion angles in the RNA-A and DNA-B conformations deduced from fiber diffraction data differ by ∿70°.[11]

Duplex Dissociation Rates

The inosine H-8, cytidine H-6 and downfield H-1′ resonances exhibit a temperature dependent line width of >35 Hz in the premelting region below 45°C, shift as average peaks during the melting transition of poly(I-C) in 0.1 M phosphate and narrow to their value in the strand state above 55°C (Figure 3). By contrast the inosine H-2, cytidine H-5 and upfield H-1′ resonances which exhibit $\Delta\delta > 0.5$ ppm (Figure 2, Table III) exhibit uncertainty broadening contributions in the fast exchange region during the melting transition (Figure 1). The excess line width contributions yield a dissociation rate constant of ∿2.5 ± 0.5 × 10^3 sec^{-1} for poly(I-C) in 0.1 M phosphate at the midpoint of the transition. By contrast, the slow exchange between duplex and strand states for poly(dI-dC) through the melting transition (Figure 2) requires that the duplex dissociation rate constant be <0.1 × 10^3 sec^{-1} at the transition midpoint in 0.1 M phosphate solution. This suggests that the average length of the branched regions are longer for poly(dI-dC) compared to poly(I-C) since the dissociation rate constants are faster for the latter duplexes.

Premelting Transition[31-34]

We observe premelting structural transitions in the NMR spectra of poly(I-C) and poly(dI-dC) in 0.1 M phosphate solution (Figure 2). The inosine H-2 resonance (located towards the center of the duplex) shifts upfield while the inosine H-8 and cytidine H-6 resonances (directed towards the backbone phosphates) shift downfield on lowering the temperature in the premelting region for both poly(I-C) and poly(dI-dC) duplexes (Figure 2). These results suggest that the structural changes corresponding to the premelting transition are common to both synthetic RNA and DNA duplexes. It appears that this premelting transition in alternating purine-pyrimidine sequences may reflect conversion of a highly branched duplex structure to a less branched duplex structure on lowering the temperature.

Sugar Pucker

The pucker of the sugar rings in poly(I-C) and poly(dI-dC) can be estimated from the vicinal proton-proton coupling constants between the H-2′ (and H-2″) proton(s) and the H-1′ proton as monitored at the latter protons. The sugar H-1′ resonances are broad in the duplex state (~50 Hz) and narrow during the cooperative melting transition to a value of a few Hz in the strand state (Figure 3). Thus, the multiplicity of the H-1′ resonances can be measured in the strand state and at very low values of

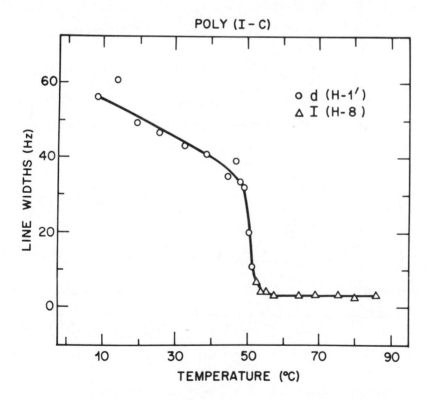

Figure 3 The temperature dependence (5° to 95°C) of the line width of the downfield (H-1′) (o) and inosine H-8 (Δ) resonances of poly(I-C) in 0.1 M phosphate, 1 mM EDTA, ^2H_2O solution.

the fraction of duplex state, prior to broadening out of the resonance and loss of the coupling information on decreasing the temperature further. The relationship between vicinal coupling constants and dihedral angles predicts coupling constants $H_{1'}H_{2'}$ ~10 Hz and $H_{1'}H_{2'}$ + $H_{1'}H_{2''}$ ~16 Hz for C2' *endo* and $H_{1'}H_{2'}$ ~0 Hz and $H_{1'}H_{2'}$ + $H_{1'}H_{2''}$ ~7 Hz for C3' *endo* type geometries in solution.[35]

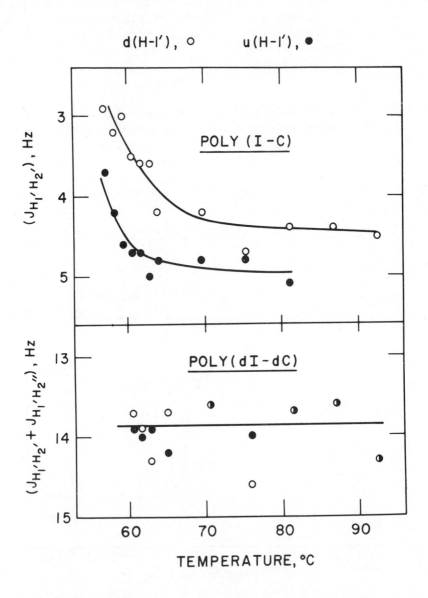

Figure 4. The temperature dependence of the $H_{1'}H_{2'}$ coupling constant (57° to 95°C) for poly(I-C) and the $H_{1'}H_{2'}$ + $H_{1'}H_{2''}$ coupling constant sum (60° to 95°C) for poly(dI-dC) in 0.1 M phosphate, 1 mM EDTA, ^2H_2O solution.

The $H_{1'}H_{2'}$ coupling constant values are 4 to 5 Hz for the poly(I-C) sugar resonances between 60° to 95°C (Figure 4), indicative of an equilibrium between C3' *endo* and C2' *endo* type pucker geometries for both sugars in the postmelting transition region. These coupling constants decrease on lowering the temperature below 60°C (Figure 4), indicative of a shift in the equilibrium towards the C3' *endo* type pucker for both sugars on poly(I-C) duplex formation ($t_{1/2}$ = 50.1°C). These studies on a synthetic RNA are supported by earlier conclusions on the sugar pucker geometries in a self-complementary ribohexanucleotide duplex.[6]

The $H_{1'}H_{2'}$ + $H_{1'}H_{2''}$ coupling constant sum of $\sim$14 Hz for both sugars resonances in

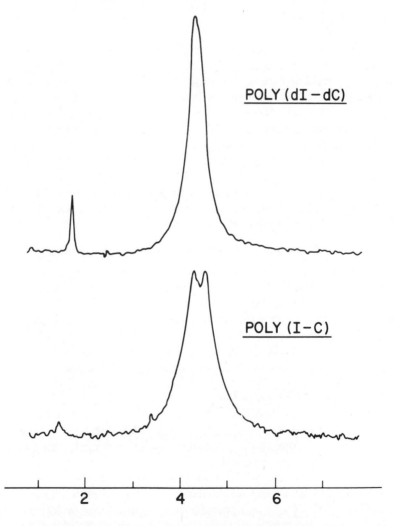

Figure 5. The 145.7 MHz proton noise decoupled ^{31}P NMR spectra (1 to 7 ppm upfield from standard trimethylphosphate) of 21.5 mM poly(I-C), $s_{20,w}$ = 10.2, and 24 mM poly(dI-dC), $s_{20,w}$ = 6.9, in 10 mM NaCl, 1 mM cacodylate, 1 mM EDTA, ^{2}H$_2$O, pH $\sim$6.9, 35.5°C.

poly(dI-dC) between 65° and 95°C (Figure 4) are indicative of a predominant C2′ *endo* type sugar pucker in the postmelting transition temperature range. The coupling sum remains unchanged on lowering the temperature below 65° to 60°C (Figure 4) which suggests that the predominant C2′ *endo* type sugar pucker in the strand state may be maintained on poly(dI-dC) duplex formation ($t_{1/2}$ = 53.5°C). These studies on a synthetic DNA are supported by earlier NMR investigations at the self-complementary deoxyoligonucleotide level which demonstrated that the sugar pucker was predominately C2′ *endo* in the duplex and strand states.[2] For a detailed discussion of this aspect read the section by Sarma on oligonucleotide conformation in solution in this volume.[132]

Backbone Phosphates

The internucleotide phosphodiester linkages in alternating purine-pyrimidine polynucleotides are either of the pyrimidine(3′-5′)purine or the purine(3′-5′)pyrimidine type and one may expect to observe two resonances in the ^{31}P NMR spectrum of the synthetic DNA and RNA in solution.

Experimentally, the phosphorus NMR spectrum of poly(dI-dC) exhibits a resonance at 4.37 ppm (upfield from standard trimethylphosphate) with a shoulder at higher field while the corresponding spectrum of poly(I-C) exhibits two partially resolved resonances of equal area at 4.485 and 4.252 ppm at 35.5°C (Figure 5). This observation of resolved resonances in the synthetic RNA but not the synthetic DNA was confirmed by parallel studies on the corresponding alternating adenosine-uridine and guanosine-cytidine sequences.

The magnitude of the splitting of the poly(I-C) resonances (0.24 ppm, 35.5°C) and the poly(G-C) resonances (0.5 ppm, 30°C)[21] is similar to the range of ^{31}P shifts in oligonucleotide duplexes. Thus, the internucleotide phosphates in the self-complementary octanucleotide dG-dG-dA-dA-dT-dT-dC-dC resonate between 3.9 and 4.5 ppm upfield from standard trimethylphosphate in the duplex state.[10] This suggests that the observation of two resonances in the synthetic RNA's reflect the contributions from the base sequence and small variations in the phosphodiester O3′-P and P-O5′ torsion angles about the *gauche⁻, gauche⁻* region.

"Alternating DNA" Conformation

The structural and dynamic studies of synthetic DNA's outlined in the previous section are extended to alternating guanosine-cytidine polynucleotides which contain three Watson-Crick hydrogen bonds.

The individual base and sugar nonexchangeable protons of dC-dG-dC-dG, dC-dG-dC-dG-dC-dG, (dG-dC)₁₀₋₁₅ and poly(dG-dC) in low salt can be followed as average resonances during the duplex to strand transition with large shifts observed at the cytidine H-5 and H-6 protons and the two sugar H-1′ protons (Figure 6). The chemical shifts in the duplex state are summarized in Table VI with the base protons

Table VI
Chemical Shift, ppm, Duplex State[1]

	(dC-dG)$_2$	(dC-dG)$_3$	(dG-dC)$_{10-15}$	poly(dG-dC)
G(H-8)		7.905	7.840	7.775
C(H-6)	7.370	7.335	7.240	7.200
C(H-5)	5.480	5.440	5.330	5.280

[1]The chemical shifts in the duplex state are measured at 0°C for (dC-dG)$_2$, 25°C for (dC-dG)$_3$, 50°C for (dG-dC)$_{10-15}$ and 80°C for poly(dG-dC). The tabulated chemical shifts are for the internal base pair protons at the oligomer level.

shifting to higher field with increasing chain length. These differences between the short duplexes and the synthetic DNA may reflect the contributions from fraying at the ends of the duplex at the oligomer level. The magnitude and direction of the base and sugar H-1′ chemical shift changes associated with the duplex to strand transition of poly(dG-dC) in low salt (10 mM NaCl, 1 mM phosphate) (Table VII) are similar to the corresponding values for the melting transition of poly(dA-dT). This suggests that the low salt poly(dG-dC) conformation and the salt independent conformation of poly(dA-dT) exhibit similar base pairs overlaps (monitored at the base protons) and glycosidic torsion angles (monitored at the sugar H-1′ protons).

The calculated ring current shifts[30] based on the overlap geometries of the A and B conformations[11] of poly(dG-dC) are summarized in Table VIII. There is agreement between the experimental upfield shifts on formation of the synthetic DNA duplex and the calculated shifts based on the B-DNA overlap geometry at the guanosine H-1 proton (located in the center of the base pair) and the cytidine H-5 proton (located in the major groove) (Tables VII and VIII). A similar agreement was observed for the purine H-2 proton (located in the minor groove) in the poly(dA-dU) and poly(dI-dC) duplexes (Tables III - V). By contrast, the calculated ring current shifts underestimate the experimental upfield shifts on poly(dG-dC) duplex formation (Tables VII and VIII) at the cytidine H-6 and guanosine H-8 protons (which are directed towards the sugar-phosphate backbone). The origin of this discrepancy is not understood. Pullman and coworkers have emphasized that atomic diamagnetic anisotropy and polarisability effects make non-negligible contributions to the calculated shifts.[36] Further efforts are being directed towards a reliable estimation of the contributions of these latter effects and also an evaluation of the selective upfield shifts observed at the sugar H-1' protons on duplex formation.

DNA Conformational Transition

The recognition by repressors, polymerases, and restriction enzymes of specific nucleic acid sequences in DNA suggest that conformations other than the regular B-

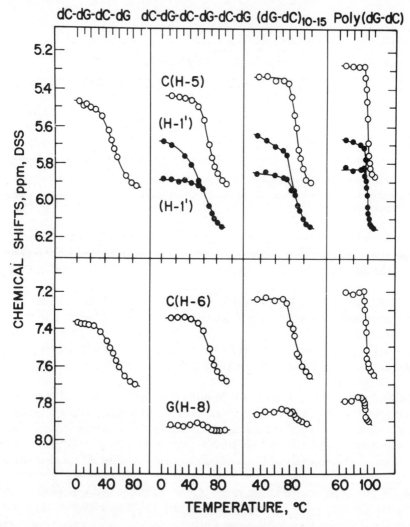

Figure 6. The temperature dependence of the base and sugar H-1' proton chemical shifts of dC-dG-dC-dG and dC-dG-dC-dG-dC-dG in 0.1 M cacodylate, 1 mM EDTA, ^{2}H$_2$O, pH 6.25, and (dG-dC)$_{10\text{-}15}$ and poly(dG-dC) in 0.1 M NaCl, 1 mM phosphate, 1 mM EDTA, ^{2}H$_2$O solution. The chemical shifts of the internal base pairs are plotted at the oligomer level.

Table VII
Experimental Upfield Chemical Shifts on
Poly(dG-dC) Duplex Formation

	Duplex State[1] δ_d,ppm	Strand State[2] δ_s,ppm	Upfield Shift' $\Delta\delta$,ppm
G(H-1)	12.90	13.70	0.80
G(H-8)	7.775	7.96	0.205
C(H-5)	5.28	5.94	0.66
C(H-6)	7.20	7.70	0.50
d(H-1')	5.83	6.15	0.32
u(H-1')	5.68	6.15	0.47

[1]Chemical shift in the duplex state, δ_d, of poly(dG-dC) at 80°C.

[2]Chemical shift in the strand state, δ_s, of oligo(dG-dC) and poly(dG-dC) at ∼110°C.

DNA structure may play a fundamental role in protein-nucleic acid interactions.[11/37-40]

A striking conformational transition has been reported by Pohl and Jovin for (dG-dC) oligomers and polymers on addition of high salt.[41] They observed a cooperative and reversible inversion of the circular dichroism (C.D.) spectrum of these alternating sequence duplexes with the transition midpoint occurring at 2.56 M NaCl and 0.66 M MgCl$_2$.[41] A similar inversion of the C.D. spectrum of poly(dG-dC) was observed in 60% ethanol solution[42] and in the mitomycin complexes of this synthetic

Table VIII
Computed Upfield Ring Current Shifts, ppm
on Alternating Guanosine-Cytidine
Polynucleotide Duplex Formation[1]

	A-RNA/DNA	B-DNA
G(H-1)	0.67	0.74
G(H-8)	0.15	0.05
C(H-6)	0.26	0.07
C(H-5)	1.02	0.51

[1]The computations are based on nearest neighbor, next-nearest neighbor and cross-strand ring current calculations tabulated by Arter and Schmidt.[30]

DNA.[43] This transition was specific for alternating (dG-dC) sequences since neither the alternating polymers poly(dA-dT) and poly(dI-dC) nor the complementary polymer poly dG•poly dC exhibited this conformational change.[41] The transition was also not observed for the corresponding ribo sequence poly(G-C).[41]

The low and high salt forms of (dG-dC)$_n$ are separated by a 22 KCal barrier in both directions with the trypanocidal agent ethidium bromide binding more tightly to the low salt form.[44] Crystals of dC-dG-dC-dG grown in low and high salt solutions exhibit different space groups and it has been possible to reversibly interconvert between the two crystalline states.[45] The structure of dC-dG-dC-dG-dC-dG crystals grown from a solution containing Na, Mg and spermine has been solved recently. (Wang, A., Quigley, G., and Rich, A., private communication.) For a summary of this recently solved structure read the section by Seeman[134] on crystallographic investigation of oligonucleotide structure in this volume.

Our experience with NMR studies of alternating sequence polynucleotides[16-19] prompted us to undertake an NMR investigation of the salt dependent conformations of $(dG-dC)_n$ at the oligomer and polymer level. The NMR line widths are narrower at the oligomer duplex level and we report below on spectral data on $(dG-dC)_n$ of chain length 16 (n = 8) and 20-30 (n = 10-15).[46] Parallel investigations on the salt dependent conformation of poly(dG-dC) in solution are presented in the companion volume of this Symposium.[21]

Hydrogen Bonding

The guanosine H-1 proton of $(dG-dC)_8$ is observed at 13.0 ppm in low salt (0.2 M NaCl) and 13.4 ppm in high salt (4.0 M NaCl) solution indicating that duplex structures are generated in both conditions. The hydrogen bonding is most likely to be of the Watson-Crick type in both states.

Base Pair Overlaps

The chemical shifts of the base protons in oligo(dG-dC) in low (10 mM NaCl) and high (4.0 M NaCl) salt are compared in Figure 7. The 22 KCal barrier results in slow exchange between salt dependent conformations so that the resonances in the two states cannot be correlated by mixing experiments. The tentative assignments of some of the resonances in the high salt conformation are being currently checked by specific labelling experiments.

The guanosine H-8 and cytidine H-6 resonances of oligo(dG-dC) in the duplex state exhibit similar chemical shifts in low and high salt (Figure 7). These resonances (which are directed towards the backbone phosphates) are sensitive to perturbations in the sugar-phosphate backbone as well as the base pair overlap geometries. By contrast, the cytidine H-5 proton of oligo(dG-dC) in the duplex state shifts 0.15 ppm to higher field in the high salt conformation (Figure 7).

The experimental data on the exchangeable and nonexchangeable base protons suggest that the base pair overlap geometry for the oligo(dG-dC) duplex changes on addition of high salt such that there is considerably less overlap with adjacent base pairs (0.4 ppm downfield shift) at the guanosine H-1 position in the center of the base pair and somewhat greater overlap with adjacent base pairs (0.15 ppm upfield shift) at the cytidine H-5 position located in the major groove.

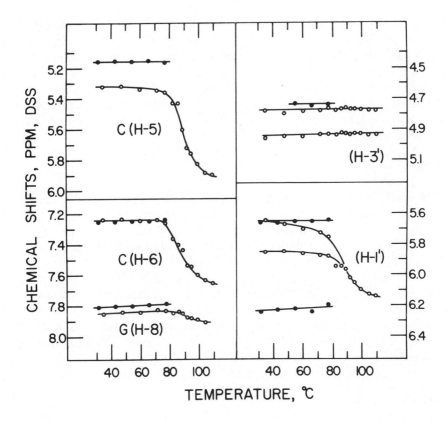

Figure 7. The temperature dependence of the nonexchangeable base and sugar protons of $(dG\text{-}dC)_{10\text{-}15}$ in 10 mM NaCl, (o), and $(dG\text{-}dC)_8$ in 4.0 M NaCl, (•), in 1 mM cacodylate, 0.1 mM EDTA buffer solution.

Glycosidic Torsion Angles

The chemical shifts of the sugar H-1′ protons in the duplex state of oligo(dG-dC) in low and high salt are compared in Figure 7. The H-1′ resonances cannot be assigned to the guanosine and cytidine sugars in either conformation at this time. However, the chemical shift data demonstrate that one of the two sugar H-1′ protons shifts downfield by 0.5 ± 0.1 ppm in high salt solution (Figure 7). The sugar H-1′ chemical shift is sensitive primarily to changes in the glycosidic torsion angle[47] and the experimental data suggest that either the cytidine or the guanosine glycosidic torsion angle in oligo(dG-dC) changes dramatically during the salt induced conformational transition.

Sugar Pucker and/or Orientation

The sugar H-3′ protons of oligo(dG-dC) are separated by 0.16 ppm in low salt and their chemical shifts remain unchanged during the duplex to strand transition (Figure 7). By contrast, the sugar H-3′ protons are superimposable for oligo(dG-dC) in high salt solution so that there is a selective upfield shift of ~0.2 ppm at one of the two sugar H-3' protons on proceeding from low to high salt (Figure 7). The H-3' proton is located on the C3'-C4' backbone bond and its chemical shift may be sensitive to changes in the sugar pucker and/or changes in the orientation of the sugar ring relative to the helix axis. If this interpretation of the H-3' chemical shifts is valid, one tentatively concludes that either the cytidine or the guanosine sugar pucker and/or sugar ring orientation in oligo(dG-dC) changes during the salt induced conformational transition.

Phosphodiester Linkage

The ^{31}P NMR spectrum of (dG-dC)$_8$ in low salt (0.2 M NaCl) exhibits a single resonance 4.2 ppm upfield from standard trimethylphosphate. By contrast, two ^{31}P resonances of approximately equal area with chemical shifts of 2.85 and 4.34 ppm are observed for (dG-dC)$_8$ in high salt (4.0 M NaCl) solution (Figure 8).

The experimental data suggest that the oligo(dG-dC) phosphodiester linkages dC(3'-5')dG and dG(3'-5')dC are similar in low salt and that the 1.5 ppm downfield shift reflects a change in conformation[48-50] of one of these phosphodiester linkages in high salt. We are unable to assign the two well resolved resonances in oligo(dG-dC) in high salt at this time and therefore cannot determine which phosphodiester linkage adopts a conformation different from that observed in B-DNA.

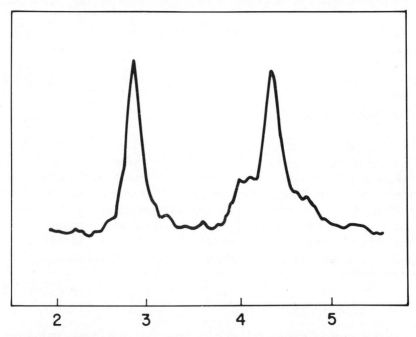

2	3	4	5

Figure 8. The 145.7 MHz proton noise decoupled ^{31}P NMR spectra of (dG-dC)$_8$ in 4.0 M NaCl, 1 mM cacodylate, 0.1 mM EDTA solution, pH 6.0, 27°C. The shift scale is upfield from standard trimethylphosphate.

Crystallographic Data and Model Building

The crystal structures of the alternating sequences pdA-dT-dA-dT[51] (ammonium salt) and dC-dG-dC-dG-dC-dG (crystals grown from solution containing Na, Mg and spermine; Wang, A., Quigley, G., and Rich, A., private communication) have been solved recently. The tetranucleotide forms an unusual right handed duplex structure where dinucleotide segments on one strand form Watson-Crick hydrogen bonds with complementary sequences on separate partner strands.[51] The dA glycosidic torsion angle and the dT(3′-5′)dA phosphodiester torsion angle(s) adopts values significantly different from those observed in the B-DNA conformation. These conformational features have been incorporated into an "alternating B-DNA" model for poly(dA-dT).[52] The hexanucleotide forms an unusual left-handed self-complementary duplex with six base pairs stabilized by Watson-Crick hydrogen bonds. The dG glycosidic torsion angle is in the *syn* conformation and the dG(3′-5′)dC phosphodiester torsion angles adopt values different from the normal *gauche*⁻, *gauche*⁻ orientations. The structural parameters can readily be used to generate the Z-DNA conformation for poly-(dG-dC). It should be noted that the "alternating B-DNA" conformation of Klug, *et al.* predicts an altered phosphodiester conformation at the pyrimidine(3′-5′)purine linkage while the Z-DNA structure of Rich, *et al.* observes an altered phosphodiester conformation at the purine(3′-5′)pyrimidine linkage.

Such structures (where the symmetry units repeats every two base pairs) may explain the selective cleavage of alternating purine-pyrimidine polynucleotides at the purine(3′-5′)pyrimidine site by the enzyme DNAse I[53] in contrast to cleavage at all sites in regular B-DNA.

The NMR chemical shift parameters demonstrate that every other glycosidic torsion angle, phosphodiester linkage and sugar pucker/orientation for the (dG-dC)$_n$ duplex in high salt adopts a different conformation from that observed in B-DNA. These results demonstrate that the symmetry unit repeats every base pair for (dG-dC)$_n$ in low salt but every two base pairs for (dG-dC)$_n$ in high salt solution. A further characterization of this novel conformation in solution must await the result of selective labelling studies to determine which glycosidic torsion angle (pyrimidine or purine) and which phosphodiester linkage [dG(3′-5′)dC or dC(3′-5′)dG] adopts a conformation different from that observed in B-DNA.

Repressor-Operator Interactions in lac System

The *lac* operator[54/55] contains two hexanucleotide segments with the sequence dT-

```
        1    5       10      15      20       25      30      35
5'  T-G-T-G-T-G-G-A-A-T-T-G-T-G-A-G-C-G̶-G-A-T-A-A-C-A-A-T-T-T-C-A-C-A-C-A   3'

3'  A-C-A-C-A-C-C-T-T-A-A-C-A-C-T-C-G-C-C-T-A-T-T-G-T-T-A-A-A-G-T-G-T-G-T   5'
```

dG-dT-dG-dT-dG on either side of a 23 base pair region that forms contacts with the *lac* repressor.[56-58]

We propose that these alternating pyrimidine-purine hexanucleotide segments can adopt a DNA structure where the symmetry unit repeats every two base pairs as discussed above. The alternating nature of the sugar-phosphate backbone of regions 1 to 6 and 30 to 35 would distinguish the intervening *lac* operator segment 7 to 29 from bulk DNA and serve as a recognition site for the *lac* repressor.

5' | T-G-T-G-T-G | G-A-A-T-T-G-T-G-A-G-C-G-G-A-T-A-A-C-A-A-T-T-T | C-A-C-A-C-A | 3'

3' | A-C-A-C-A-C | C-T-T-A-A-C-A-C-T-C-G-C-C-T-A-T-T-G-T-T-A-A-A | G-T-G-T-G-T | 5'

| ALTERNATING | REGULAR | ALTERNATING |
| DNA | B-DNA | DNA |

This proposal suggests that irregularities in the DNA conformation on either side of regulatory protein binding site may account for the specificity of protein-nucleic acid interactions.

Phenanthridine Antibiotic-Nucleic Acid Complexes

The study of antibiotic-DNA complexes are of great interest since a fundamental understanding of the complementary surfaces and the role of symmetry in these interactions would provide a molecular basis for the design of clinically useful pharmacological agents that function at the DNA level.

X-ray crystallographic information is available at atomic resolution for a series of antibiotics intercalated into miniature RNA and DNA duplexes.[59-61] (See also sections in this volume by Sobell, Berman and Neidle and Rich et al.) Current efforts are directed towards extending such studies to complexes at the oligonucleotide level.

The binding of a series of antibiotics to stable oligonucleotide duplexes has been monitored by NMR spectroscopy,[17] and the published examples are compiled in Table IX.[62-69] Our laboratory has undertaken parallel NMR investigations on a series of intercalating antibiotics complexed to poly(dA-dT) to evaluate the potential of NMR spectroscopy to study drug-nucleic acid interactions at the synthetic DNA level. This section outlines and interprets the NMR parameters for the complex of the trypanocidal antibiotic ethidium bromide (EtdBr) with nucleic acids in solution.

Ethidium Bromide

EtdBr is a trypanocidal antiobiotic which interferes with the replication and transcription process related to nucleic acid synthesis.[70] A variety of physical techni-

Table IX
Summary of Antibiotic•Oligonucleotide
Complexes Investigated by NMR Spectroscopy

Antibiotic	Sequence	Reference
Actinomycin D	dA-dT-dG-dC-dA-dT	9
	dC-dG-dC-dG	2
	$(dC_{15}-dA_{15})•(dT_{15}-dG_{15})$	7
Ethidium Bromide	dC-dG-dC-dG	63
	tRNA	64,65
Miracil	$(A_{20})•(U_{20})$	5
Proflavine	dC-dC-dG-dG	66
	dG-dG-dC-dC	66
9-Aminoacridine	dA-dT-dG-dC-dA-dT	67
Daunomycin	dG-dC-dG-dC	68
Netropsin	dG-dG-dA-dA-dT-dT-dC-dC	69

ques including viscosity,[71] flow dichroism,[71] and transient electric dichroism[72] on linear DNA, sedimentation studies on supercoiled circular DNA[73/74] and X-ray diffraction[75] and electron microscopy[76] studies on DNA fibers have demonstrated that the strong binding of ethidium to nucleic acids involves an intercalation complex of the type first proposed by Lerman.[77]

Strong support for the intercalation model came from X-ray studies of Sobell and coworkers on the 2:2 crystalline complex of ethidium and the dinucleoside phosphates iodoC-G[78] and iodoU-A.[79] The phenanthridine ring of one ethidium bromide intercalates between Watson-Crick base pairs of a miniature dinucleoside duplex with its phenyl and ethyl side chains exposed to solvent in the minor groove. Structural changes in the nucleic acid include an unwinding of the duplex by 26° and change of sugar puckers (C3′ *endo* at pyrimidine and C2′ *endo* at purine) and glycosidic torsion angles ($\chi \sim 25°$ for pyrimidine and $\sim 100°$ for purine) on complex formation. The crystallographic information on the ethidium•miniature RNA duplex served as a starting point for the generation of a neighbor exclusion ethidium•DNA complex in which ethidium intercalates at every other base pair and the DNA structure is characterized by alternating sugar puckers and glycosidic torsion angles.[80] By contrast, Sundaralingam and coworkers demonstrated that ethidium bromide binds at a single nonintercalative site in complexes of the trypanocidal agent diffused into crystals of yeast tRNA[Phe].[81]

Considerable effort has been directed towards the characterization of the ethidium•nucleic acid complexes in solution by fluorescence, temperature jump and nuclear magnetic resonance spectroscopic techniques. These investigations include complexes at the dinucleotide,[82-88] stable tetranucleotide,[63] transfer RNA[64/65/89] and DNA[44/90/91] level in solution.

We are primarily interested in understanding the intercalation process and hence the NMR studies of the EtdBr•poly(dA-dT) complex were undertaken at high Nuc/D ratios ($=25$ and 15) in 1 M NaCl solution. The NMR resonances broaden on complex formation at the synthetic DNA level[91] and it was necessary to undertake parallel studies on the complex with shorter oligomer duplexes whose narrow resonances could be more readily monitored through the helix-coil transition. These latter investigations have been undertaken on the self-complementary tetranucleotide dC-dG-dC-dG in 0.1 M buffer solution.[63]

Spectral Resolution

The NMR spectra of the nonexchangeable protons (5.5 to 8.5 ppm) in the Nuc/D $\simeq$ 25 EtdBr complexes with dC-dG-dC-dG and with poly(dA-dT) at temperatures near the midpoint of the melting transition are presented in Figures 9A and 9B respectively. The drug resonances are partially resolved from the nucleic acid base and sugar resonances so that each component can be monitored independently of each other during the dissociation of the complex with temperature. The EtdBr resonances are quite broad in the complex at the synthetic DNA level (Figure 9B) but can be readily

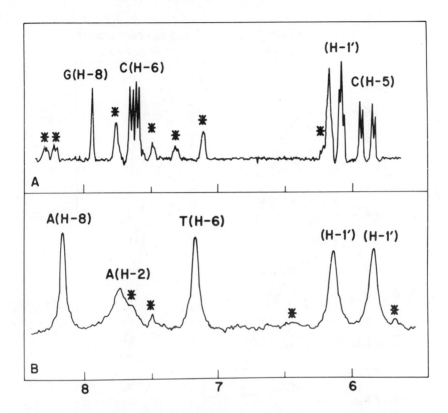

Figure 9. The 360 MHz proton NMR spectra of (A) EtdBr•dC-dG-dC-dG Complex, Nuc/D = 24, in 0.1 M phosphate, 1 mM EDTA, ^2H_2O, pH 6.74, 76.5°C ($t_{1/2}$ of complex = 74 ± 2°C) and (B) the EtdBr•poly(dA-dT) complex, Nuc/D = 25, in 1 M NaCl, 10 mM cacodylate, 0.1 mM EDTA, ^2H_2O at 78.6°C ($t_{1/2}$ of complex = 81.5°C). The EtdBr resonances in complex are designated by asterisks. The resolution of spectrum A was improved by convolution difference techniques while the signal to noise of spectrum B was improved by adding a line broadening contribution of 3 Hz.

observed at the tetramer level (Figure 9A) in the presence of excess nucleic acid.

Drug Resonances

The temperature dependence of the EtdBr resonances in the Nuc/D = 25 complexes with dC-dG-dC-dG (in 0.1 M phosphate buffer) and with poly(dA-dT) (in 1 M NaCl, 10 mM cacodylate buffer) are plotted in Figures 10A and 10B respectively. The antibiotic resonances of the complex in the presence of excess nucleic acid (Nuc/D = 25) reflect the chemical shift of free EtdBr at high temperature (≥ 100°C) and EtdBr bound to the nucleic acid at low temperature (≤ 60°C).

The protons on the phenanthridine ring shift upfield between 0.7 to 1.0 ppm on complex formation while the protons on the phenyl and ethyl side chains undergo much smaller shifts (≤ 0.2 ppm) to high and low fields on complex formation (Figures 10A and 10B). The large upfield shifts at the phenanthridine protons on

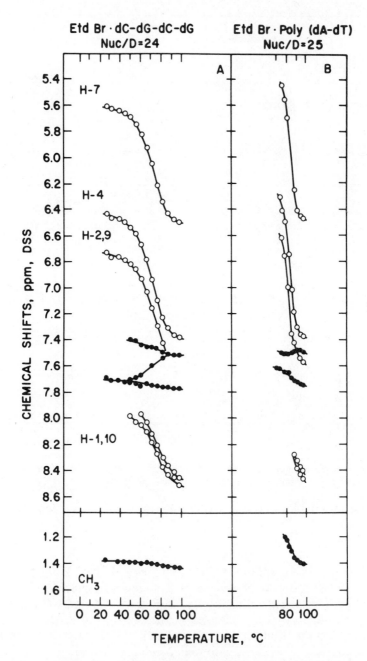

Figure 10. A comparison of the temperature dependent chemical shifts of the antibiotic in EtdBr•nucleic acid complexes at the oligonucleotide duplex and synthetic DNA level. (A) The EtdBr•dC-dG-dC-dG complex, Nuc/D = 24, in 0.1 M phosphate, 1 mM EDTA, ^2H_2O, pH 6.74. (B) The EtdBr•poly(dA-dT) complex, Nuc/D = 25, in 1 M NaCl, 10 mM cacodylate, 0.1 mM EDTA, ^2H_2O. The tetranucleotide concentration was 40 mM in nucleotides and the synthetic DNA concentration was 26.3 mM in nucleotides. The phenanthridine ring protons and side chain protons are represented by (o) and (•) respectively.

Table X
Experimental and Calculated Upfield
Phenanthridine Complexation Shifts ($\Delta\delta$, ppm)
in the EtdBr•dC-dG-dC-dG Complex

| | Free EtdBr[1] | dC-dG-dC-dG Complex[2] | | Intercalation at dC-dG Site[3] |
	δ	δ	$\Delta\delta$	$\Delta\delta$
H-1/10	8.57	7.86	0.71	0.5
H-1/10	8.63	7.96	0.67	0.5
H-2/9	7.65	6.75	0.90	0.7
H-2/9	7.52	6.44	1.08	0.7
H-4	7.52	6.44	1.08	0.75
H-7	6.60	5.63	0.97	0.75

[1]Chemical shift of 1 mM EtdBr in 0.1 M phosphate, 1 mM EDTA, ^2H_2O at 90°C.

[2]EtdBr•dC-dG-dC-dG complex, Nuc/D = 24, in 0.1 M phosphate, 1 mM EDTA, ^2H_2O, pH 6.74. Chemical shift (δ, ppm) values in the duplex state at 25°C. The experimental upfield complexation shift ($\Delta\delta$, ppm) is relative to free EtdBr at 90°C.

[3]The calculations are based on ring current and atomic diamagnetic anisotropy contributions[108] based on the intercalation overlap geometry depicted in the text.

complex formation (Tables X and XI) demonstrate intercalation of the chromophore between base pairs so that its proton experience upfield ring current contributions from adjacent base pairs[29/30] at the intercalation site. By contrast, the

Table XI
Experimental and Calculated Upfield
Phenanthridine Complexation Shifts ($\Delta\delta$, ppm)
in the EtdBr•Poly(dA-dT) Complex

| | Free EtdBr[1] | Poly(dA-dT) Complex[2] | | Intercalation at dT-dA Site[3] |
	δ	δ	$\Delta\delta$	$\Delta\delta$
H-1/10	8.57			0.65
H-1/10	8.63			0.65
H-2/9	7.65	≤ 6.61	≥ 1.04	0.7
H-2/9	7.52	≤ 6.41	≥ 1.11	0.7
H-4	7.52	≤ 6.41	≥ 1.11	0.85
H-7	6.60	≤ 5.44	≥ 1.16	0.85

[1]Chemical shift of 1 mM EtdBr in 0.1 M phosphate, 1 mM EDTA, ^2H_2O at 90°C.

[2]EtdBr•poly(dA-dT) complex, Nuc/D = 25, in 1 M NaCl, 10 mM cacodylate, 0.1 mM EDTA, ^2H_2O. Chemical shift (δ, ppm) values in the duplex state at <81°C. The experimental upfield complexation shift ($\Delta\delta$, ppm) is relative to free EtdBr at 90°C.

[3]The calculations are based on ring current and atomic diamagnetic anisotropy contributions[108] based on the intercalation overlap geometry depicted in the text.

phenyl and ethyl side chains are exposed to solvent and their protons are minimally perturbed by the ring currents of the base pairs.

Nucleic Acid Resonances

The base and sugar proton chemical shifts monitor an average of EtdBr-free and - bound regions of the complex in the presence of excess nucleic acid.

The thymidine H-3 proton is observed in the EtdBr•poly(dA-dT) complex, Nuc/D = 25, 43.5°C between 12.5 to 13.1 ppm (Figure 11A) which demonstrates that the base pairs are intact in the complex. The temperature dependence of this imino proton in poly(dA-dT) and its Nuc/D = 25 and 15 EtdBr complexes are plotted in

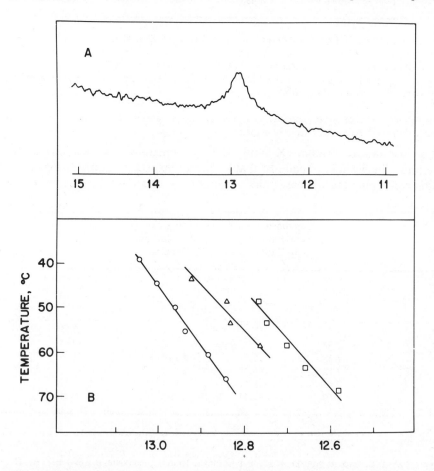

Figure 11. (A) The 360 MHz continuous wave proton NMR spectrum (11 to 15 ppm downfield from DSS) of the EtdBr•poly(dA-dT) complex, Nuc/D = 25, in 1 M NaCl, 10 mM cacodylate, 0.1 mM EDTA, H₂O, 43.5°C. (B) The temperature dependence of the thymidine H-3 resonance in poly(dA-dT), (o), and the EtdBr•poly(dA-dT) complexes, Nuc/D = 25, (Δ), and Nuc/D = 15, (□), in 1 M NaCl, 10 mM cacodylate solution.

Figure 11B with the resonance shifting to high field with increasing drug concentration.

The chemical shifts of the base and sugar non-exchangeable protons of poly(dA-dT) and its EtdBr complexes in high salt are plotted in Figure 12. The transition midpoint increases by 9.0°C and 13.5°C on formation of the Nuc/D = 25 and 15 complexes, respectively, demonstrating stabilization of the duplex by bound ethidium bromide. The nucleic acid resonances shift as average peaks during the melting transition so that the dissociation of the complex (k_d = 6.1 × 10^3 sec^{-1} for Nuc/D = 25 complex at 80.5°C, $t_{1/2}$ = 81.5°C; k_d = 5.5 × 10^{3}sec^{-1} for Nuc/D = 15 complex at 86°C, $t_{1/2}$ = 86°C) remains fast on the NMR time scale.

The base proton chemical shifts are either unaffected (adenosine H-8 and thymidine CH$_3$-5 resonances, Figure 12) or shift upfield (adenosine H-2 resonance, Figure 12,

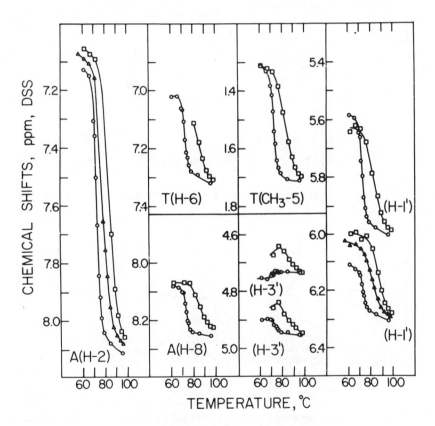

Figure 12. The temperature dependence of the chemical shifts of the nonexchangeable nucleic acid base and sugar protons in poly(dA-dT), (o), and the EtdBr•poly(dA-dT) complex, Nuc/D = 15, (□), in 1 M NaCl, 10 mM cacodylate, 0.1 mM EDTA, ^{2}H$_2$O solution. Plots of the adenosine H-2 and downfield H-1′ resonances for the Nuc/D = 25 complex, (△), are also included for comparison purposes. The synthetic DNA concentration was 26.3 mM in nucleotides.

thymidine H-3 resonance, Figure 11) on formation of the Nuc/D = 15 complex. This reflects the somewhat larger upfield ring current contributions from the phenanthridine ring[92] compared to the dA•dT base pair[29/30] it displaces following intercalation.

The sugar H-1′ protons resonate at ~5.6 and ~6.1 ppm in the duplex state of poly(dA-dT) and undergo duplex to strand transition shifts of 0.4 and 0.2 ppm, respectively (Figure 12). The 6.1 ppm resonance in the synthetic DNA shifts upfield on addition of EtdBr and resonates at 6.0 ppm for the Nuc/D = 15 complex. By contrast, the resonance at 5.6 ppm shifts downfield to a much smaller extent on addition of the antibiotic (Figure 12). The selective perturbation of one of the sugar H-

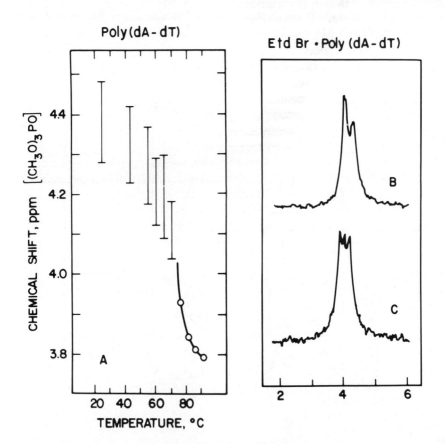

Figure 13. (Left, A) The temperature dependence (20° to 90°C) of the ^{31}P chemical shifts of poly(dA-dT) in 1 M NaCl, 10 mM cacodylate, 10 mM EDTA, ^2H_2O, pH 6.2. The spectra are an envelope of broad overlapping resonances between 20° and 70°C with a shift in intensity from the upfield to the downfield components on going from 55° to 70°C. (Right) The proton noise decoupled 145.7 MHz ^{31}P NMR spectra of the EtdBr•poly(dA-dT) complex, Nuc/D = 25 at 66°C (B) and Nuc/D = 15 at 71°C (C) in 1 M NaCl, 10 mM cacodylate, 1 mM EDTA, ^2H_2O solution. The chemical shifts are upfield from standard trimethylphosphate.

1′ proton chemical shifts on addition of EtdBr to poly(dA-dT) suggests a change in the corresponding glycosidic torsion angle[47] following intercalation of ethidium bromide into the duplex.

Phosphodiester Linkage

The chemical shifts of the internucleotide phosphates of poly(dA-dT) in high salt (1 M NaCl) are plotted as a function of temperature in Figure 13A. An unresolved envelope of resonances was observed in the duplex state and the center of the envelope shifts downfield from 4.4 ppm to 4.2 ppm on raising the temperature from 20° to 65°C in the premelting transition region (Figure 13A). A more pronounced downfield shift is associated with the melting transition ($t_{1/2}$ = 72.5°C) with the envelope shifting from 4.2 ppm at 65°C to narrow resonances at 3.85 ppm at 80°C (Figure 13A). Further downfield shifts are observed with increasing temperature in the post-melting transition range (>80°C) conditions under which the stacked strand-unstacked strand equilibrium shifts towards the latter. These downfield shifts probably reflect changes in the O-P rotation angles of the phosphodiester linkage during the transition from stacked states to unstacked strands.[2/93-95]

The proton noise decoupled 145 MHz ³¹P NMR spectra of the EtdBr•poly(dA-dT) Nuc/D = 25 complex (66°C) and Nuc/D = 15 complex (71°C) in high salt at temperatures corresponding to the duplex state are presented in Figures 13B and 13C respectively. No additional resonances were observed within 6 ppm to low and high field of the main envelope. The poly(dA-dT) envelope between 4.1 to 4.3 ppm at 65°C ($t_{1/2}$ = 72.5°C) shifts downfield to resolved peaks between 3.9 to 4.15 ppm at 71°C ($t_{1/2}$ = 86°C) for the Nuc/D = 15 complex. Since the nucleic acid is in excess under these conditions, the downfield complexation shifts represent an average of antibiotic-free and -bound nucleic acid states. It is not clear why more than two resolved resonances are observed in the EtdBr•poly(dA-dT) complex (Figures 13B and 13C) for a polynucleotide duplex containing only two types of phosphodiester linkages (dApdT and dTpdA).

Small downfield shifts have been previously reported for EtdBr•nucleic acid complexes at the dinucleotide[88] and stable tetranucleotide duplex[63] level. These solution results parallel the crystallographic data where the 03′-P and P-05′ dinucleoside monophosphate backbone torsion angles retain their *gauche⁻*, *gauche⁻* configuration on formation of the EtdBr miniature intercalation complexes.[78/79] The ³¹P chemical shift changes associated with the intercalation of EtdBr into nucleic acids (Figure 13) are much smaller than the large shifts of 1.5 ppm and 2.5 ppm associated with the intercalation of actinomycin D into oligonucleotide duplexes.[2/9] These results suggest that the conformation of the phosphodiester linkages are different in the generation of EtdBr and actinomycin D intercalation sites. By contrast, proposed models of the actinomycin-nucleic acid complex retain the *gauche⁻*, *gauche⁻* phosphodiester linkage at the intercalation site.[96/97]

Sugar Pucker Geometry

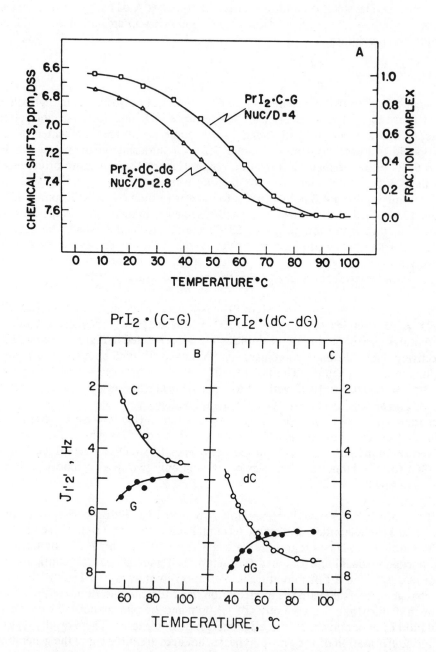

Figure 14. (Top) The temperature dependence of the phenanthridine H-2/H-9 proton in the $PrI_2 \bullet C\text{-}G$ complex, Nuc/D = 4,(□), and the $PrI_2 \bullet dC\text{-}dG$ complex, Nuc/D = 2.8, (△), in 0.1 M phosphate, 1 mM EDTA, ^2H_2O, pH 6.56. The dinucleotide concentration was 33 mM in nucleotides (Bottom) The temperature dependence of the sugar $J_{1'2'}$ proton vicinal coupling constant in the $PrI_2 \bullet$ miniature dinucleotide duplex complexes with C-G (B) and dC-dG (C).

The sugar rings exhibit a C3′*endo* (3′-5′)C2′*endo* pucker geometry in the 2:2 EtdBr•iodoC-G[78] and EtdBr•iodoU-A[79] complexes in the crystalline state. We have undertaken parallel NMR studies in solution to evaluate the sugar pucker in phenanthridine intercalating antibiotic complexes with miniature RNA and DNA duplexes.[85] The approach is based on the correlation which relates the vicinal sugar proton-proton coupling constant $J_{1'2'}$ monitored at the H-1′ proton with the pucker of the sugar ring.[35] The coupling constant is $\sim$0 Hz for C3′*endo* (C2′*exo*) sugar pucker and $\sim$10 Hz for C2′*endo* (C3′*exo*) sugar pucker.[35] The coupling information (splitting of $\leq$10 Hz) is lost under the broad ($\geq$50 Hz) proton resonances at the synthetic DNA duplex level. We have therefore undertaken NMR studies on miniature duplexes containing the intercalating agent propidium diiodide PrI_2 (for solubility considerations) and the self-complementary C-G and dC-dG sequences (due to the greater stability of G•C compared to A•T base pairs) as a function of Nuc/D ratio and salt concentration.[85]

The chemical shift changes in the drug resonances of PrI_2 with C-G and dC-dG as a function of temperature[85] parallel those observed in the EtdBr•dC-dG-dC-dG[63] and EtdBr•poly(dA-dT)[91] complexes (Figure 10). This establishes that the phenanthridine ring intercalates between base pairs and the phenyl and propyl side chains are exposed to solvent in the miniature complexes. The experimental data for the temperature dependence of the chemical shift of the H-2/H-9 proton in the PrI_2•C-G complex (Nuc/D = 4) and the PrI_2•dG-dG complex (Nuc/D = 2.8) in 0.1 M buffer are presented in Figure 14A. The transitions are quite broad with complex formation approaching completion at low temperature.

Experimentally, the sugar H-1′ resonances in the PrI_2•dinucleotide mixtures are narrow at high temperature and broaden progressively on lowering the temperature towards the midpoint of the transition. This limits the coupling constant information to the range f_c (fraction of complex) = 0.0 to $\sim$0.5 and trends in the coupling constant changes are used to estimate the values in the fully complexed state ($f_c = 1$) at low temperature.

The temperature dependence of the vicinal $J_{1'2'}$ coupling constants for the cytidine and guanosine residues in the PrI_2•C-G complex, Nuc/D = 4, and the PrI_2•dC-dG complex, Nuc/D = 2.8, in 0.1 M buffer are plotted in Figure 14B and 14C respectively. For the miniature RNA system, the $J_{1'2'}$ coupling constants are 4.5 to 5.0 Hz above 80°C (Figure 14B). The cytidine $J_{1'2'}$ coupling decreases to $\sim$2 Hz while the guanosine $J_{1'2'}$ increases to $\sim$6 Hz on lowering the temperature of the complex to $\sim$50°C (Figure 14B) conditions under which f_c = 0.55 (Figure 14A). The coupling constant trend with decreasing temperature requires that $J_{1'2'}$ for the cytidine residue should decrease towards 0 Hz (C3′*endo* pucker) and the $J_{1'2'}$ for the guanosine residue should increase towards 7 Hz (predominant C2′*endo* pucker) in the region below 10°C where f_c approaches unity. The experimental results demonstrates that C-G exhibits a mixture of C3′*endo* and C2′*endo* sugar puckers at each residue when free in solution at high temperature but switches to a C3′*endo*(3′-5′) predominant-C2′*endo* sugar pucker on intercalation of PrI_2 into a miniature RNA duplex at low temperature.

The coupling constant trends for the miniature DNA complex are such that $J_{1'2'}$ for the dC residue decreases from 7.5 Hz ($f_c = 0$) to ~2.0 Hz ($f_c = 1$) and $J_{1'2'}$ for the dG residue increases from 6.5 Hz ($f_c = 0$) to ~9.5 Hz ($f_c = 1$) (Figure 14C). The experimental results demonstrate that dC-dG exhibits a predominant C2′endo sugar pucker at each residue when free in solution at high temperature but rearranges to a predominant-C3′endo(3′-5′)C2′endo pucker geometry on intercalation of PrI₂ into a miniature DNA duplex at low temperatures.

Lee and Tinoco have reported on a seminal study of EtdBr complexes with miniature RNA duplexes containing bulge loops.[98] Their NMR contributions[98] along with our own results[85] demonstrate a predominant mixed sugar pucker geometry at the intercalation site of phenanthridine antibiotics into miniature RNA and DNA duplexes in solution.

Sequence Specificity

The sequence specificity of intercalative drug binding to nucleic acids was first probed at the dinucleoside monophosphate level with the drug acting as a template on which the nucleic acid forms a miniature duplex.[99/100] These studies demonstrated that actinomycin D exhibits a specificity for purine(3′-5′)pyrimidine sites[99-102] while ethidium bromide exhibits a specificity for pyrimidine(3′-5′)purine sites.[83/84/87]

Since dG•dC containing tetranucleotide sequences form stable duplexes at mM concentrations at low temperature in the absence of drugs[2/68/103] they serve as excellent models for the investigation of drug binding to stable nucleic acid duplexes. The dC-dC-dG-dG duplex contains a central dC-dG site (but no dG-dC site) while the dG-dG-dC-dC duplex contains a central dG-dC site (but no dC-dG site) so that these sequences are excellent models[63/66] for differentiating pyrimidine(3′-5′)purine specificity from purine(3′-5′)pyrimidine specificity associated with drug complexation.

The absorbance change in the 480 nm band of ethidium bromide has been monitored on gradual addition of the tetranucleotides dC-dC-dG-dG and dG-dG-dC-dC in 0.1 M phosphate at 1° and 21.5°C (Figure 15). A comparison of the tetranucleotide concentrations corresponding to half-maximal change demonstrates stronger binding of EtdBr to dC-dC-dG-dG compared to dG-dG-dC-dC at both temperatures. This establishes a relative sequence specificity for complex formation of EtdBr at the pyrimidine(3′-5′)purine sites at the stable duplex level in solution.[63/66]

By contrast, binding studies of actinomycin D with self-complementary tetranucleotides demonstrates that this antibiotic exhibits a purine(3′-5′)pyrimidine specificity at the stable duplex level since stronger binding is observed with dG-dG-dC-dC compared to dC-dC-dG-dG in solution.[66] These binding studies have been subsequently repeated in another laboratory.[104/105]

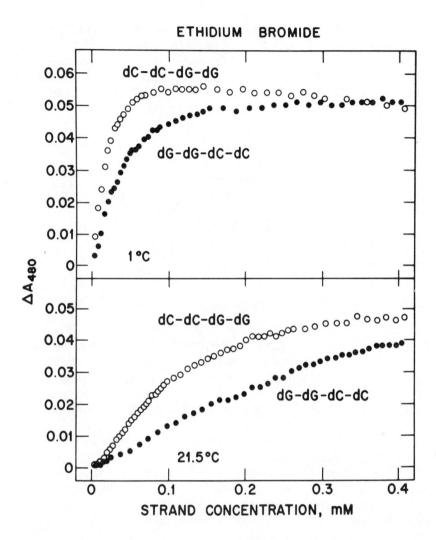

Figure 15. Changes in the 480 nm absorbance of 0.02 mM ethidium bromide on addition of dC-dC-dG-dG (o) and dG-dG-dC-dC (•) in 0.1 M phosphate solution, pH 7.0, at 1°C and at 21.5°C. The concentrations are based on extinction coefficients of ϵ_{480} = 5600 M^{-1}cm^{-1} for ethidium bromide, 2.90 × 10^4 M^{-1}cm^{-1} for dC-dC-dG-dG (in strands, no added salt, 70°C) and 3.05 × 10^4 M^{-1}cm^{-1} for dG-dG-dC-dC (in strands, no added salt, 70°C).

Overlap Geometry at Intercalation Site

The phenanthridine proton complexation chemical shift changes following intercalation into C-G and dC-dG duplexes are approximately equivalent[85] so that a similar overlap geometry between the phenanthridine ring and adjacent base pairs is predicted for the miniature RNA and DNA complexes. A parallel comparison in the crystalline state cannot be undertaken since crystals of EtdBr intercalated into a miniature DNA duplex have not been successfully grown to date.

A schematic representation of a proposed overlap geometry for EtdBr intercalated into a deoxy pyrimidine(3′-5′)purine site is presented below with the (o) symbols representing the location of the phenanthridine ring protons. The mutual overlap of the two base pairs at the intercalation site involves features derived in a linked atom conformational calculation of the intercalation site of the proflavine•DNA complex[106] and observed in the crystal of a platinum metallo-intercalator•miniature DNA complex.[107]

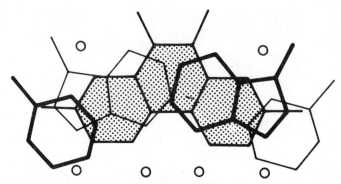

The experimental upfield phenanthridine proton complexation shifts for the EtdBr•dC-dG-dC-dG complex and the EtdBr•poly(dA-dT) complex in the presence of excess nucleic acid are presented in Tables X and XI respectively. The calculated upfield shifts of the phenanthridine protons due to ring currents and atomic diamagnetic anisotropy contributions[108] from adjacent base pairs for the overlap geometry presented above are listed for intercalation at dC-dG (Table X) and dT-dA (Table XI) sites.

The calculated upfield shifts are somewhat smaller than the experimental complexation shifts at the phenanthridine protons in the complexes at the oligonucleotide (Table X) and polynucleotide (Table XI) level. This difference may reflect in part the contributions from next-nearest neighbor base pairs which are not included in the evaluation of the calculated phenanthridine upfield shifts.

Intercalating Agent•Poly(dA-dT) Complexes

The NMR studies on the ethidium bromide•nucleic acid complexes in the previous section demonstrated that the magnitude of the upfield chemical shifts of the antibiotic protons on complex formation could be approximately correlated with the overlap geometry of the phenanthridine ring with adjacent base pairs at the intercalation site. This section briefly extends these studies to the poly(dA-dT) intercalation complexes of a platinum metallointercalator, the mutagen proflavine and the antitumor antibiotic daunomycin in high salt (1 M NaCl) solution.

Platinum Metallointercalator

Lippard and coworkers have introduced a series of platinum metallointercalating agents[109/110] and have extensively studied their complexes with RNA,[111] linear DNA[112] and circular DNA[113] in solution and in the fiber state.[114]

Current interest is focused on 2-hydroxyethane-thiolato-2,2′,2″-terpyridineplatinum II (TPH)[112] and highlighted by the solution to atomic resolution of a 2:2 complex of TPH and dC-dG.[107] (Also see reference 60 and the section by Rich *et al.* in this volume.) A molecule of TPH is intercalated into the miniature DNA duplex formed by two Watson-Crick dG•dC base pairs which are unwound by 23°.[107] The glycosidic torsion angles and sugar puckers differ dramatically for the dC ($\chi = 33°$, C3′ *endo* pucker) and the dG ($\chi = 115°$, C2′ *endo* pucker) residues in the complex.[107]

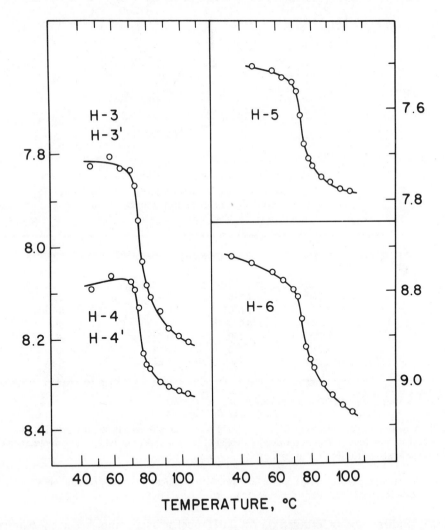

Figure 16. The temperature dependence of the chemical shifts of the TPH•poly(dA-dT) complex, Nuc/D = 15, in 1 M NaCl, 10 mM cacodylate, 0.1 M EDTA, ²H₂O. The poly(dA-dT) concentration was 26 mM in nucleotides.

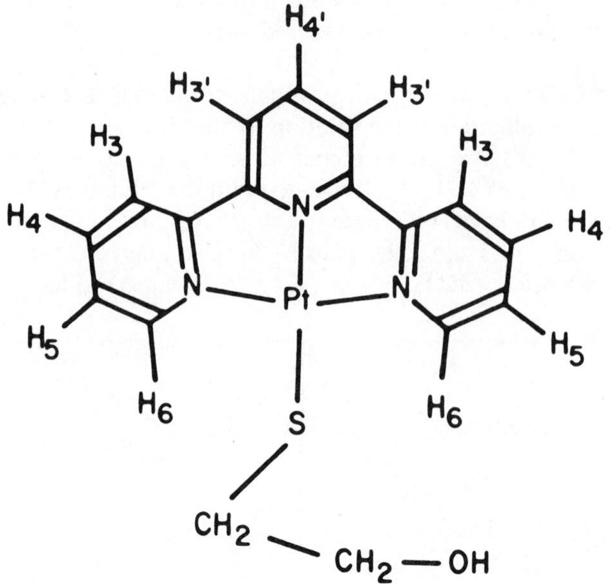

Table XII
Experimental and Calculated Upfield
Terpyridyl Complexation Shifts ($\Delta\delta$, ppm)
in the TPH•poly(dA-dT) Complex

| | Free TPH[1] | Poly(dA-dT) Complex[2] | | Intercalation at dT-dA Site[3] |
	δ	δ	$\Delta\delta$	$\Delta\delta$
H-3′	8.260	7.810	0.45	0.65
H-3	8.230	7.810	0.42	0.75
H-4′	8.420	8.080	0.34	0.3
H-4	8.350	8.080	0.27	0.5
H-5	7.810	7.510	0.30	0.5
H-6	9.170	8.730	0.44	0.55

[1]Chemical Shift of 1 mM TPH in 0.1 M phosphate at 90°C.

[2]TPH•poly(dA-dT) complex, Nuc/D = 15, in 1M NaCl, 10 mM cacodylate, 0.1 mM EDTA, ^2H_2O. Chemical shift (δ, ppm) values in the duplex state at 45°C. The experimental upfield complexation shift ($\Delta\delta$, ppm) is relative to free TPH at 90°C.

[3]The calculations are based on ring current and atomic diamagnetic anisotropy contributions[108] based on the intercalation overlap geometry depicted in the text.

NMR studies were undertaken on the TPH•nucleic acid complexes at the synthetic DNA level to deduce the overlap geometry at the intercalation site in solution. The temperature dependent chemical shifts of the partially resolved terpyridyl protons during the dissociation of the TPH•poly(dA-dT) complex, Nuc/D = 15, in high salt (1 M NaCl) solution are plotted in Figure 16. The chemical shifts of the terpyridyl protons in the poly(dA-dT) complex at 45 °C ($t_{1/2}$ of complex = 75.1 °C) and the experimental upfield complexation shifts relative to the value of free TPH at high temperature are summarized in Table XII.

The X-ray structure of the TPH•dC-dG complex demonstrates that TPH intercalates into the miniature duplex with the 2-hydroxyethanethiolato side chain directed into the major groove. A schematic drawing of the overlap geometry based on the X-ray structure is shown below with (o) designating the location of the terpyridyl protons.

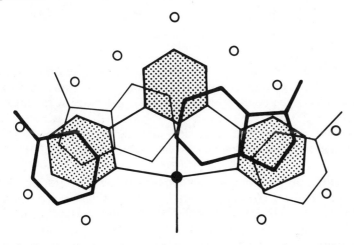

The calculated upfield shifts of the terpyridyl protons due to ring currents and atomic diamagnetic anisotropy effects of adjacent base pairs for this overlap geometry are summarized in Table XII. The calculated upfield shifts are somewhat larger than the experimental upfield shifts for TPH protons on complex formation at the synthetic DNA level (Table XII).

Proflavine

The cationic acridine dye proflavine binds to DNA and causes frame shift mutations.[115] The structure of the 3:2[116] and 2:2[117] proflavine:C-G complexes have been solved by X-ray crystallographic techniques. Proflavine intercalates into the miniature RNA duplex in these heavily hydrated crystals and the structure provides striking verification for Lerman's seminal proposal for the intercalation of planar dyes into nucleic acid double helices.[77] The conformational characteristics of possible proflavine intercalation sites in DNA[106] and RNA[118] have been probed by linked atom conformational calculations.

The mutagen and nucleic acid protons are well resolved in the proflavine•poly(dA-dT) complex in 1 M NaCl solution in the presence of excess nucleic acid.[119] Both sets of resonances shift as average peaks during the dissociation of the complex and the temperature dependent chemical shifts of the four types of nonexchangeable mutagen resonances in the Nuc/D = 8 proflavine•poly(dA-dT) complex are plotted in Figure 17A. The four resonances shift ~0.9 ppm to high field on complex formation ($t_{1/2}$ = 84.3°C in 1 M NaCl solution) (Table XIII).

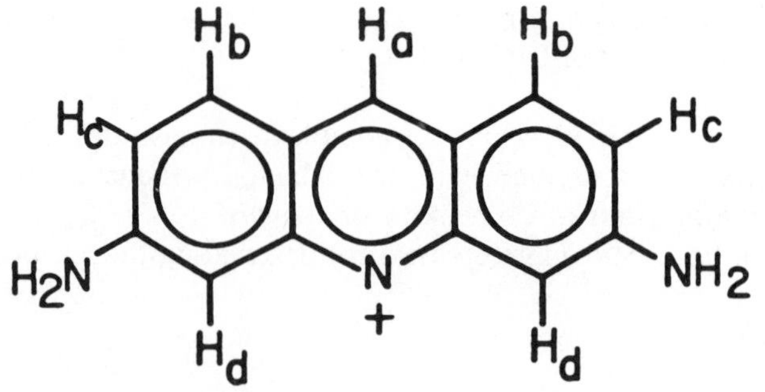

Proflavine, ethidium bromide and the platinum metallointercalator unwind closed circular DNA to the same extent following intercalation between base pairs.[73/113] The base pair geometry at the pyrimidine(3′-5′)purine intercalation site was kept the same as in the examples discussed earlier and the overlap of the proflavine ring with adjacent base pairs varied until there was approximate agreement between the experimental and calculated shifts. The overlap geometry shown below has the long axis of the proflavine ring colinear with the direction of the Watson-Crick hydrogen bonds and there is significant overlap between the mutagen ring system and both purine rings of adjacent base pairs.

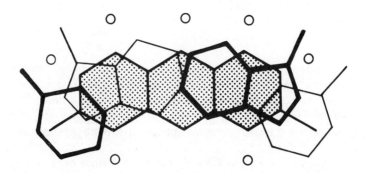

The calculated upfield complexation shifts at the mutagen protons from the shielding contributions of adjacent base pairs are listed in Table XIII and are somewhat smaller than the corresponding experimental values in the proflavine•poly(dA-dT) complex.

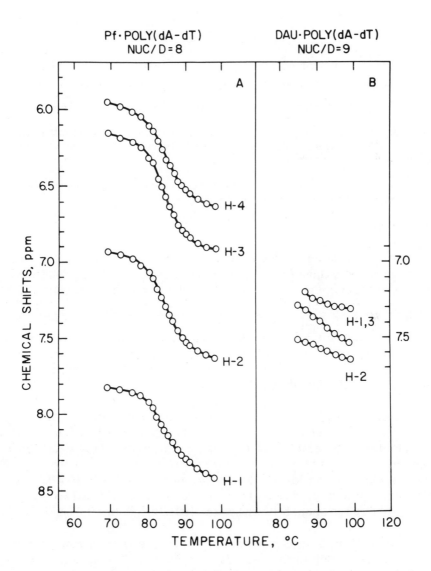

Figure 17. A comparison of the temperature dependence of the (A) acridine protons in the proflavine•poly(dA-dT) complex, Nuc/D = 8, in 1 M NaCl, 10 mM cacodylate, 10 mM EDTA, ^2H_2O, pH 7 and (B) the anthracycline ring D protons in the daunomycin•poly(dA-dT) complex, Nuc/D = 9, in 1 M NaCl, 10 mM cacodylate, 1 mM EDTA, ^2H_2O, pH 6.5.

Table XIII
Experimental and Calculated Upfield
Acridine Complexation Shifts ($\Delta\delta$, ppm)
in the Proflavine•Poly(dA-dT) Complex

	Free Prf[1] δ	Poly(dA-dT) Complex δ	$\Delta\delta$	Intercalation at dT-dA site $\Delta\delta$
H-1	8.740	7.820	0.92	0.85
H-2	7.880	6.930	0.95	0.8
H-3	7.060	6.155	0.905	0.65
H-4	6.840	5.955	0.885	0.7

[1]Chemical shift of 1.3 mM proflavine in 0.1 M phosphate, 1 mM EDTA, ^2H_2O, pH 6.6 at 100°C.

[2]Prf•poly(dA-dT) complex, Nuc/D = 8, in 1 M NaCl, 10 mM cacodylate, 10 mM EDTA, ^2H_2O, pH 7. Chemical shift (δ, ppm) values in the duplex state at 69°C. The experimental upfield complexation shift ($\Delta\delta$, ppm) is relative to free Prf at 90°C.

[3]The calculations are based on ring current and atomic diamagnetic anisotropy contributions[108] based on the intercalation overlap geometry depicted in the text.

Daunomycin

The anthracycline antibiotic daunomycin[120] and its analogs adriamycin[121] and car-minomycin[122] are in clinical use as potent antitumor agents. The available evidence suggests that the antitumor properties are associated with the intercalation of the antibiotic into DNA and subsequent blocking of RNA synthesis.[123/124] Though crystallographic information is available on daunomycin[125] and carminomycin,[126] attempts to generate a miniature duplex intercalation complex have been unsuccessful to date. Fiber diffraction patterns[127] and physico-chemical measurements[128/129] on the daunomycin-DNA complex have been interpreted in terms of intercalation into the nucleic acid.

We have investigated the NMR parameters for the daunomycin•poly(dA-dT) complex as a function of the Nuc/D ratio (50, 25, 9 and 5) in 1 M NaCl solution.[130] There are non-exchangeable protons on anthracycline ring D of daunomycin and hence only this part of the intercalating chromophore can be monitored on complex formation with nucleic acids. The temperature dependence of the H-1, H-2, and H-3 anthracycline protons of the Nuc/D = 9 daunomycin•poly(dA-dT) complex ($t_{1/2}$ ~ 90°C in 1 M NaCl solution) between 85° and 100°C are plotted in Figure 17B.

A comparison of the upfield chemical shifts of the proflavine protons (Figure 17A) with the corresponding shifts of the daunomycin ring D protons (Figure 17B) on complex formation with poly(dA-dT) demonstrates much smaller complexation shifts with the latter antibiotic. The experimental data for daunomycin complexes at the oligonucleotide and polynucleotide level at high Nuc/D ratios are summarized in Table XIV. The small magnitude of the anthracycline ring D proton upfield com-

Table XIV
Experimental Upfield Anthracycline
Complexation Shifts ($\Delta\delta$, ppm) in
Daunomycin•Nucleic Acid Complexes

	Free Dau[1]	dG-dC-dG-dC Complex[2]		Poly(dA-dT) Complex[3]	
	δ	δ	$\Delta\delta$	δ	$\Delta\delta$
H-1/3	7.680	7.465	0.215	7.330	0.350
H-1/3	7.465	7.275	0.190		
H-2	7.755			7.570	0.185
OCH$_3$-4	3.965	3.810	0.155		

[1]Chemical shift of 2 mM daunomycin in 0.1 M phosphate, 1 mM EDTA, ^2H_2O, at 90°C.

[2]Dau•dG-dC-dG-dC complex, Nuc/D = 28 in 0.1 M phosphate, 1 mM EDTA, ^2H_2O, pH 6.4. Chemical shift (δ, ppm) in the duplex state between 40° and 60°C. The experimental upfield shift ($\Delta\delta$, ppm) is relative to free Dau at 90°C.

[3]Dau•poly(dA-dT) complex, Nuc/D = 25 in 1 M NaCl, 10 mM cacodylate, 1 mM EDTA, ^2H_2O, pH 6.5. Chemical shift (δ, ppm) in the duplex state at 80°C. The experimental upfield shift ($\Delta\delta$, ppm) is relative to free Dau at 90°C.

plexation shifts suggest that these protons project onto the periphery of the shielding contours of adjacent nucleic acid base pairs at the intercalation site. This result suggests that either ring B and/or C of the anthracycline chromophore overlap with the nucleic acid base pairs in the intercalation complex of daunomycin with synthetic DNA in high salt solution. The stereodynamics of the interaction of the anthracycline drugs with nucleic acids is discussed by Krugh et al[135] elsewhere in this volume.

Comments on Calculated Shifts

The calculated upfield shift contributions from adjacent stacked base pairs underestimates somewhat the experimental shifts observed at the phenanthridine protons of EtdBr (Tables X and XI) and the acridine protons of proflavine (Table XIII) and overestimates the experimental shifts observed at the terpyridyl protons of TPH (Table XII) in these intercalating drug-nucleic acid complexes. The calculated shifts are very sensitive to the choice of overlap geometry at the intercalation sites and the various shielding contribution(s) from stacked adjacent base pairs.

It is not clear at this time whether the intercalation site conformations observed in the miniature RNA and DNA dinucleoside duplex-drug complexes in the crystalline state can be extrapolated to the polymer duplex level in solution. Recent transient electric dichroism studies on intercalating drug-DNA complexes in solution suggest that the drugs are tilted from perpendicularity by 21 ± 7° from their long helix axis[131] (see also Crothers *et al*, this volume) in contrast to the observation of no pronounced tilt for the miniature intercalation complexes observed in the crystal.[59-61] The static view of the intercalation site interactions presented above may need modification since planar drug like proflavine may oscillate within the intercalation site about a mean orientation.

The original shielding contours used to estimate the calculated shifts were based on ring currents only[29/30] and were later refined to include atomic diamagnetic anisotropy contributions and polarizability effects.[36] The relative importance of these separate contributions to the total calculated shift needs further elucidation based on simpler well-defined systems. The calculated shifts listed in Tables X to XIV were based on ring current and atomic diamagnetic anisotropy shielding contours at a distance of 3.4Å from the base pair planes. Sarma in one of the sections in this volume[133] discusses the derivation of shielding contributions from ring current, diamagnetic and paramagnetic anisotropy from x, y, z coordinates and such an approach should resolve some of the limitations associated with the methodology used in the present work.

Summary

The section on synthetic RNA and DNA duplexes demonstrates the application of high resolution NMR spectroscopy to investigate the helix-coil transition of alternating purine-pyrimidine polynucleotides in solution. The premelting, melting and postmelting transitions can be monitored at markers distributed throughout the base pairs and sugar-phosphate backbone. Qualitative estimates of changes in the base pair overlaps, the glycosidic torsion angles and the phosphodiester linkages can be evaluated from the variations in the chemical shift of the base protons, the sugar protons and the backbone phosphates, respectively. We observe several disagreements between the experimental and calculated ring current upfield shifts on formation of the poly(purine-pyrimidine) duplexes. This suggests that either the overlap geometries for these alternating synthetic RNA and DNA polynucleotides differ from the classical A and B conformations respectively in solution or that contributions other than ring current effects contribute to the calculated shifts.[132]

We have probed structural aspects of the salt dependent poly(dG-dC) transition by NMR spectroscopy. The chemical shift parameters demonstrate Watson-Crick base pairing for the high salt poly(dG-dC) duplex with the symmetry unit repeating every two base pairs resulting in an alternating conformation in solution. Specifically, every other phosphodiester linkage and glycosidic torsion angle adopts values different from that observed in B-DNA. The same alternating feature is tentatively assigned to either the pucker and/or orientation of the sugar-ring. The poly(dG-dC) transition from low to high salt results in considerably less overlap of the Watson-Crick imino H-1 proton and somewhat greater overlap of the cytidine H-5 proton with adjacent base pairs. The X-ray structure of dC-dG-dC-dG-dC-dG has been successfully solved by Rich and coworkers (Private communication) and this structure most likely corresponds to the high salt conformation based on the crystallization medium (contains Na, Mg and spermine). Each of the structural aspects of the high salt conformation of $(dG-dC)_n$ deduced from the NMR parameters[21/46] has been observed in the hexanucleotide duplex in the crystalline state at atomic resolution. The details available from the X-ray structure of dC-dG-dC-dG-dC-dG in turn will permit a more quantitative correlation of the observed NMR chemical shifts with torsion angle changes about specific bonds in the polynucleotide backbone.

The application of NMR spectroscopy to investigate the interaction of ethidium bromide with nucleic acids in solution has been presented above in considerable detail. The drug and nucleic acid resonances can be monitored independently of each other during the temperature dependent dissociation of the complex. The large upfield chemical shifts of the ethidium bromide resonances on complex formation demonstrates that the phenanthridine ring intercalates between base pairs and their relative magnitude defines an approximate overlap geometry between the phenanthridine ring and adjacent base pairs. We observe mixed sugar pucker and glycosidic torsion angles and an unperturbed phosphodiester linkage at the intercalation site in ethidium•deoxy nucleic acid duplexes similar to the earlier observation of these features in the ethidium ribodinucleoside miniature duplex in the crystal. The relative specificity of ethidium for pyrimidine(3'-5')purine sites at the dinucleoside monophosphate level has been extended to the stable tetranucleotide duplex level.

These studies are extended to derive the NMR parameters for the poly(dA-dT) complexes with the intercalating agents terpyridineplatinum, proflavine and daunomycin in the presence of excess nucleic acid. The approximate overlap geometries between the planar intercalating chromophore and adjacent base pairs at the intercalation site are proposed for the nucleic acid complexes with the metallointercalator and the mutagen in solution.

References and Footnotes

1. Cross, A.D., and Crothers, D.M., *Biochemistry, 10*, 4015-4023 (1971).
2. Patel, D.J., *Biopolymers, 15*, 533-558 (1976).
3. Kallenbach, N.R., Daniel, W.E., Jr., and Kaminker, M.A., *Biochemistry, 15*, 1218-1224 (1976).
4. Arter, D.B., Walker, G.C., Uhlenbeck, O.C., and Schmidt, P.G., *Biochem. Biophy. Res. Commun., 61*, 1089-1094 (1974).
5. Heller, M.J., Tu, A.T., and Maciel, G.E., *Biochemistry, 13*, 1623-1631 (1974).
6. Borer, P.N., Kan, L.S., and T'so, P.O.P., *Biochemistry, 14*, 4847-4863 (1975).
7. Early, T.A., Kearns, D.R., Burd, J.F., Larson, J.E., and Wells, R.D., *Biochemistry, 16*, 541-551 (1977).
8. Selsing, E., Wells, R.D., Early, T.A., and Kearns, D.R., *Nature, 275*, 249-250 (1978).
9. Patel, D.J., *Biochemistry, 13*, 2396-2402 (1974).
10. Patel, D.J., and Canuel, L.L., *Eur. J. Biochem., 96*, 267-276 (1979).
11. Arnott, S., Chandrasekaran, R., and Selsing, E., in *Structure and Conformation of Nucleic Acids and Protein-Nucleic Acid Interactions*, eds. Sundaralingam, M., and Rao, S.T., University Park Press, Baltimore, 577-596 (1975).
12. Brunner, W.C., and Maestre, M.F., *Biopolymers, 13*, 343-357 (1975).
13. Pilet, J., Blicharski, J., and Brahms, J., *Biochemistry, 14*, 1869-1876 (1975).
14. Wells, R. D., Larson, J. E., Grant, R.C., Shortle, B.E., and Cantor, C.R., *J. Mol. Biol., 54*, 465-497 (1970).
15. Ivanov, V.I., Minchenkova, L.E., Schyolkina, A.K., and Polatayev, A.I., *Biopolymers, 12*, 89-110 (1973).
16. Patel, D.J., and Canuel, L.L., *Proc. Natl. Acad. Sci. U.S.A., 73*, 674-678 (1976).
17. Kearns, D.R., *Progr. Nucleic Acids Res. and Mol. Biol., 6*, 477-523 (1977).
18. Patel, D.J., *Eur. J. Biochem., 83*, 453-464 (1978).
19. Patel, D.J., *J. Polym. Sci., Polym. Symp., 62*, 117-141 (1978).
20. Early, T.A., Olmstead, J., Kearns, D.R., and Lewis, A.G., *Nucleic Acids Res., 5*, 1955-1970 (1978).
21. Patel, D.J., in *Stereodynamics of Molecular Systems*, Ed. Sarma, R.H., Pergamon Press, New York, Oxford, 397-422 (1979).
22. Spatz, C., and Baldwin, R.L., *J. Mol. Biol., 11*, 213-222 (1965).
23. Baldwin, R.L., in *Molecular Associations in Biology*, Academic Press, New York, 145-162 (1968).
24. Patel, D.J., *Acc. Chem. Res., 12*, 118-125 (1979).
25. Kearns, D.R., Patel, D.J., and Shulman, R.G., *Nature, 229*, 338-340 (1971).
26. Crothers, D.M., Hilbers, C.W., and Shulman, R.G., *Proc. Natl. Acad. Scs. U.S.A., 70*, 2899-2901 (1973).
27. Patel, D.J., and Tonelli, A.E., *Biopolymers, 13*, 1943-1964 (1974).
28. Hilbers, C.W., in *NMR in Biology*, ed. Shulman, R.G., Academic Press, in press.
29. Gressner-Prettre, C., Pullman, B., Borer, P.N., Kan, L.S., and T'so, P.O.P., *Biopolymers, 15*, 2277-2286 (1976).
30. Arter, D.B., and Schmidt, P.G., *Nucleic Acids Research, 3*, 1437-1447 (1976).
31. Gennis, R.B., and Cantor, C.R., *J. Mol. Biol., 65*, 381-399 (1972).
32. Brahms, S., Brahms, J., and VanHolde, K.E., *Proc. Natl. Acad. Scs. U.S.A., 73*, 3453-3457 (1976).
33. Greve, J., Maestre, M.F., and Levin, A., *Biopolymers, 16*, 1489-1504 (1977).
34. Paleck, E., *Progr. Nucl. Acid. Res. Mol. Biol., 18*, 151-213 (1976).
35. Altona, C., and Sundaralingam, M., *J. Am. Chem. Soc., 95*, 2333-2344 (1973).
36. Giessner-Prettre, C., Pullman, B., and Caillet, J., *Nucleic Acids Research, 4*, 99-116 (1977).
37. Wells, R.D., Blakesley, R.W., Burd, J.F., Chan, H.W., Dodgson, J.B., Hardies S.C., Horn, G.T., Jensen, K.F., Larson, J.E., Nes, I.F., Selsing, E., and Wartell, R.M., *Critical Reviews in Biochemistry, 4*, 305-340 (1977).

38. Crick, F.H.C., and Klug, A., *Nature (London), 255,* 530-533 (1975).
39. Sobell, H.M., Reddy, B.S., Bhandary, K., Jain, S.C., Sakore, T.D., and Seshadri, T.P., *Cold Spring Harbor Symp. Quant. Biol., 42,* 87-102 (1977).
40. Selsing, E., Wells, R.D., Alden, C.J., and Arnott, S., *J. Biol. Chem., 254,* 5417-5422 (1979).
41. Pohl, F.M., and Jovin, T.M., *J. Mol. Biol., 67,* 375-396 (1972).
42. Pohl, F.M., *Nature, 260,* 365-366 (1976).
43. Mercardo, C.M., and Tomasz, M., *Biochemistry, 16,* 2039-2046 (1977).
44. Pohl, F.M., Jovin, T.M., Baehr, W., and Holbrook, J.J., *Proc. Natl. Acad. Scs. U.S.A., 69,* 3805-3809 (1972).
45. Drew, H.R., Dickerson, R.E., and Itakura, K., *J. Mol. Biol., 25,* 535-543 (1978).
46. Patel, D.J., Canuel, L.L., and Pohl, F.M., *Proc. Natl. Acad. Scs. U.S.A., 76,* 2508-2511 (1979).
47. Giessner-Prettre, C., and Pullman, B., *J. Theor. Biol., 65,* 171-188 (1977).
48. Patel, D.J., *Biochemistry, 13,* 2388-2395 (1974).
49. Gueron, M., and Shulman, R.G., *Proc. Natl. Acad. Scs. U.S.A., 72,* 3482-3485 (1975).
50. Gorenstein, D.G., Findlay, J.B., Momii, R.K., Luxon, B.A., and Kar, D., *Biochemistry, 15,* 3796-3803 (1976).
51. Viswamitra, M.A., Kennard, O., Shakked, Z., Jones, P. G., Sheldrick, G.M., Salisbury, S., and Falvello, L., *Nature, 173,* 687-688 (1978).
52. Klug, A., Jack, A., Viswamitra, M.A., Kennard, O., Shakked, Z., and Steitz, T.A., *J. Mol. Biol., 131,* 669-680 (1979).
53. Scheffler, I.E., Elson, E.L., and Baldwin, R.L., *J. Mol. Biol., 36,* 291-304 (1968).
54. Gilbert, W., and Maxam, A., *Proc. Natl. Acad. Scs. U.S.A., 70,* 3581-3584 (1973).
55. Dickson, R.C., Abelson, S., Barnes, W.M., and Reznikoff, W.S., *Science, 187,* 27-35 (1975).
56. Jobe, A., Sadler, J.R., and Bougeois, S., *J. Mol. Biol., 85,* 321-348 (1974).
57. Ogata, R., and Gilbert, W., *Proc. Natl. Acad. Scs. U.S.A., 74,* 4973-4976 (1977).
58. Goeddel, D.V., Yansura, D.G., and Caruthers, M.H., *Proc. Natl. Acad. Scs. U.S.A., 75,* 3578-3582 (1978).
59. Lozansky, E.D., Sobell, H.M., and Lessen, M., in *Stereodynamics of Molecular Systems,* Ed. Sarma, R.H., Pergamon Press, New York, Oxford, 265-270 (1979).
60. Rich, A., Quigley, G.J., and Wang, A.H.J., in *Stereodynamics of Molecular Systems,* Ed. Sarma, R.H., pergamon Press, New York, 315-330 (1979).
61. Berman, H.M., and Neidle, S., in *Stereodynamics of Molecular Systems,* Ed. Sarma, R.H., Pergamon Press, New York, Oxford, 367-382 (1979).
62. Kallenbach, N., and Berman, H.M., *Quart. Revs. Biophys., 10,* 138-236 (1977).
63. Patel, D.J., and Canuel, L.L., *Proc. Natl. Acad. Scs. U.S.A., 73,* 3343-3347 (1976).
64. Jones, C.R. and Kearns, D.R., *Biochemistry, 14,* 2660-2665 (1975).
65. Jones, C.R., Bolton, P.H., and Kearns, D.R., *Biochemistry, 17,* 601-607 (1978).
66. Patel, D.J., and Caunuel, L.L., *Proc. Natl. Acad. Scs. U.S.A., 74,* 2624-2628 (1977).
67. Reuben, J.H., Baker, B.M., and Kallenbach, N.R., *Biochemistry, 17,* 2915-2919 (1978).
68. Patel, D.J., *Biopolymers, 17,* 553-569 (1979).
69. Patel, D.J., *Eur. J. Biochem., 99,* 369-378 (1979).
70. Waring, M., in *Antibiotics III,* eds. Corcoran, J.W., and Hahn, F.E., Springer-Verlag, New York, 141-165 (1975).
71. LePecq, J.B., and Paoletti, C.J., *J. Mol. Biol., 27,* 87-106 (1967).
72. Hogan, M., Dattagupta, N., and Crothers, D.M., *Biochemistry, 17,* 280-288 (1979).
73. Crawford, L.V., and Waring, M.J., *J. Mol. Biol., 25,* 23-30 (1967).
74. Bauer, W., and Vinograd, J., *J. Mol. Biol., 33,* 141-172 (1968).
75. Fuller, W., and Waring, M.J., *Ber. Bunsenges, Physik. Chem., 68,* 805 (1964).
76. Freifelder, D., *J. Mol. Biol., 60,* 401-403 (1971).
77. Lerman, L.S., *J. Mol. Biol., 3,* 18-30 (1961).
78. Tsai, C.C., Jain, S.C., and Sobell, H.M., *J. Mol. Biol., 114,* 301-315 (1977).
79. Jain, S.C., Tsai, C.C., and Sobell, H.M., *J. Mol. Biol., 114,* 317-331 (1977).
80. Sobell, H.M., Tsai, C.C., Jain, S.C., and Gilbert, S.G., *J. Mol. Biol., 114,* 333-365 (1977).
81. Liebman, M., Rubin, J., and Sundaralingam, M., *Proc. Natl. Acad. Scs. U.S.A., 74,* 4821-4825 (1977).

82. Kreishman, G.P., Chan, S.I., and Bauer, W., *J. Mol. Biol., 61*, 45-58 (1971).
83. Krugh, T.R., Wittlin, F.K., and Cramer, S.P., *Biopolymers, 14*, 197-210 (1975).
84. Krugh, T.R., and Reinhardt, C.G., *J. Mol. Biol., 97*, 133-162 (1975).
85. Patel, D.J., and Shen, C., *Proc. Natl. Acad. Scs. U.S.A., 75*, 2553-2557 (1978).
86. Krugh, T.R., and Nuss, M.E., in *Biological Applications of Magnetic Resonance*, ed., Shulman, R.G., in press.
87. Davanloo, P., and Crothers, D.M., *Biochemistry, 15*, 5299-5305 (1976).
88. Reinhardt, C.G., and Krugh, T.R., *Biochemistry, 17*, 4845-4854 (1978).
89. Wells, B.D., and Cantor, C.R., *Nucleic Acids Research, 4*, 1667-1680 (1977).
90. Bresloff, J.F., and Crothers, D.M., *J. Mol. Biol., 95*, 103-123 (1975).
91. Patel, D.J., and Canuel, L.L., *Biopolymers, 16*, 857-873 (1977).
92. Giessner-Prettre, C., and Pullman, B., *C. R. Acad. Sci. Paris., 283D*, 675-677 (1976).
93. Tewari, R., Nanda, R.K. and Govil, G., *Biopolymers, 13*, 2015-2035 (1974).
94. Olson, W.K., *Biopolymers, 14*, 1797-1810 (1975).
95. Yathindra, N., and Sundaralingam, M., *Proc. Natl. Acad. Sci., U.S.A., 71*, 3325-3328 (1974).
96. Sobell, H.M., *Scientific American, 231*, 82-91 (1974).
97. Sobell, H.M., Tsai, C.C., Jain, S.C., and Gilbert, S.G., *J. Mol. Biol., 114*, 333-365 (1977).
98. Lee, C-H., and Tinoco, I., Jr., *Nature, 274*, 609-610 (1978).
99. Schara, R., and Muller, W., *Eur. J. Biochem., 29*, 210-216 (1972).
100. Krugh, T.R., *Proc. Natl. Acad. Sci. U.S.A., 69*, 1911-1914 (1972).
101. Davanloo, P., and Crothers, D.M., *Biochemistry, 15*, 4433-4438 (1976).
102. Patel, D.J., *Biochem. Biophys. Acta., 442*, 98-108 (1976).
103. Patel, D.J., *Biopolymers, 16*, 1635-1656 (1977).
104. Kastrup, R.V., Young, M.A., and Krugh, T.R., *Biochemistry, 17*, 4855-4865 (1978).
105. Chiao, Y-C., Gurudath Rao, K., Hook, III, J. W., Krugh, T.R., and Sengupta, S.K., *Biopolymers, 18*, 1749-1762 (1979).
106. Alden, C.J., and Arnott, S., *Nucleic Acids Research, 2*, 1701-1717 (1975).
107. Wang, A.H.J., Nathans, J., van der Marel, G., van Boom, J.H., and Rich, A., *Nature, 276*, 471-474 (1978).
108. Giessner-Prettre, C., and Pullman, B., *Biochem. Biophys. Res. Commun., 70*, 578-581 (1976).
109. Jennette, K.W., Lippard, S.J., Vassiliades, G.A., and Bauer, W.R., *Proc. Natl. Acad. Sci. U.S.A., 71*, 3839-3843 (1974).
110. Jennette, K.W., Gill, J.T., Sadownick, J.A., and Lippard, S.J., *J. Am. Chem. Soc., 98*, 6159-6168 (1976).
111. Barton, J.K., and Lippard, S.J., *J. Am. Chem. Soc., 18*, 2661-2668 (1979).
112. Lippard, S.J., *Acc. Chem. Res., 11*, 211-217 (1978).
113. Howe-Grant, M., Wu, K.C., Bauer, W.R., and Lippard, S.J., *Biochemistry, 15*, 4339-4356 (1976).
114. Bond, P.J., Langridge, R., Jennette, K.W., and Lippard, S.J., *Proc. Natl. Acad. Sci., U.S.A., 72*, 4825-4829 (1975).
115. Brenner, S., Barnett, L., Crick, F.H.C., and Orgel, A., *J. Mol. Biol., 3*, 121-124 (1961).
116. Neidle, S., Achari, A., Taylor, G.L., Berman, H.M., Carrell, H.L., Glusker, J.P., and Stallings, W.C., *Nature, 269*, 304-307 (1977).
117. Sobell, H.M., in *International Symposium on Biomolecular Structure, Conformation, Function and Evolution*, ed., Srinivasan, R., Pergamon Press, Oxford, in press.
118. Alden, C.J. and Arnott, S., *Nucleic Acids Research, 4*, 3855-3861 (1977).
119. Patel, D.J., *Biopolymers, 16*, 2739-2754 (1977).
120. DiMarco, A., Gaetani, M., Orezzi, P., Scarpinato, B.M., Silvestrini, R., Soldati, M., Dasdia, J., and Valentini, L., *Nature, 201*, 706-707 (1964).
121. Arcamone, F., Cassinelli, G., Fatini, G., Grein, A., Orozzi, P., Pol, C., and Spalla, C., *Biotechnol. Bioeng., 11*, 1101-1105 (1969).
122. Gauze, G.F., Sveshikova, M.A., Ukholina, R.S., Gavrilina, G.V., Filicheva, V.A., and Gladkikh, E.G., *Antibiotiki, 18*, 675-678 (1973).
123. DiMarco, A., Arcamone, F., and Zunino, F., *Antibiotics*, Vol. 3, eds. Corcoran, J.W., and Hahn, F.E., Springer-Verlag, Berlin, 101-128 (1974).

124. Neidle, S., *Topics in Antibiotic Chemistry*, ed. Sammes, P., Ellis Horwood Ltd., Chichester, Vol. 2, Part D, 241-278 (1978).
125. Neidle, S., and Taylor, G., *Biochem. Biophys. Acta., 479*, 450-459 (1977).
126. VonDreele, R.B., and Einck, J.J., *Acta. Crystallogr. Sec. B, 33*, 3283-3288 (1977).
127. Pigram, W.J., Fuller, W., Hamilton, L.D., *Nature New Biology, 235*, 17-19 (1977).
128. Zunino, F., Gambetta, R., DiMarco, A., Luoni, G., and Zaccara, A., *Biochem. Biophys. Res. Commun., 69*, 744-750 (1976).
129. Gabbay, E.J., Grier, D., Fingerle, R.E., Reimer, R., Levy, R., Pearce, S.W., and Wilson, W.D., *Biochemistry, 15*, 2062-2070 (1976).
130. Patel, D.J., and Canuel, L.L., *Eur. J. Biochem., 90*, 247-254 (1978).
131. Crother, D.M., Dattagupta, N., and Hogan, M., in *Stereodynamics of Molecular Systems*, ed., Sarma, R.H., Pergamon Press, New York, Oxford, p. 383 (1979).
132. Sarma, R.H., in *Nucleic Acid Geometry and Dynamics*, ed., Sarma, R.H., Pergamon Press, New York, Oxford, pp. 143-184 (1980).
133. Sarma, R.H., in *Nucleic Acid Geometry and Dynamics*, ed., Sarma, R.H., Pergamon Press, New York, Oxford, pp. 1-46 (1980).
134. Seeman, N.C., in *Nucleic Acid Geometry and Dynamics*, ed., Sarma, R.H., Pergamon Press, New York, Oxford, pp. 109-142 (1980).
135. Krugh, T.R., Hook III, J.W., Balakrishnan, M.S., and Chen, F-M., in *Nucleic Acid Geometry and Dynamics*, ed., Sarma, R.H., Pergamon Press, New York, Oxford pp. 351-366 (1980).

Structure, Fluctuations and Interactions of Base Pairs in Nucleic Acids Monitored by NMR Tritium Exchange and Stopped-Flow Hydrogen Deuterium Exchange Measurements

N.R. Kallenbach,* C. Mandal† and S.W. Englander†
Department of Biology,* Biochemistry and Biophysics†
University of Pennsylvania
Philadelphia, PA 19104

Introduction

The dynamics of nucleic acid helices in solution are of increasing interest in connection with attempts to define in molecular terms mechanisms of mutagenesis, regulation and formation of nucleoprotein complexes. Structural details of nucleic acids have been derived from X-ray diffraction analysis of fibers and crystals of base paired systems.[1/53] A corresponding body of information about the energetics of base-pairing has been developed from experiments on denaturation of nucleic acids by extremes of temperature, pH, solvent and so on.[2] However the nature, rates and energetics of fluctuational processes that mediate access of reagents or ligands to the bases within double helices are not well understood. Here we present a summary of the equilibrium and kinetic characteristics of ordered nucleic acids from the viewpoint of denaturation or helix-coil transition experiments. This furnishes a reference for discussing exchange reactions as probes of the base-pair opening reactions in oligonucleotides and tRNA molecules, in which the predominant pathway for opening appears to coincide with denaturation. Exchange reactions in a high molecular duplex system are then considered, and finally a section on dynamics of a ligand interaction in DNA investigated by exchange is presented.

Energetics of Base Pairing in Solution

The association of complementary base pairs, while demonstratable in mononucleosides,[3] requires several base-pairs in order for stable complexes to form in aqueous solution.[1] This is because much of the free energy of association between bases in an oligomeric duplex is derived from stacking interactions.[2/4] The extreme cooperativity characteristic of the denaturation of high molecular weight DNA for example suggests that configurations of the duplex in which adjacent pairs stack are more stable than ones in which solvent molecules intervene.[5] This tendency of helical segments to condense introduces a long-range order into nucleic acids systems.

A quantitative description of the cooperative transitions between native and denatured states of nucleic acids[6] requires a minimum of three elemental equilibrium constants corresponding to microscopic states of base pairs in a duplex.

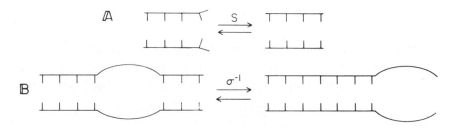

Figure 1. Equilibrium constants corresponding to fundamental processes in nucleic acids. A. Addition of a base pair at an end of a previously nucleated duplex. B. Condensation of two independently nucleated duplex regions to form one longer region of identical number of base-pairs.

1. Pairing of A•T(rA•U) at the ends of a nucleated helix is assigned the equilibrium constant S_A, corresponding to the reaction diagrammed in Figure 1a. This equilibrium constant depends on T, ionic strength, pH and other experimental conditions. Note that, as illustrated, S_A includes a contribution from stacking an A•U pair upon a previously formed one.

2. The greater stability of G•C pairs under most solvent conditions is accounted for by assigning a stability constant S_G ($>S_A$) for similarly adding an terminal G•C pair to a duplex.

3. Cooperativity between sequences of paired bases is introduced by assigning the equilibrium constant $\sigma < 1$ to each independent sequence of adjacent paired bases. As illustrated in Figure 1b, σ measures the difficulty of initiating a sequence of paired bases rather than adding them onto a preexisting duplex.

The free energy of base-pair formation (A•T or G•C) at the end of a double helix corresponds to

$$\Delta G^\circ_{A,G} = \Delta H^\circ_{A,G} - T\Delta S^\circ_{A,G} \tag{1}$$

Calorimetric data permits measurement[7][8] of ΔH°—for A•U pairing ΔH° is -7 kcal/mole, for G•C -10 kcal/mole. The entropy terms can be obtained from the relation that $\Delta G^\circ_{A,G} = 0$ at $T = T_m$ of an appropriate homopolymer:

$$\Delta S^\circ_{A,G} = \Delta H^\circ_{A,G}/T_m \tag{2}$$

The additional constant σ can be evaluated from the dependence of the slope of a transition profile of a homopolymer on temperature; the more cooperative the profile,[5] the smaller the σ. More recently, σ has been measured by fitting differential thermal transition data on DNA restriction fragments of known sequence.[9]

Values of σ of 10^{-4} to 10^{-5} imply a correlation length of helical regions of $\sigma^{-1/2} \geq 100$ base pairs, consistent with the long range of cooperative interactions observed. The fact that a DNA molecule undergoing denaturation contains alternating helices and "bubbles" means that the dependence of the free energy of a loop or bubble of unpaired bases on length has to be taken into account. Since we consider here only minimal opening processes, this factor is not discussed further. It should be noted that the scheme presented can be extended to oligonucleotides, in which case finer details of neighboring base sequence and single strand interaction effects come into play.[4/10]

In a series of homologous base-paired oligonucleotides, the nucleation constant conventionally[11] designated β proves to be about 10^{-3} M^{-1}. Thus in a minimal duplex of four or so base pairs in length, the dissociation of the two strands tends to approximate an all-or-none process. In longer oligonucleotides the dominant opening pathway will consist of progressive opening from the ends of the duplex, since the free energy for internal bubbling measured by the equilibrium constant σ is highly unfavorable. Sequence might be expected to modulate this effect to some extent, because internal blocks of A•T pairs are less stable than G•C. For oligonucleotides, the major effect of end melting prevails even in fairly long chains.[12/13] In high molecular weight molecules sequences of A•T pairs clearly form bubbles at lower temperatures than do G•C runs; this forms the basis of the mapping procedure due to Inman and Schnos,[14] in which A•T rich sequences react faster with formaldehyde than others.

Formaldehyde is an example of a reagent that requires opening of a duplex for complete reaction with imino and amino groups on the bases.[15] One estimate of the availability of bases for initial reaction with formaldehyde at internal sites, for DNA outside any detectable denaturing transition region, is derived by calculating

$$K_{op} = \sigma \cdot (S_A)^{-1}$$

at the appropriate temperature.[15] At room temperature in high salt, for example, this predicts a value of 10^{-5}, for an A•T site or region. For G•C, the value would be smaller, by about a factor of 10. On the other hand, the rate of reaction at an end would be a factor of σ faster for either species. It is shown below that this estimate is far too low to account for the open state responsible for hydrogen exchange.

H-exchange in Oligonucleotides and tRNA Molecules

The slow exchange of hydrogen bonded ring protons (N_1-H of G and N_3-H of T or U) is strikingly demonstrated by the presence of resonances (at extremely low field) in H_2O solutions of self-complementary oligonucleotides and tRNA. Figure 2 illustrates the behavior of these ring protons for an octanucleotide duplex.[16] Due to the symmetry of this duplex, there are four protons detected in the low field (i.e., below 10 ppm from standard DSS) region. No resonances appear in this part of the spectrum in the presence of D_2O or at higher temperature (Figure 2); hence these protons can exchange with solvent.

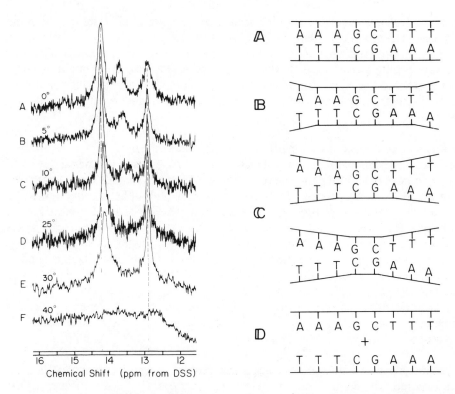

Figure 2. 220 MHz ^{1}H NMR spectrum of 9 mM d-A $_3$GCT$_3$ in H$_2$O solution with 1.5 M NaCl, .01M Tris HCl, 5 × 10^{-4}M EDTA pH 7 at several different temperatures. The protons represent the ring protons G N$_1$-H and T N$_3$-H in the hydrogen bonded self-complementary duplex, which has T$_m$ above 50°C under these conditions.

Figure 3. Scheme illustrating the major opening pathway in the self-complementary octanucleotide d-A $_3$GCT$_3$ duplex (A). Initially (B) the terminal base-pair undergoes rapid opening-closing, followed successively by the two remaining A•T base pairs, (C) leaving a stable G•C core which finally leads to separation of the strands.

The temperature dependence of the chemical shifts directly permits one to assign the protons in the spectrum. First, the three peaks at 0°C (initially 1:2:1 in intensity) are seen to reduce to two of roughly equal intensity by 25°C. The conditions of this experiment are 1.5 M NaCl, pH 7, favoring the duplex which exhibits a T$_m$ of 50°C under these conditions. According to the anticipated major opening pathway for this duplex, denaturation should follow the progressive sequence illustrated in Figure 3. Second, two of the resonances broaden out with minor change in shifts (Figure 4), whereas the position of the third resonance has already shifted considerably by 0°C. Similar behavior is clearly shown in a study by Patel[17] of the self-complementary hexanucleotide dA-dT-dC-dC-dA-dT, in which the resonance of the terminal T N$_3$-H is found to continue shifting to -30°C (with added methanol).

To understand these two types of line-broadening the behavior of an exchangeable proton in the duplex must be considered in more detail. It is supposed that such a proton cannot exchange at all in an intact base pair, but requires unpairing in order

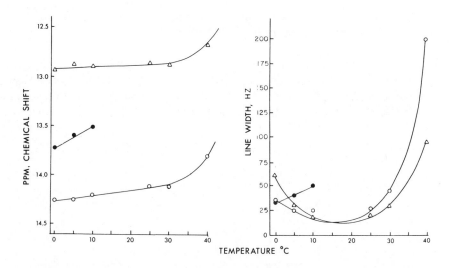

Figure 4. Graph of the chemical shifts and line widths of ^{1}H resonances of ring protons of base-paired G and T in the d-A$_3$G•CT$_3$ octanucleotide. This graph summarizes the behavior of the resonances shown in Figure 2; the symbols are: •, the T N$_3$-H resonances from the terminal base pairs, $\triangle$, the G N$_1$-H resonances from the G•C core; o, the T N$_3$-H of the remaining A•T base- pairs.

to permit solvent or base exchange (see Equation 5, below). Thus from the NMR viewpoint line broadening is determined by the rates of the opening-closing and the chemical pathways in the scheme. Some limiting cases for this general situation have been described by Crothers, *et al.*[18]

If the internal ring protons of the duplex exchange (in the presence of buffer, as is the case for the data of Figure 4) at rates limited by the dissociation rate of the duplex, the line width of the resonances measure the rate of dissociation itself - Equation 7a. The case of the fraying termini represents a situation in which k_{op} and k_{cl} both faster than the transfer rate, so that several rounds of opening-closing precede exchange. The chemical shift thus corresponds to an intermediate frequency

$$W = f_p \, Wp + f_u \, Wu$$

where Wp is the frequency corresponding to paired termini, Wu that for unpaired, and f_p, f_u the fractional quantities of paired or unpaired termini. That a terminal base pair can open with a rate above 10^5 sec^{-1} is expected from relaxation experiments.[19]

The structure of intermediates such as C in Figure 3 presumably also involves additional sets of protons that display rapid fluctuation between paired and unpaired states, by analogy with the terminal pair at lower temperature. However, perhaps because of residual base stacking the chemical shifts of the two forms are similar and only a small change in shift results. Similar interpretations of the opening of the hexanucleotide dA-dT-dG-dC-dA-dT from the ends have been presented from the point of view of non-exchanging protons in the hexamer.[20/21]

Exchange of the ring hydrogens in tRNA has been intensively investigated by proton NMR. These studies have focussed principally on defining the total number of slowly exchanging N-H in order to correlate the structure of tRNA in solution with that of yeast tRNA[phe] in crystals.[22] Depending on the sequence present in a particular

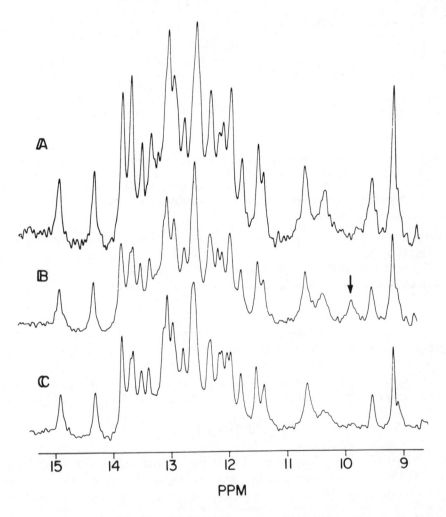

Figure 5. Low field ^{1}H NMR at 360 MHz of ring protons in *E. coli* tRNA$_1^{val}$, in H_2O, 0.1 M NaCl, .01 M MgCl$_2$, 0.1 M sodium cacodylate. Each spectrum represents 800 transients of 0.8 sec using correlation spectroscopy. *A.* Temperature 30°C, pH 7.2. *B.* Temperature 30°C, pH 6.1. *C.* Temperature 40°C, pH 6.1. The resonance at 9.8 ppm indicated by an arrow in B is clearly absent in A and C. As was described in Ref. 30, this is accompanied by an equilibrium structural change in the tRNA, and represents the slow exchange of some group (perhaps A$_9$) in the molecule which is in rapid exchange at neutral pH or higher temperature at pH 6.

molecule, the low field spectrum of a tRNA can consist of more than twenty distinct peaks (Figure 5). The presence of so many resonances makes assignment of each peak a formidable task, particularly because of resonances due to the teriary structure of tRNA. By systematic studies of fragments of different molecules, spin-labelling selective bases[23] and utilizing saturation transfer methods described below, fairly secure assignments have recently been obtained.[24-27] Once these are at hand, the course of thermal or solvent induced unfolding of a particular molecule can be followed as in simpler self-complementary duplexes. The sensitivity of the low field spectrum as a monitor of fine structural changes in tRNA is illustrated in Figure 5, which shows the appearance of a resonance at 9.8 ppm at pH 6, 30°C (panel B).[28] The resonance broadens at 40° or pH 7, 30° as shown in panels (A) and (C). Direct tritium exchange measurements on tRNA suggest that at 0°C in the presence of Mg^{2+} in excess of 120 slowly exchanging hydrogens per molecule can be detected.[29/30] (Figure 6). This implies i) that all the base paired imino and amino hydrogens in tRNA exchange slowly as they do in synthetic poly(rG•rC) or poly(dA-dT)•poly(dA-dT) (see next section) and ii) that the total of these hydrogens cannot explain the number of hydrogen bonding interactions in the clover leaf secondary structure. On this basis, it was argued that[29] a tertiary structure involving additional hydrogen bonding interactions exists in the presence of Mg^{2+}. It is now clear that the ring proton spectrum of tRNA includes resonances from tertiary base pair interactions as well as the secondary structure.[24-27] Moreover, by using Fourier

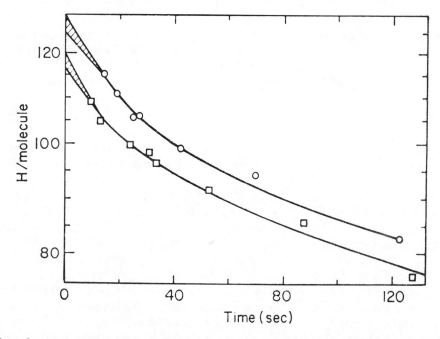

Figure 6. Demonstration by direct tritium exchange experiments using Sephadex gel filtration to resolve the macromolecular tRNA from ^3H_2O that two amino acid specific tRNA molecules contain base-paired structure in excess of that possible for clover leaf configurations.[31] □, *E. coli* tRNA$_F^{Met}$o, *E. coli* tRNA^{Tyr}.

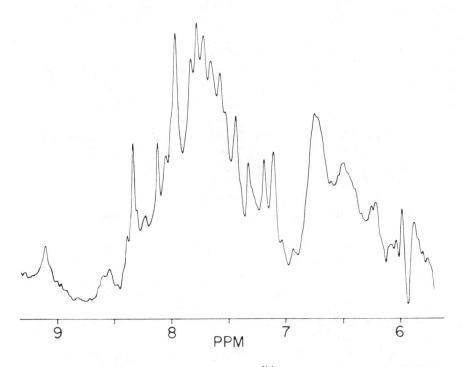

Figure 7. ¹H NMR spectrum at 360 MHz of *E. coli* tRNA$_1^{Val}$ at 30°C, pH 7 (H₂O), showing the region from 6 to 9 ppm downfield from DSS. The resonances in this region include a large number of amino protons as well as aromatic protons of the bases. A small temperature increase leads to diminution of the exchangeable protons, but not the aromatic protons.

ᵗransform NMR with a special pluse sequence that makes it possible to detect resonances of a solute in the presence of a vast excess of solvent, Johnston and Redfield[31/32] have been able to demonstrate that at higher temperature, saturation transfer from H₂O to low-field tRNA resonances is dominated by exchange. This is not the case for saturation transfer experiments at low temperature however. The rapidity of signal accumulation by Fourier transform permits direct H-D rates to be measured for individual resonances in the low-field region.[33] In yeast tRNA^{phe}, there is extreme heterogeneity in rates of exchange among different ring protons. One remarkably slow group with a half-life of 465 minutes at 15°C is found; retardation of this magnitude might well involve hydrogens of bases that are immobilized by Mg²⁺ binding.[33] The ability of ¹H NMR in aqueous (H₂O) solution to monitor the resonances of amino groups in tRNA is illustrated in Figure 7, for *E. coli* tRNA. The sharp peaks between 7 and 8.5 ppm represent aromatic protons of the bases; a substantial number of amino ¹H protons occur in this region as well as between 6 and 7 ppm.

These examples illustrate how exchange measurements can yield information both about the dynamics of structural changes in solution as well as the extent of structure present under specified conditions. We now turn to measurements of exchange from high molecular weight nucleic acids.

Exchange in Polynucleotides

We have seen that in oligonucleotide duplexes the rate of exchange of nucleic acid protons is slower than in free bases or denatured chains. In polynucleotides, slow exchange of both ring and amino protons can be monitored by ^{3}H-exchange using Sephadex chromatography to remove labelled solvent from the macromolecule.[29/34/35] The kinetics of ^{3}H exchange-out from two synthetic polynucleotides-poly(dA-dT)•poly(dA-dT) and poly(rG)•poly(rC), at 0°C are presented in Figures 8 and 9. In both cases, it is apparent that extrapolation of the

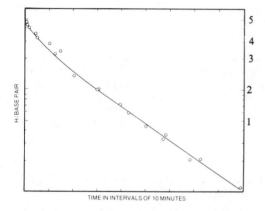

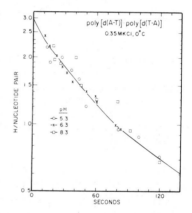

Figure 8. Tritium exchange experimental data for poly (rG)•poly (rC); pH 7, 0°C in 1 mM Mg²⁺, 0.15M KCl pH 7.4. The ordinate is H/base pair, each interval of the abscissa represents 10 minutes. This shows that all N-H protons of the bases exchange slowly.

Figure 9. Demonstration by direct ³H → ¹H exchange that all three N-H protons in poly(dA-dT)•poly(dA-dT) at 0°C, 0.35M KCl, exchange slowly with solvent. The fact that the exchange rate is pH independent over the range shown is discussed in detail in the text.

trace to zero time yields more than the number of hydrogens involved in Watson-Crick pairing.[29] This proves to be a consequence of the chemical events in the exchange reaction[34] as is discussed below. It has recently been reported that H-D exchange in nucleic acids can be monitored by a kinetic difference spectrum in the UV.[36-39] The method enormously enhances the time resolution of exchange experiments, and has the added advantage of introducing a selectivity by appropriate choice of chromophore. Figure 10 illustrates a stopped flow measurement of exchangeable base protons in a high molecular weight synthetic duplex polynucleotide, poly(rA)•poly(rU).[40] This measurement provides an approach to the dynamics of an important structural opening process in this helix, some characteristics of which can be inferred from analysis of the salt, pH and temperature dependence of the H- exchange rates. In addition, measurement of H- exchange rate by absorbance can yield information about the extent of base pairing in larger partially ordered molecules such as rRNA, and potentially can even furnish details about nucleic acid structure

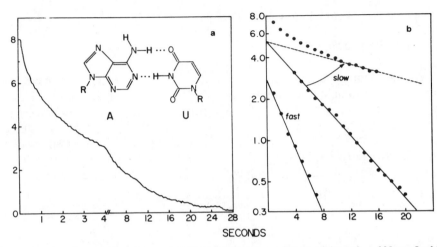

Figure 10. HD exchange of poly(rA)•poly(rU) detected by stopped-flow mixing at λ = 285 nm. In this experiment, a sample of poly•(rA)•poly(rU) in 0.1M phosphate buffer, D₂O at pD7, 20°C has been mixed with a ten-fold excess of H₂O containing salt and buffer at the same concentration. Final concentration of polymer was 0.22 mM in phosphorus. The fast and slow exchange processes are assigned to U-N₃H and A-NH₂ respectively as described in the text. (a) Kinetic trace of transmittance recorded on two time scales. (b) Semi-logarithmic plots of the trace: curve (1) shows the slow reaction, and curve (2) the early time data on a two-fold expanded scale.

in the presence of large amounts of protein as in ribosomal particles, chromatin, and viruses. The experiment in Figure 10 was performed by monitoring the long wavelength UV transmittance change upon rapidly diluting into H₂O a solution of poly(rA)•poly(rU) in D₂O using a stopped-flow spectrophotometer. Exchange of deuterons for protons is accompanied by a detectable difference spectrum near 290 nm.

Two kinetic processes are revealed in poly (rA)•poly(rU). The slower proves to represent the H-D exchange of the two exocyclic amino protons of A in the duplex. These two hydrogens exchange at identical rates despite the obvious structural difference between them: one is a member of an A-NH to U-C=O hydrogen bond while the second is freely exposed to solvent. The identity in rates results from the exchange of both from a conformational fluctuation of the double helix which severs the Watson-Crick H-bonds and makes both A-NH₂ protons equally vulnerable to exchange. This fluctuation is referred to as the H- exchange open state. The faster process in Figure 10 registers the exchange of the N₃-H ring proton of U. Despite its apparent inaccessibility due to the intramolecular U-N₃-H to A-N₁ hydrogen bond and the bases stacked above and below it, this hydrogen exchanges at a rate equal to the rate of the base pair separation reaction just described.

The basis for these assertions is presented below, with a discussion of the possible nature of the H- exchange open state in poly(rA)•poly(rU) and in other nucleic acid duplexes. Finally, a new application of H- exchange is presented in which details of the intercalation of the phenanthridine dye ethidium bromide (EB) into DNA can be studied owing to the slowing of the free exchange rate of at least one of the amino protons of the dye molecule. The H- exchange slowing seems to indicate an intermoleculr EB-DNA hydrogen bond in the complex.

Exchange Chemistry

Any interpretation of the exchange data must rest on an understanding of the exchange behavior of the bases in the absence of duplex structure. Figure 11 summarizes the exchange of the U N_3-H proton as a function of pH.[40] The two branches of the profile correspond to an H⁺ catalyzed process below pH 4 and an OH⁻-catalyzed domain above pH 4. In the alkaline region the U-N_3-H proton is catalyzed by OH⁻ with a rate constant near the diffusion limited value, 1.5×10^{10} M⁻¹ sec⁻¹. The chemical step in this case appears straightforward near neutral pH:

$$\text{HO}^{\ominus} + \text{D-N} \longrightarrow \text{HOD} + {}^{\ominus}\text{N} \tag{3}$$

Catalysis of the exchange of the U-N_3-H proton by buffers such as imidazole confirms this, since the measured rate in the presence of imidazole base as acceptor is closely predicted by the difference in pK between imidazole (7.1) and uracil (9.5), i.e., $10^{10} \times 10^{-2.4}$ or near 10^7 M⁻¹ sec⁻¹.

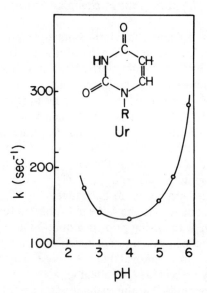

Figure 11. Exchange of the ring proton N_3-H in uridine at different pH determined by proton NMR in H_2O 0.1M solution in the absence of buffer. The rate is determined by the excess line width of the ring proton (near 11 ppm downfield from DSS), at 20°C.

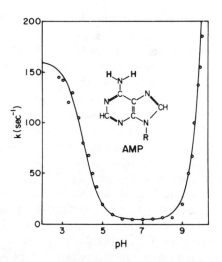

Figure 12. Exchange of the amino protons of AMP measured by stopped-flow H-D exchange in the ultraviolet.[36] Near neutral pH, hydroxide is less effective than the pathway summarized in reaction (2) in which the ring N_1 position first protonates followed by removal of the now favorable amino protons by OH⁻ (or any buffer base).

By contrast, the exchange behavior of adenosine is more complex, as seen in Figure 12. Near neutral pH, the amino proton of adenine exchanges by a pH independent mechanism involving two proton transfer steps-protonation of N_1 on the ring which reduces a highly unfavorable pK for deprotonating the amino group, facilitating removal of H from this group by OH⁻ or other bases;[34/36/41]

$$(4)$$

The pK for the first step is 4; the reduction in pK of the amino group following protonation is seen in the analogous case of N_1-methyl adenosine, for which the effective pK is near 8, rather than an estimated value near 20 in the case of adenosine itself.[34] The pathway in Equation 4 involves the product of (H⁺) and (OH⁻), which is just the constant Kw. Hence the independence of the rate with pH. Added general base catalysts can greatly increase the A-NH₂ exchange rate.[36] The pH independent exchange rate in neutral poly(rA) is found to be slower by a factor of three than in AMP.[40]

Evidence for An Open State in Poly (A•U) And Identity of the Exchanging Protons

From previous tritium exchange experiments, it is found that all three nitrogen protons of the bases exchange in A•T and A•U duplexes (and all five in the case of G•C duplexes, Figures 8 and 9). A number of observations demonstrate that a transient breaking of base pairs in the duplex A•U structure acts as an intermediate state in the exchange reactions. i) Despite the relatively greater protection of ring protons from solvent, especially in the case of the central G-N_1-H to C-N_3 hydrogen bond of G•C pairs, these exchange readily under conditions in which duplex structure is stable and unperturbed by denaturaton. In fact, they exchange faster than the more exposed amino protons. ii) Both A-NH₂ protons exchange at the same rate in poly(rA•rU), poly(rA-rU)•poly(rA-rU) and in poly(dA-dT)•poly(dA-dT),[34] despite the involvement of only one in hydrogen bonded structure. iii) Exchange of A-NH₂ protons in A•U or A•T duplexes is pH-independent over the same pH range as for free A, implying that the pathway in Equation 4, requiring prior protonation at N_1 of the adenine ring, is operative. This site is blocked by an internal hydrogen bond in the duplex. iv) The similar pathway involving proton removal by buffer base, which also requires a pre-protonation at the normally protected N_1 of adenine, goes just as well in the double helix as in free AMP. We conclude that the exchange from stable hydrogen bonded duplexes takes place via a transient fluctuational base unpairing within the double helix.

A minimal description of an opening-dependent pathway, involving three states of an exchangeable proton, can be written as follows (see, for example, 34):

$$\text{closed} \underset{k_{cl}}{\overset{k_{op}}{\rightleftharpoons}} \text{open} \overset{k_{tr}}{\rightarrow} \text{exchanged} \qquad (5)$$

The rate constants k_{op}, k_{cl} pertain to the conformational process, for which the equilibrium constant $K_{eq} = k_{op}/k_{cl}$, and k_{tr} represents the chemical transfer rates discussed in the previous section. The overall exchange rate corresponding to scheme (5) is given by:

$$k_{ex} = \frac{k_{op}k_{tr}}{k_{op} + k_{cl} + k_{tr}} \qquad (6)$$

As has often been noted before, Equation 6 admits of two simple limiting situations: (1) if $k_{tr} \geq\geq k_{cl}$, so that the transfer is rapid, exchange becomes opening-limited and:

$$k_{ex} = k_{op} \qquad (7a)$$

(2) conversely if $k_{tr} \leq\leq k_{cl}$,

$$k_{ex} = K_{eq}/(1 + K_{eq}) k_{tr} \sim K_{eq}k_{tr} \qquad (7b)$$

the preequilibrium limit. Equation 6 also has a simple Lineweaver-Burke type of reciprocal form:

$$1/k_{ex} = 1/k_{op} + [(1 + K_{eq})/K_{eq}] 1/k_{tr} \qquad (8)$$

Assignment of the slower rate class in poly $(rA \bullet rU)$ to the two A-NH_2 protons is on the basis of their pH independence over the same range as in AMP, and on their identical response to buffer catalysts as the NH_2 protons in AMP.[40] Most critically for the mechanism we have postulated as well as for the identity of the faster rate class, the experiment summarized in Figure 13 makes it clear that the faster process is in fact insensitive to general base catalysis as predicted by Equation 7a. The base used in this experiment, trifluoroethylamine (TFEA) would accelerate the free U-N_3-H exchange rate by orders of magnitude over the concentration range shown. But U-N_3-H, already exchanging at the opening-limited rate (Equation 7a), is unaffected. Only the slower process, representing the A-NH_2 protons, exchanging via a pre-equilibrium opening pathway (Equation 7b), is accelerated by TFEA. The slower rate class asymptotically approaches the same value of k_{op} inferred from Equation 7a, just as required by Equation 8.

The importance of this experiment is hard to exaggerate. Not only does it furnish a quantitative test of the concepts which lead to Equation 8, but it also establishes for the first time that both classes of hydrogens in poly$(rA \bullet rU)$ exchange from the same

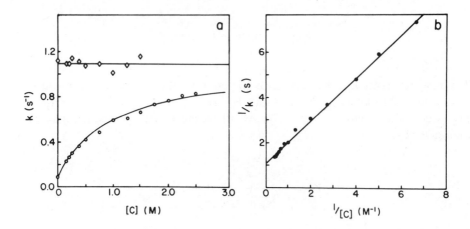

Figure 13. Test of open limiting exchange in poly(rA)•poly(rU) by response of exchange rates for the fast and slow processes in the polymer to a general base catalyst. The base used in this case was trifluoroethylamine (TFEA) (a). It is seen that the faster rate does not respond to buffer base over the range indicated, under which conditions free U-N₃-H would experience an approximate 10^5 increase in rate. On the other hand, the slower process asymptotically approaches the faster rate if catalyst is added. In (b) the reciprocal-plot of Equation 8 is illustrated, from which both k_{ex} and k_{op} can be evaluated. The data are those for the slower process in panel (a).

open state. If we accept the notion that the unperturbed closed form of the double helix is in equilibrium with a variety of fluctuational open states varying in energy from minimal openings through gross denaturation, this result implies that the H-exchange open state is the lowest in energy of these states and also displays the fastest rate at which the duplex can open. As such, the structure of this opening is of some interest.

The facts which any model for the H- exchange open state must reconcile are these: i) $\Delta G°$ for opening is about $+2$ kcal/mole at 25°C, with $\Delta H°$ near $+5$ kcal/mole and $\Delta S°$ 10 e.u. ii) The opening rate at 20°C is 1 sec⁻¹, the closing rate 20 sec⁻¹, with an energy of activation of $+15$ kcal/mole for opening. iii) The open state does not exhibit significant features of a bulk denaturation process of the helix: in particular, neither salt nor divalent ion concentrations, which affect Tm appreciably, have much effect on exchange rates.

For these reasons, the opening discussed above that entails denaturation of a base pair within a duplex measured by $(\sigma S_A)^{-1}$, does not seem tenable. The structure we presently lean toward represents a minimal opening in which an internal A•U pair ruptures and the U swings out, leaving the A more or less stacked. The extreme slowness of this process may be due to two important factors: i) The high activation barrier required to rupture the internal hydrogen bonds without simultaneously transferring the U-N₃-H and A-N₁ groups to water, as occurs in breaking base pairs at the ends of a duplex. ii) The difficulty of swinging out one member of a base pair from an internal bonded position because of steric contacts with adjacent base pairs.

Most flexibility in nucleotides is associated with the O-P torsional angles, rotation about which is highly restricted in this situation. The $\Delta G°$ for unpairing a U base and moving it out of the helix seems within range of $+2$ kcal/mole, based on model A•U duplexes incorporating non-complementing U residues.[42] It is more difficult to rationalize the large activation energy corresponding to the open state.

In this connection, it should be noted that the enthalpy corresponding to A•U hydrogen bonding in anhydrous chloroform is about -6 kcal/mole,[43] while that for transfer of U from its crystal to water is about as large.[44] The observed activation energy for opening might include both kinds of contribution. Finally we should indicate that the possibility of coupling the opening to internal fluctuations of the double helix cannot be eliminated, and experiments with short oligonucleotide duplexes as well as solvents of much higher viscosity are planned to investigate this.

The existence of this open state raises a number of questions concerning its relevance to protein-nucleic acid recognition and the dynamics of the DNA duplex in chromatin. If G•C base pairs are refractory both to melting and opening in the sense we have described,[35] one can imagine that A•T sequences provide a more rapid path to melting of DNA by proteins such as the RNA polymerase holoenzyme for example. At sites defined by a sequence including the Pribnow box[45/46] RNA polymerase forms an open complex with DNA in which a short region of DNA melts.[47/48] The highest efficiency promoters in *E. coli* and a number of phages correspond to a version of this sequence with six adjacent A•T pairs, while lower efficiency promoters include one or more G•C pairs at this site. Is this due to more rapid access of enzyme

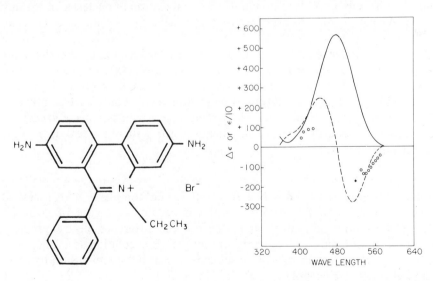

Figure 14. *Left* Structure of the phenanthridim dye ethidium bromide. **Figure 15.** *Right* Absorbance and differential absorbance of ethidium bromide in H_2O and D_2O. The figure shows the visible absorbance spectrum in H_2O reduced twenty-fold as a solid line, the equilibrium difference spectrum (H_2O-D_2O) as a dashed line, and the exchange derived (kinetic) difference spectrum as individual points.

to the open state(s) of the A•T pairs? To answer this we need to define the open state in DNA carefully, in order to determine whether there are neighbor base pair effects for instance that influence opening of individual pairs in a heterogeneous environment.

A second line of enquiry concerns the dynamics of the DNA molecule within the nucleosome core structure in chromatin.[49] In this case, we can measure the open state and its kinetic properties for isolated core-particles released from chromatin by nuclease digestion. The particle from calf thymus consists of about 145 b.p. of DNA with a protein complement involving 26 tyrosines but no tryptophanes.[50] Hence the signal near 290 nm should register the nucleic acid exchange as in free DNA. In this case, it is conceivable that changes in the dynamics might prove interesting, regardless of the magnitude of difference in equilibrium opening present. The point is that the open state defines both an equilibrium and dynamic aspect of nucleic acid behavior, and should thus be significant.

New Applications of Exchange Methods

The advantage of H-D exchange measurements using stopped-flow spectrophotometry over tritium exchange methods lies both in the rapidity and selectivity that can be achieved. While tritium-exchange experiments cannot directly resolve protein from nucleic acid exchange in protein-nucleic acid complexes, the isotope effect on absorption spectra of the base chromophores may well make this possible. As pointed out above, the requirements are simply that the protein not possess sufficient aromatic amino acids to dominate the contribution of nucleic acid around 280-290 nm. We believe that the exchange of nucleic acids in the presence of considerable amounts of protein can be followed by this technique. Englander *et al.*[51] have recently shown that exchange of the peptide group itself can be monitored in the vicinity of 230 nm, and it may therefore be possible to detect exchange of proteins separately from nucleic acid in this region. Obviously, individual protons of protein and nucleic acid can be readily resolved using proton NMR, but it is frequently difficult to obtain high resolution spectra on high molecular weight systems, and chemical exchange is not necessarily the most significant factor in line-broadening or saturation transfer experiments.[52]

We have recently carried out an exchange study of protons in the intercalating dye ethidium bromide (EB), both free in solution and complexes with excess DNA. There are two sets of -NH$_2$ protons in this molecule (see Figure 14), which has a strong absorption band in the visible. Spectral and difference spectral properties of EB are shown in Figure 15. Exchange of the free EB in solution as a function of pH (Figure 16) shows that in this case exchange proceeds by a diffusion-limited acid-catalyzed pathway below pH 9, according to the scheme:

$$\text{>}-ND_2 + H_3O^{\oplus} \longrightarrow \text{>}-\overset{\oplus}{N}HD_2 + H_2O \qquad (9)$$

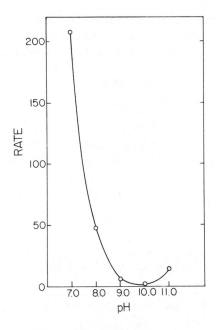

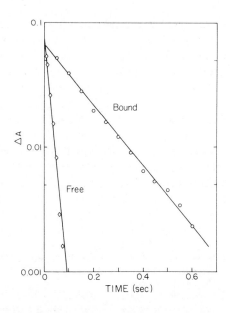

Figure 16. Dependence of exchange rate of amino protons of ethidium bromide on pH. The rates were determined using the difference spectrum of λ = 540 nm, and a stopped-flow spectrophotometer, diluting dye in D_2O into H_2O. Each solution contained 1mM phosphate to stabilize pH.

Figure 17. Comparison of the HD exchange of free ethidium bromide (0.5mM) with that of the dye complexed in the presence of 3mM calf thymus DNA. The retardation is a factor of seven, leading to an estimated dissociation constant for dye from complex of 5×10^{-4} M. So far we cannot discriminate which bound state the immobilized amino protons are retarded by; it seems likely to be the fully intercalated state.

Exchange of EB from the intercalated state proves to be slowed down significantly with respect to the free state at the same pH (Figure 17). This behavior can be understood in terms of participation of the $-NH_2$ drug protons in the complex:

$$EB(ND_2) + DNA \underset{k_{off}}{\overset{k_{on}}{\rightleftharpoons}} DNA\text{-}EB(\text{-}ND_2)$$

$$\downarrow k_{ex}^{free} \qquad\qquad\qquad\qquad \downarrow k_{ex}^{bound}$$

$$EB(NH_2) \qquad\qquad\qquad DNA\text{-}EB(\text{-}NH_2) \qquad (10)$$

The complexing of dye to DNA proceeds by a minimal two stage kinetic process;[19] the rate constants k_{on} and k_{off} apply to that step involved in restricting exchange of dye protons. If $k_{off} \geq\geq k_{ex}^{free}$ exchange of dye in the complex will approximate a preequilibrium pathway as in Equation 4b, with the equilibrium constant now corresponding to dissociation of EB from DNA. The calculated value of the dissociation constant for drug at pH 9 is 5×10^{-4} M, in fair agreement with the value determined from binding isotherms measured using fluorescence. Since the rate k_{ex}^{free} can be accelerated by acid catalysis at fixed pH or by shifting pH below 7, a situation in which $k_{ex}^{free} \geq k_{off}$ can also be achieved permitting direct determination of the off rate for EB from the complex. Whether or not the slowing in exchange of dye amino protons is due specifically to their participation in hydrogen bonds to DNA groups or to some indirect effect of intercalation remains to be established.

Acknowledgements

This research was supported by grants HL21757, AM11295 and 1RO1-CA24101-01 from the National Institutes of Health and PCM 77-26740 from the National Science Foundation.

References and Footnotes

1. Kallenbach, N.R. and Berman, H.M. *Quart. Revs. Biophysics 10*, 138-236 (1977).
2. Bloomfield, V.A., Crothers, D.M., and Tinoco, I., Jr. *Physical Chemistry of Nucleic Acids*, Harper and Row (1974).
3. Raszka, M. and Kaplan, N.O. *Proc. Natl. Acad. Sci. U.S.A. 69*, 2025-2029 (1972).
4. Appleby, D.W. and Kallenbach, N.R. *Biopolymers 12*, 2093-2120 (1973).
5. Crothers, D.M. and Zimm, B.H. *J. Mol. Biol. 9*, 1-14 (1964).
6. Zimm, B.H. *J. Chem. Phys. 33*, 1349-1356 (1960).
7. Filimonov, V.V. and Privalov, P.L. *J. Mol. Biol. 122*, 465-470 (1978).
8. Breslauer, K.J. and Sturtevant, J.M. *Biophys. Chem. 7*, 205-209 (1977).
9. Lyubchenko, Y.L., Vologodskii, A.V., Frank-Kamenetskii, M.D. *Nature 271*, 28-31 (1978).
10. Tinoco, I. Jr., Uhlenbeck, O.C., and Levine, M.D. *Nature 230*, 362-367 (1971).
11. Applequist, J. and Damle, V. *J. Am. Chem. Soc. 88*, 3895-3900 (1966).
12. Crothers, D.M., Kallenbach, N.R., and Zimm, B.H. *J. Mol. Biol. 11*, 802-820 (1965).
13. Kallenbach, N.R. *J. Mol. Biol. 37*, 445-466 (1968).
14. Inman, R.B. and Schnos, M. *J. Mol. Biol. 49*, 93-98 (1970).
15. McGhee, J.D. and von Hippel, P.H. *Biochemistry 16*, 3276-3293 (1977).
16. Kallenbach, N.R., Daniel, W.E., Jr., and Kaminker, M.A. *Biochemistry 15*, 1218-1224 (1976).
17. Patel, D.J. in *Nucleic Acid Geometry and Dynamics*, Ed., Sarma, R.H., Pergamon Press, New York, Oxford (1980).
18. Crothers, D.M., Cole, P.E., Hilbers, C.W., and Shulman, R.G. *J. Mol. Biol. 87*, 63-88 (1974).
19. Porschke, D. *Biophys. Chem. 2*, 97-101 (1974).
20. Patel, D.J. *Biochemistry 14*, 3984-3989 (1975).
21. Patel, D.J. and Hilbers, C.W. *Biochemistry 14*, 2651-2656 (1975).
22. Holbrook, S.R., Sussman, J.L., Warrant, R.W., and Kim, S.H. *J. Mol. Biol. 123*, 631-660 (1978).
23. Daniel, W.E., Jr., and Cohn, M. *Proc. Natl. Acad. Sci. U.S.A. 72*, 2582-2586 (1975).
24. Reid, B.R., McCollum, L., Ribeiro, N.S., Abbate, J., and Hurd, R.E. *Biochemistry 18*, 3996-4005 (1979).
25. Hurd, R.E. and Reid, B.R. *Biochemistry 18*, 4005-4011 (1979).
26. Hurd, R.E., Azhderian, E., and Reid, B.R. *Biochemistry 18*, 4012-4017 (1979).
27. Hurd, R.E. and Reid, B.R. *Biochemistry 18*, 4017-4024 (1979).
28. Steinmetz-Kayne, M., Benigno, R., and Kallenbach, N.R. *Biochemistry 16*, 2064-2073 (1977).
29. Englander, J.J., Kallenbach, N.R. and Englander, S.W. *J. Mol. Biol. 63*, 153-169 (1972).
30. Goldstein, R.N., Stefanovic, S., and Kallenbach, N.R. (1972). *J. Mol. Biol. 69*, 217-236 (1972).
31. Johnston, P.D. and Redfield, A.G. *Nuc. Acids Res. 4*, 3599-3615 (1972).
32. Johnston, P.D. and Redfield, A.G. *Nuc. Acids. Res. 5*, 3913-3927 (1978).
33. Johnston, P.D., Figueroa, N., and Redfield, A.G. *Proc. Natl. Acad. Sci U.S.A. 76*, 3130-3134 (1979).

34. Teitelbaum, H. and Englander, S.W. *J. Mol. Biol. 92*, 55-78 (1975).
35. Teitelbaum, H. and Englander, S.W. *J. Mol. Biol. 92*, 79-92 (1975).
36. Cross, D.G. (1975). *Biochemistry 14*, 357-362 (1975).
37. Cross, D.G., Brown, A., and Fisher, H.F. *Biochemistry 14*, 2754-2749 (1975).
38. Nakanishi, M. and Tsuboi, M. *J. Mol. Biol. 124*, 61-71 (1978).
39. Nakanishi, M., Nakamura, H., Hirakawa, A.Y., Tsuboi, A., Nagamura, T., and Saijo, Y. *J. Am. Chem. Soc. 100*, 272-276 (1978).
40. Mandal, C., Kallenbach, N.R. and Englander, S.W. *J. Mol. Biol. 133* (in press).
41. McConnell, B. *Biochemistry 13*, 4516-4523 (1974).
42. Lomant, A.J. and Fresco, J.R. *Prog. Nuc. Acid Research and Mol. Biol. 15*, 185-216 (1975).
43. Binford, J.S. and Holloway, D.M. *J. Mol. Biol. 31*, 91-99 (1963).
44. Scruggs, R.L., Achter, E.K. and Ross, P.D. (1972). *Biopolymers, 11*, 1961-1972 (1972).
45. Pribnow, D. *Proc. Natl. Acad. Sci. U.S.A. 72*, 784 (1975).
46. Pribnow, D. *J. Mol. Biol. 99*, 419 (1975).
47. Hinkle, D.C. and Chamberlin, M.J. *J. Mol. Biol. 70*, 187-195 (1972).
48. Chamberlin, M., in *DNA Polymerase*, Eds. Losick, R. and Chamberlin, M., Cold Spring Harbor, 159-192 (1976).
49. Felsenfeld, G. *Nature 271*, 115-121 (1978).
50. Bostock, C.J. and Summer, A.T. *The Eukaryotic Chromosome*, North-Holland Publishing Co., Amsterdam, New York, 1978. Ch. 4.
51. Englander, J.J., Calhoun, D.B. and Englander, S.W. *Anal. Biochem. 92*, 517-524 (1979).
52. Bresloff, J. and Crothers, D.M. *J. Mol. Biol. 95*, 103-123 (1975).
53. Seeman, N.C. in *Nucleic Acid Geometry and Dynamics*, Ed., Sarma, R.H., Pergamon Press, New York, Oxford (1980).

Torsional Flexibility of DNA As Determined By Electron Paramagnetic Resonance

Ian Hurley, B.H. Robinson, C.P. Scholes and L.S. Lerman

Center for Biological Macromolecules
State University of New York at Albany
Albany, NY 12222

Introduction

In this chapter, we will discuss some experiments in which electron paramagnetic resonance (EPR) has been used to study the rotational dynamics of the DNA double helix. We will assume that the reader is familiar with the use of EPR as a tool for studying molecular structure and motion. Those lacking such a background might find a text on magnetic resonance[1,2] helpful in understanding what is to follow.

First, let us explain why we think a study of the rotational dynamics of DNA is relevant to biology. There are a number of interesting events or processes in which local untwisting of the double helix takes place. Some examples are: (1) nucleotide reactions involving an out-of-plane attack, such as covalent alteration by mutagens; (2) breathing modes; (3) premelting (or melting zones); (4) strand interchange as seen in branch point migration; and (5) the mechanics of recombination, transcription and replacement. At the temperatures at which these processes occur, some of the internal energy of motion of the DNA double helix must be in the form of torsional vibrations, that is, oscillation of base pairs in a plane perpendicular to the local helix axis. These torsional vibrations can contribute to the unwinding process. We believe that we can use a knowledge of the rotational dynamics of DNA in devising and evaluating physical models of processes and events which involve unwinding.

Next we describe our experiments in rough outline, so that the reader will have some idea of what is to come. Since we cannot follow the motion of the DNA directly as it is not paramagnetic and therefore cannot give rise to an EPR signal, we attached small paramagnetic probes to the DNA. We recorded the EPR signals from these probes under conditions which we will discuss later. From these spectra we obtained information about the geometry of attachment of our probes to the DNA and the role of rotational motion of the probes. We devised four physical models of DNA which might account for rotational motion and inferred a rate of rotational motion for each model. We then compared the predictions of our models with our observations.

There is an implicit assumption in the analysis of our experiments: that the motion of the probe mirrors the motion of interest of the DNA. This is such an important point that we will discuss it in some detail before going any further. There are at least three conditions which a good probe should fulfill.

I. A probe should not be capable of any movement relative to the local DNA helix.

II. Attachment of the probe should not significantly alter the elastic properties of the helix.

III. The geometry of the attachment of the probe should have at least helical symmetry.

The reason for the third condition may not be quite so obvious as for the first two. If the probes were attached in some random manner, different probe molecules or ions on the same piece of DNA would be in different electrical and motional environments. The overall EPR spectrum we would observe would be some combination of EPR signals characteristic of these. This composite signal would probably be too complex to sort out. With helical symmetry, the geometry of the probe attachment would be the same as the fundamental geometry of the double helix we wanted to study.

We decided to use intercalating agents modified by attaching a paramagnetic substituent as our probes. We reasoned that intercalators ought to fulfill the third condition at least approximately. How well they fulfilled it would depend upon how tightly it would be possible to constrain the position of the unpaired spin in a fixed orientation to the part of the probe which intercalated. In addition we knew that intercalation was the only type of bonding for which a structural relation has been well established. (However, see Crothers et al[35] in this volume). Unfortunately, intercalation must alter the elasticity of the double helix to which it is attached in the vicinity of its attachment, so condition (II) cannot be exactly fulfilled. This was clear in the earliest configurational study.[3] However, we expected that the motion of the DNA double helix as a whole would be determined by lengths of DNA much longer than the intercalation site, so the stiffening of two or four base pairs might not invalidate our results. As to condition (I), we had no idea when we began this project whether or not it could be met by our intercalators. Most serious from our point of view was oscillation of the probe relative to DNA in the plane of the intercalation site. Certainly, we could not see anything obvious to prevent such oscillation when we examined the fit between space-filling models of our intercalators and of short lengths of double helical DNA.

We have set out the structures of the probes we have used in Figure 1 and Table I. In each case an intercalator, either an acridine derivative or a phenanthridinium derivative, is attached by an amide linkage to pyrroline ring containing a nitroxyl radical. The pyrroline nitroxyl radical belongs to a class of substituents called "spin labels" which have been widely used in recent years. Readers who are interested in spin labels and spin labeling should refer to the books by Rozantzev[4] and Berliner[5/6]

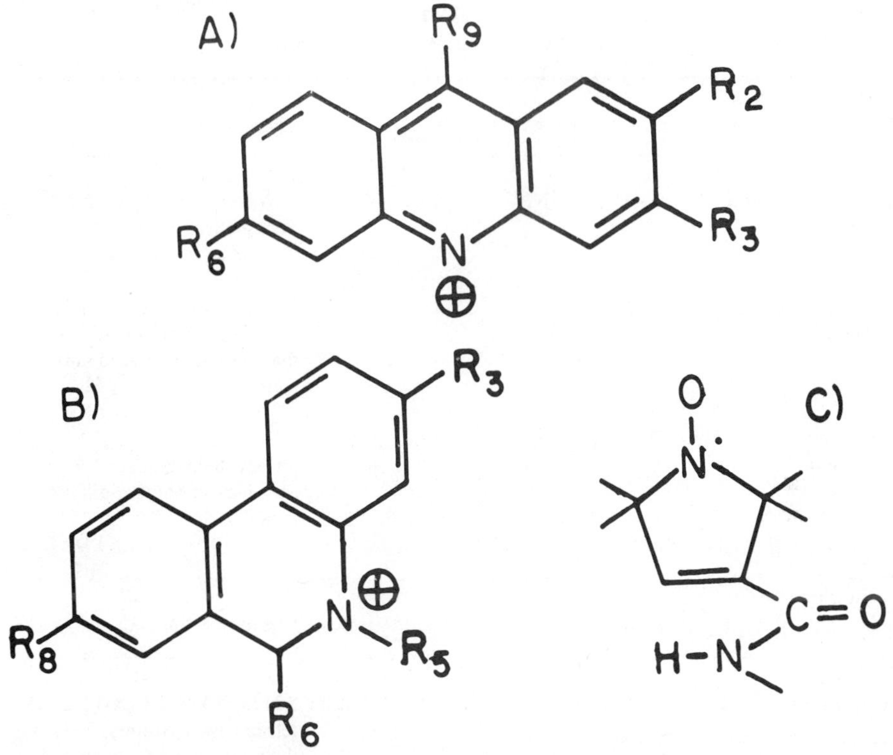

Figure 1. Structural formulae of a) the acridinium and b) the phenanthridinium systems, showing the numbering of substituents R_i (see Table I), and c) the moiety containing the nitroxyl radical which is attached to the heterocyclic ring system.

which are listed in the bibliography. Spin labels have two advantages over other paramagnetic substituents which we might have used instead. First, they are very stable. We have prepared dilute aqueous solutions containing spin labels which lost less than 2 percent of their unpaired spins in four weeks at 4°C. Second, their spectra are very sensitive to motional processes in the 10^{-11} to 10^{-7} second ranges. Motion on this time scale corresponds to high frequency motion for a stiff worm-like chain such as double-stranded DNA. Amide linkages were used between spin label and intercalator because these are relatively resistant to hydrolysis in solutions of near neutral pH and because we expected that the combination of the conjugated electron distribution between both ring systems and steric repulsion between the amide oxygen and the aromatic ring would prevent rotation of the spin label relative to the intercalating ring, helping to fulfill condition (III).

Fiber Studies

In one set of experiments[21] we took EPR spectra of oriented fibers of chicken erythrocyte DNA which had been allowed to equilibrate with acridine probes. These experiments had two purposes: (1) to demonstrate that the condition (III) discussed

Table I
List of Substituents on Various Probes

Acridinium Probes Simple Name	R_2	R_3	R_6	R_9	Chemical Name
6-substituted-3-amino acridinium probe	-H	-NH₂	SpLb	-H	6-N-(3-carbonyl-l-oxyl-2,2,5,5-tetramethylpyrroline)-3,6-diaminoacridinium cation
6-substituted-9-amino acridinium probe	-OC₂H₅	-H	SpLb	-NH₂	6-N-(3-carbonyl-1-oxyl-2,2,5,-tetramethylpyrroline)-6,9-diamino-2-ethoxyacridinium cation
9-substituted acridinium probe	-OC₂H₅	-H	-NH₂	SpLb	9-N-(3-carbonyl-1-oxyl-2,2,5,5-tetramethylpyrroline)-6,9-diamino-2-ethoxyacridinium cation

Phenanthridinium Probes Simple Name	R_3	R_5	R_6	R_8	Chemical Name
ethidium probe	-NH₂	-C₂H₅	-C₆H₅	SpLb	8-N-(3-carbonyl-1-oxyl-2,2,5,5-tetramethylpyrroline)3,8-diamino-5-ethyl-6-phenyl-phenanthridinium cation
propidium probe	-NH₂	R*	-C₆H₅	SpLb	8-N-(3-carbonyl-1-oxyl-2,2,5,5-tetramethylpyrroline)-3,8,-diamino-5(3-propyldiethyl-methylammonio)-6-phenyl-phenanthridinium cation

R* = -C₃H₆N⁺ Et₂Me

in the introduction was met; and (2) to provide spectra which we could match to computer simulations. Representative EPR spectra of these systems appear in Figure 2. These show a marked orientation dependence. We found the probes to be ordered with respect to the helix in a fashion characteristic of the probe's structure. The spectrum of the 6-spin labeled acridine bound to a fiber oriented perpendicular to the external magnetic field corresponded closely to the spectrum of the 9-substituted probe bound to a fiber parallel to the field, and vice versa. Our simulations of these spectra, which we will discuss below, suggest that the spin orbital of the 6-substituted probes lies at about 40°. The basis of this strong tilt away from the plane of intercalation is shown in the photographs of a space-filling model of a 6-substituted probe in Figure 3. These reveal that van der Waals contact between the oxygen of amide and the hydrogens at the 5- and 7- positions of acridine prevents coplanarity. Further, the crystal structure of a simple analogous compound,

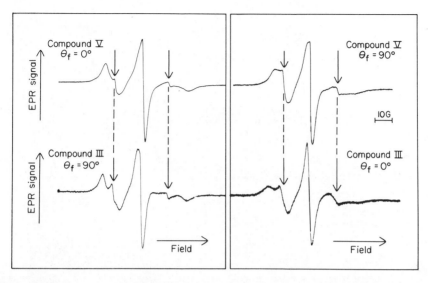

Figure 2. EPR spectra of acridinium probes bound to DNA fibers as a function of fiber orientation and the position of the nitroxyl- containing substituent on the acridinium ring. Compound V is 6-substituted-3-amino acridinium and compound III is 9-substituted-6-amino acridinium. θ_f is the angle between the fiber axis and the magnetic field. Note that similar orientations are diagonally arranged in the figure. Arrows mark the cross-over points of the freely tumbling (unbound) intercalator.

acetanilide, shows a similar degree of tilt.[7] These facts suggest that the spin orbital ought to be oriented at 40° to the acridine ring. This in turn implies that the plane of the acridine must be nearly perpendicular to the helix axis, as is indeed expected in intercalation. The larger tilt observed in the instance of the 9-substituted acridine may result from the even closer proximity of the 1- and 8- hydrogens of acridine to the amide oxygen attached to the nitrogen at the 9- position of the ring.

Our computer simulations of these spectra were made using a slightly modified program written by Balasubramanian and Dalton.[8] Our program contained the following assumptions: (1) that every molecule of the probe is bound to the double helix in the same way and that there is a uniform angle between the local helix and the spin orbital; (2) that the disorder in the packing of helices within the macroscopic DNA fiber corresponds to the sum of two components, one with a Gaussian distribution in the angle between the axes of the helices and the fiber axis, and the other involving a small fraction of randomly oriented helices;[9] and (3) that the only motion was rotation about a single axis. The calculation proceeded in two steps. The first step involved calculation of the spectra of an array of nitroxyls incorporating the molecular and fiber geometry at the rapid motional limit.[10/11] The second step involved incorporating the motion as a convolution of the rapid limit spectrum using a broadening function that was calibrated against isotropic spectra of known correlation times between 0.01 and 5 nsec.[12] The magnetic parameters we used to describe the pyrroline nitroxyl radical were gxx = 2.0088, gyy = 2.0066, gzz = 2.0031, Axx = 5.0 G, Ayy = 9.0 G, Azz = 34 G. The relaxation times we used were $T_2 = 2.5 \times 10^{-8}$ sec and $T_1 = 3.0 \times 10^{-6}$ sec. We took account of unresolved hyperfine coupling of the methyl protons by convoluting each spin packet with a Gaussian of 1.5 G

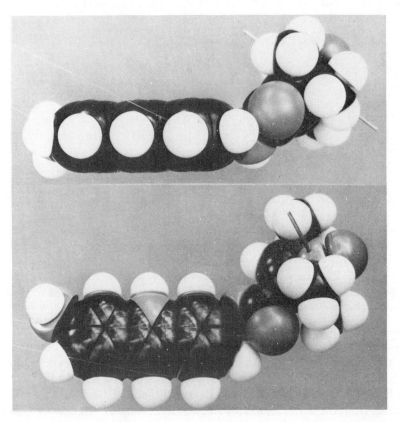

Figure 3. Photographs of a space-filling model of 6-substituted acridinium probe viewed a) in the plane of the acridinium ring and b) perpendicular to the acridinium ring. The metal rod through the nitrogen atom of the pyrroline ring shows the orientation of the spin orbital. Contact between the amide oxygen and the 5-hydrogen of acridinium prevents the pyrroline ring from becoming coplanar with acridinium.

width.[13] Figure 4 shows how well the calculated spectrum of a representative system compares with our actual spectrum for three different orientations of the fiber.

We are not the first to study the EPR spectra of intercalating probes in the presence of oriented DNA with respect to geometry. This was done by Onishi and McConnell[14] in 1965 using a chlorpromazinium probe. Concurrently with our efforts, Hong and Piette,[15] working with a probe very similar to our spin ethidium, found the same sort of orientation dependence as in the case of our 6-substituted acridine probe. Unfortunately, these authors did not attempt computer simulations, so we cannot make any direct comparison of their reported orientation dependence with our own.

Solution Studies

These experiments fall naturally into two groups, those involving the acridinium

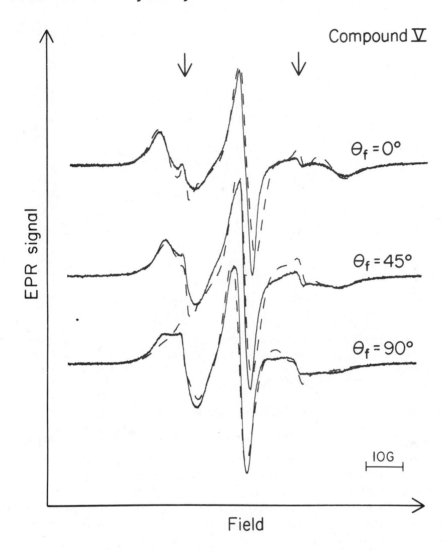

Figure 4. EPR spectra of 6-substituted-3-amino acridinium in oriented fibers in standard buffer with 0.2 M NaCl and 100 mg/ml poly(ethylene oxide) 6000. θ_τ is the angle between the fiber axis and the magnetic field. Spectra are shown for θ_f at 0°, 45° and 90°. Dashed lines represent theoretical simulations obtained using the parameters given in the text. The arrows indicate the cross over magnetic field values of the unbound intercalator.

probes and those involving phenanthridinium probes. We will discuss the acridinium probes first.

The EPR spectrum of a 6-substituted acridinium probe equilibrated with sonically fragmented chicken erythrocyte DNA in aqueous solution appears at the top of Figure 5. At the bottom of Figure 5 is the EPR spectrum of the same probe bound to precipitated randomly oriented DNA fibers, prepared in a manner similar to our

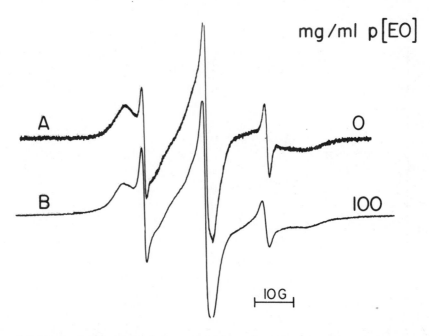

Figure 5. EPR spectra of 6-substituted-3-amino acridinium bound to DNA. Spectrum A shows this mixture in solution, Spectrum B is of an unordered pellet which was condensed by the addition of poly(ethylene oxide) to a concentration of 100 mg/ml.

oriented fibers. There are no significant differences between these spectra. This observation is rather surprising. In view of the closeness between helices in fibers prepared in this manner, whose separation has been estimated to be 44Å on the basis of X-ray diffraction studies,[16] we might expect the motion of DNA to be less rapid in the fiber than in solution. Our 9-substituted probe, however, does show a change in its spectrum upon precipitation. The outermost hyperfine maxima are closer together in EPR spectra of solutions than in EPR spectra of fibers (see Table II). The rotational correlation times corresponding to these spectra were calculated as before and appear in Table II.

The EPR spectra of our phenanthridinium probes equilibrated with aqueous solutions of chicken erythrocyte DNA appear in Figure 6. There is an important difference between these spectra and those of the acridinium derivatives. The separations of the outer hyperfine maxima of the spectra of both phenanthridinium probes are much greater than in the case of the acridinium probes. This separation is only slightly less than that reported by Hong and Piette[15] (64.3 Gauss) for a phenanthridinium probe bound to oriented fibers when the magnetic field was oriented along the fiber axis. The closeness of the 64 Gauss value to the value we obtained for fully immobilized phenanthridinium spin label (68 Gauss) from a study of viscous sucrose-containing solutions indicates that the molecular axis lies close to the fiber axis. The fact that this splitting depends upon temperature and upon DNA length, as we shall discuss later, requires that there must be residual motional effects.

Table II
Geometrical and Motional Parameters of Probes

Probe	Tilt Angle degrees	Separation* Gauss	Rotational Correlation Time nsec	DNA Form to Which Probes Are Bound
6-substituted-3-amino acridinium probe	40	47†	0.2	solution and fiber
6-substituted-9-amino acridinium probe	40	47†	0.9	solution and fiber
9-substituted acridinium probe	90	49†	2.2	solution
9-substituted acridinium probe	90	53†	9.0	fiber
propidium probe		64†	33†	solution
ethidium probe	††	63.7	28	solution

†Estimated from figures in reference 21.

*Separation between outer hyperfine maxima of EPR spectrum at 20°C. For oriented fibers this refers to the greatest separation seen as the fiber was rotated.

††Hong and Piette[15] in a study of a similar spin probe find a tilt angle of ∼0° in an oriented DNA fiber prepared by a different method.

To determine rotational correlation times for our phenanthridinium probes we have followed the simple expedient of using the procedures outlined by Freed and co-workers[17/18] and Gordon and Messenger[19] which relate the splitting of the outermost features to rotational correlation times using the model of isotropic Brownian diffusion. These correlation times appear in Table II. Correlation times determined by this method are often a good approximation to correlation times determined by spectral simulation even in the presence of a moderate anisotropy in diffusion with cylindrical symmetry.[20]

If we compare the rotational correlation times of the acridinium probes in Table II with those of the phenanthridinium probes, we see the latter are much longer. This difference suggests that the acridinium probes must have considerable freedom to move in the intercalation socket. A simple hydrodynamic model may help to account for these correlation times. All probes exhibit a correlation time of 0.05 nsec when free in solution.[21] This is the same as the value calculated for a hydrodynamically equivalent ellipsoid of revolution 3.4Å thick and 20Å in diameter in water at 20°C. If the ellipsoid is constrained to move only about its short axis[22] the calculated correlation time is 0.25 nsec, close to the value we have established for the 6-substituted acridinium probe in DNA solution. The 9-substituted acridinium probe showed a correlation time of 2.2 nsec in DNA solution and therefore must be

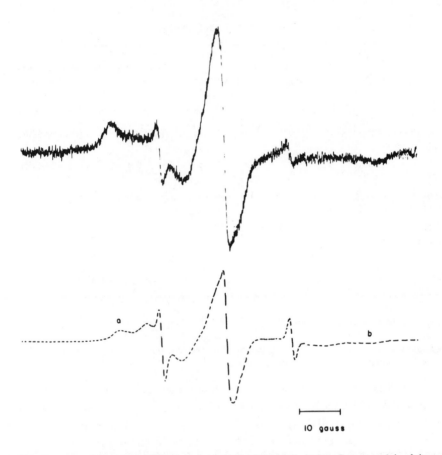

Figure 6. EPR spectra of phenanthridinium probes bound to DNA in solution. Spectrum A is of the propidium probe bound to DNA with an average length of 550 base pairs. Spectrum B is of the ethidium probe bound to DNA with an average length of 1200 base pairs.

more tightly coupled to the neighboring base pairs of DNA. Possibly this coupling is associated with the greater angle (90°) the pyrroline ring of the 9-substituted probe makes with the acridine ring as compared with the 6-substituted probe (40°). The phenanthridinium probes, which show the longest correlation times, have one substituent, phenyl, which is normally oriented at 90° to the phenacridinium ring and a second substituent, the side chain attached to the ring nitrogen, which can take up a similar orientation. Steric repulsions between these side chains and adjacent atoms of DNA may "lock" the intercalator into a fixed position relative to its neighboring base pairs.

DNA Dynamic Models

We considered four dynamic models in order to examine whether the observed rotational correlation times of the phenanthridinium probes are plausible, and consistent with other estimates of DNA properties. Diagrams illustrating these models appear in Figure 7.

Model 1. The probe is proposed to be rigidly attached to the double helix, which rotates uniformly along its length, as if it were a speedometer cable. The rotational

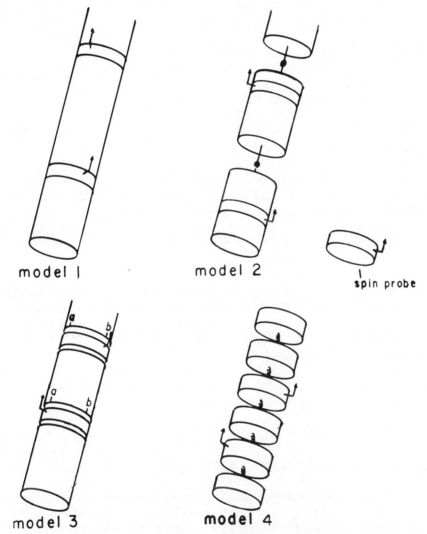

model I   model 2   spin probe

model 3   model 4

Figure 7. Diagrams illustrating 4 models of DNA rotational motion. *Model 1.* The whole length of DNA turns as a rigid rod. *Model 2.* The length of DNA is divided into lengths of 100 base pairs connected by swivels. Each length moves completely independently of its neighbors. *Model 3.* The spin probe is represented by a disc with a pointer on one side representing the unpaired electron spin. The spin probe jumps between two positions in the DNA helix. The helix itself does not move. *Model 4.* The DNA helix is made up of a stack of concentric discs (the base pairs) linked by simple torsional springs (the deoxyribose phosphate bridges). Each disc can exert a force on its neighbors through the springs. The spin probe acts as a typical disc of the stack.

correlation time we ought to observe for DNA if this model were correct would be approximately proportional to the length of the double helices. This model requires that a DNA molecule of 550 base pairs long (the average size of the sonicated DNA used in our experiments involving phenanthridinium probes) have a correlation time of 200 nsec, approximately 7 times longer than the 30 nsec we observed.[21] Therefore, either model 1 is completely unsuitable or it must be modified to allow the observed correlation time to include the effect of independent motion of the probe.

Model 2. The probe is rigidly attached to relatively short uniformly rotating segments of double helix connected by swivels in model 2. This model is compatible with our measured correlation times reported in Table II if the segment length is approximately 100 base pairs and if no coupling between adjacent segments through the swivels is permitted. While such junctions between such segments might exist if the double helix were near its melting temperature, it is hard to imagine any basis for such discontinuities at room temperature.

Model 3. The probe is hypothesized to jump rapidly between two equilibrium positions within the double helix, the positions being separated by a large potential barrier. This model requires that there be an Arrhenius temperature dependence of the rotational correlation time with the pre-exponential factor of the Arrhenius expression corresponding to the jump frequency. This pre-exponential factor we found to be several orders of magnitude too large for this requirement to be met.[21]

Model 4. Model 4 conceives of the double helix as consisting of a stack of co-axially arranged discs which are linked by simple torsional springs. This is equivalent to requiring oscillation of nucleotide pairs about the helix axis by means of a complex motion within the phosphate-deoxyribose bridges. We assume that the intercalator acts as if it were a typical disc of the stack at least to a first approximation. Our experimental correlation time can be accommodated to this model if the root mean square (rms) amplitude of rotational oscillation between adjacent base pairs is approximately $5°$.

Models 1 and 4 appear to be not inconsistent with our initial solution experiments. Before we describe the experiments we did to distinguish between the models, it is appropriate to go somewhat deeper into the theory we developed.

Theoretical analysis of the dynamics of linearly coupled rotational oscillations in a viscous medium[21/23] led to the following equations for the rotational correlation time τ_R for a double helix of N base pairs in the case of our model 4

$$\tau_R^{-1} = \tau_F^{-1} - T_{2e}^{-1} \tag{1}$$

$$\tau_F = \int_0^\infty \{\exp[-t/T_{2e}]\exp[-\Omega(g,N)t/\tau_o]\}dt \tag{2}$$

$$\Omega(g,N) = 1/N \sum_{j=1} \{1 - \exp[-g \cdot \sin^2[\pi j/N]]\}/g \cdot \sin^2[\pi j/N] \tag{3}$$

$$g = 2t/(A^2\tau_o) \tag{4}$$

where g is a composite of t, time (the variable of integration), τ_0, the correlation time of a single disc, and A, the rms oscillation amplitude of each disc. T_{2e} is the appropriate relaxation time for the magnetization of the electron, the spin-spin dephasing constant, which is about 25 nsec at 20°C, as we inferred from the spectrum of unbound probes. Equations 1 through 4 were derived under the assumption that the DNA was in the form of closed circles to avoid considering end effects. In Equation 3, Ω represents the rate at which the displacement angle of a disc loses correlation with its earlier position by mechanisms other than its own independent diffusion. At early times (g < 1), $\Omega \simeq 1$, and decorrelation occurs by free diffusion of a single disc. At intermediate times ($1 < g < N^2$) decorrelation occurs by interaction of neighboring discs. At long times (g > N^2), Ω becomes 1/N, and decorrelation is dominated by overall rotation of the whole DNA molecule.[24] Only at these longest times does Ω depend upon N. The dependence of the correlation time, τ_F, in Equation 2 upon the number of base pairs, N, is determined by the value of Ω relative to the electron relaxation rate (T_{2e}^{-1}). Since Ω contributes to the determination of τ_F only while exp($-t/T_{2e}$) is large, the dependence of τ_F upon N will be governed by the value of T_{2e}. The integral is nearly independent of Ω when t is greater than $10 T_{2e}$. If N is sufficiently small, then it is possible to have g < N^2 while t < $10 T_{2e}$. In this situation, and only in this situation, will τ_F and τ_R depend upon N. This dependence is shown in Figure 8 for two values of A at 20°C. Notice that τ_R reaches a limiting value below 400 base pairs and is independent of N for longer double helices. Dynamics equivalent to these have been developed independently by Barkley and

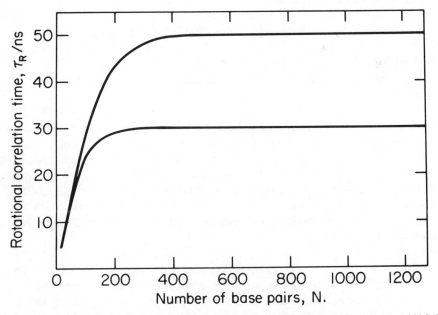

Figure 8. Calculated rotational correlation time τ_R of Model 4 as a function of DNA length at 20°C. The upper curve corresponds to an rms oscillation amplitude of 4 degrees; the lower curve to an rms oscillation amplitude of 6 degrees. The curves were calculated using equations 1-4 in the text. Equation 2 was evaluated by numerical integration over the time interval 0-1 μsec. Contributions to the integral from longer times were negligible.

Zimm[25] for the analysis of the time-dependent anisotropic fluorescence of ethidium bound to DNA.

The equation relating rotational correlation time to length of DNA for our model 1 involves a rotating rigid cylinder of radius R turning about its unique (long) axis in a medium of viscosity, η, at an absolute temperature, T,

$$\tau_R = (\pi R^2 \eta 3.4N)/kT \tag{5}$$

where π and k have their usual values as constants. This same equation was used to determine the value of τ_o in equation 4 which we employed to obtain Figure 8 by setting N = 1. This equation, 5, predicts that the rotational correlation time ought to increase uniformly with the length of DNA if model 1 is correct. It seemed reasonable, therefore, to determine the length dependence of the rotational correlation time of DNA experimentally in order to choose between model 1 and model 4.

Dependence of DNA Motion Upon Length

In these experiments,[19/25] we equilibrated ethidium probe with solutions of DNA of different lengths. The shorter lengths of DNA we made by sonicating a sample of the high molecular weight chicken erythrocyte DNA we used, and then separated the fractions by controlled precipitation from aqueous ethanol.[34] Figure 9 shows two of the EPR spectra we obtained at 20°C.

The differences between these spectra are significant and reflect the motion of the DNA to which they are bound. Free or partially free rotation of the probe within its intercalation socket could not be expected to vary with length of DNA when the length is much greater than the intercalation site. So the probe must be following the motion of the DNA closely in this case.

One important difference between the spectra in Figure 9 is that the outermost features in the spectrum shown as a solid line (DNA length $\sim$ 1200 bp) are narrower and more widely separated than the outermost features in the spectrum shown as a dotted line (DNA length $\sim$ 200 bp). Application of the semiempirical method of determining rotational correlation times we have used earlier gives the results listed in Table III. They show that correlation time increases with helix length and approaches an asymptotic limit, as expected of model 4 but not of model 1. Model 1 cannot therefore be correct.

The agreement between our experimental results and the predictions of model 4 is no more than qualitative, as can be seen by comparing our results with the curve in Figure 9. The probes' motion seems sensitive to greater lengths of DNA than predicted by the model. One way of accounting for part of this discrepancy is to alter the model so that it has open ends rather than being circular. Robinson, *et al.*[23] have done this and conclude that the mean correlation times seen for open ended molecules will correspond to the values of endless molecules half as long. The rest of

Table III
Rotational Correlation Time of DNA
As a Function of Number of Base Pairs

Solution	DNA length base pairs (mass mode)	Rotational Correlation Time, τ_R nanoseconds			Calculated length of DNA segment†† base pairs
		from splitting	from high field linewidth	from low field linewidth	
A	200	12.	6.	6.	40
B	330	16.	10.	9.	62
C	1200	28.	28.	40.	>150
D	High mw	26.			>150

*A value of 68 Gauss for the separation between outermost features in totally immobilized ethidium spin label was used in this determination. This value was obtained by determining the splitting of ethidium spin label bound to DNA in sucrose solution of various concentrations and then extrapolating a plot of viscosity⁻¹ versus splitting to the limit of infinite viscosity. A value of $\partial = 1.0$ Gauss used in the calculation was determined by finding the best fit to the low field and high field line width as recommended by Mason and Freed.[18]

††The length of a DNA circular segment which would have a rotational correlation time equal to that obtained from the measured splitting; calculated using Model 4 with $T_e = 25$ nanoseconds, $\tau_o = 0.35$ nanoseconds and A = 0.109 radians.

the discrepancy may arise from a bending motion of the ends of the DNA helix. Such a bending ought to make a larger contribution to the relaxation of probes

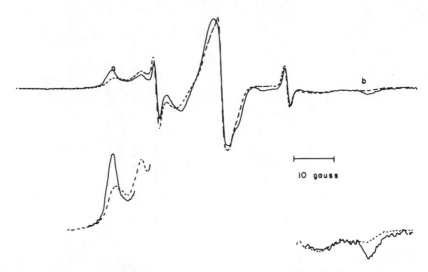

Figure 9. Room temperature electron paramagnetic resonance spectra of ethidium probe bound to chicken erythrocyte DNA.
------ Solution A - DNA length ∼ 200 base pairs.
_____ Solution C - DNA length ∼ 1200 base pairs.

bound close to the end of the piece of DNA than to probes nearer the middle and thus further reduce the size of the discrepancy between experiment and prediction.

Torsional Rigidity of DNA

We can calculate a temperature independent harmonic force constant, C, for internal twisting of a DNA helix from model 4 and our experimentally determined rotational correlation times. The force constant is related to the rms oscillation amplitude of a single disc, A, by:

$$C = lkT/A^2 \qquad (6)$$

where l is the separation of adjacent discs, .34 nm, and k is Boltzmann's constant. (A is related to the rotational correlation time, τ_R, by equations 1-4 cited above.) We find that the rotational correlation times of ethidium spin probe bound to relatively long (> 400 base pairs) DNA are fitted by an rms oscillation amplitude of 6° at 20°C and a corresponding torsional force constant of 1.2×10^{-19} erg-cm/radian². The propidium spin probe results require 4° and 3×10^{-19} erg-cm/radian² for a best fit.[21]

It is of interest to compare our results with other estimates of torsional force constant and rms oscillation amplitude, which can be inferred from two types of experiments: measurement of fluorescence depolarization, and supercoil linking number distributions, as well as from theoretical simulations of DNA structure. Assuming that the depolarization of fluorescence of ethidium bound to DNA is attributable to the rate of internal twisting, Barkley and Zimm[24] calculate a torsional force constant of 4.1 × 10^{-19} erg-cm/radian^2. This number corresponds to an rms oscillation amplitude of 3.3° at 20°C.

Two sets of investigators, Pulleyblank, et al.[27] and Depew and Wang[28] obtained estimates of the Gibbs free energy involved in forming supercoiled DNA by observing the variation in linking number of nicked circular DNA when it is restored to double stranded continuity upon incubation with ligase at various temperatures. If we assume that all this free energy is due to torsion and none to flexure[36] we can obtain the following estimates of torsional force constant and oscillation amplitude:

Pulleyblank, et al.   C = 1.5 × 10^{-19}erg-cm/radian^2 and
                         A = 5.6° at 20°C;

Depew and Wang   C = 0.64 - 1.1 × 10^{-19} erg-cm/radian^2 and
                        A = 6.5 - 8.4°.

Since some of the free energy may be due to bending, these estimates from supercoiling must represent lower limits to the force constant, and upper limits to the oscillation amplitude. Barkley and Zimm[25] have also calculated a torsional force constant from hydrodynamic and light scattering estimates of persistence length[28/29/30] under the assumption that DNA behaved as a homogenous isotropic elastic rod. They obtain C = 1.8 × 10^{-19} erg-cm/radian^2. This corresponds to A = 5.2°.

The modeling of DNA structure by Levitt[31] and by Miller[32] lead to independent estimates of the force constant from the variation in total free energy in the system upon introducing a small twist into the helix. Levitt estimated the total energy of a 20 base pair segment of straight DNA as a function of the pitch of the double helix. He used empirical energy functions which allowed for bond stretching, twisting, bending, van der Waals interactions and hydrogen bonding. However, he did constrain the terminal base pairs to remain coaxial, perpendicular to the helix axis, and at a specified net rotation. His relation between energy and pitch shows a parabolic minimum near 10.5 base pairs. If we assume the potential function fits a harmonic form for small displacement as it appears to do for his figure we obtain a torsional force constant of 20 × 10^{-19} erg-cm/radian^2 and an rms oscillation amplitude of 1.5°. Miller in his calculation specified the atomic positions of two Watson-Crick base pairs in accordance with experimental equilibrium bond lengths and angles and permitted only the conformational angles to change in order to minimize the total energy of the system. He reported a dependence of the potential energy of a C-G base pair between other C's and G's upon twist angle which implies a harmonic torsional force constant of 36 × 10^{-19} erg-cm/radian^2 and an rms oscillation amplitude of 1.1°.

The inference of the oscillation amplitude from each type of experimental observation depends on a particular set of special assumptions. It is clear that the EPR approach has not yet reached its full development in terms of rotational discrimination and accuracy. With these qualifications, it appears that the estimates from the several experimental approaches are in reasonable agreement. The estimates based on theoretical calculations from bonding parameters all imply a stiffer helix and a smaller amplitude than the experimental values. The difference suggests that the calculations impose excessive constraints or the consequences of interaction with the solvent have not been accounted for adequately.

Conclusions

In the experiments we have described, we have shown that the rotational correlation time of DNA depends upon DNA length in a manner which is predicted by our model 4, and that the torsional force constant for DNA rotation which we can derive from our data using our model 4 is consistent with other estimates made in very different ways.

Obviously, a lot can be done to strengthen the argument we have made. More and better data exhibiting the length dependence is required and presently being assembled. There is a need to improve our estimate of the phenanthridinium probe rotational correlation times by computer simulation of the spectra. In this regard, we have passage saturation transfer measurements underway to examine the contribution of anisotropic motion to relaxation, as has been exhibited in anisotropically moving phospholipid spin labels.[32] It would also be desirable to obtain more direct evidence that phenanthridinium spin labels are not moving relative to the local helix. We have some preliminary results which seem promising in this regard. We observe that when we equilibrate dilute solutions of phenanthridinium probe with high molecular weight chromatin from chicken erythrocytes the separation between outer hyperfine maxima is more than 66 Gauss at 20°C, corresponding to a rotational correlation time of more than 200 nsec assuming isotropic Brownian motion. If the mode of bonding of probe to chromatin is the same as that of probe to linear DNA, as now seems likely, and if the 200 nsec correlation time of chromatin represents independent motion of the probe, we can represent the true rotational correlation time by:

$$1/t_R(\text{true}) = 1/t_R(\text{linear DNA}) - 1/t_R(\text{chromatin})$$

which gives t_r (true) = 35 nsec, so at most 15 percent of the relaxation of the probe can be due to independent motion. Finally, we can improve our theoretical model 4 by incorporating increased stiffness in the bonds between the probe at its neighboring base pairs, and by allowing for the contributions of bending motion at the ends of short DNA molecules.

References and Footnotes

1. Feher, G. *Electron Paramagnetic Resonance with Applications to Selected Problems in Biology.* Gor-

don and Breach, New York (1970).

2. Carrington, A., and McLachlan, A.D. *Introduction to Magnetic Resonance with Applications to Chemistry and Chemical Physics.* Harper and Row, New York (1967).

3. Lerman, L.S. *J. Mol. Biol. 3*, 18 (1961).

4. Rozantsev, E.G. *Free Nitroxyl Radicals.* (Transl. by B.J. Hazzard) Plenum, New York (1970).

5. Berliner, L.J., ed. *Spin Labeling: Theory and Applications.* Academic Press, New York (1976).

6. Berliner, L.D., ed. *Spin Labeling II: Theory and Applications.* Academic Press, New York (1979).

7. Brown, C.J., and Corbridge, D.F. *Acta Cryst. 7*, 711 (1954).

8. Balasubramanian, K., and Dalton, L.R. *J. Magn. Resonance 33*, 245 (1979).

9. Auer, C. Ph.D. Dissertation (Vanderbilt University) (1978).

10. Porumb, T., and Slade, E.F. *J. Magn. Resonance 22*, 219 (1975).

11. Griffith, O.H., and Jost, P.C. in *Spin Labeling: Theory and Applications* (L.J. Berliner, ed.) Academic Press, New York, p. 454 (1976).

12. Libertini, L.J., Burke, C.A., Jost, P.C., and Griffiths, O.H. *J. Magn. Resonance 15*, 460 (1974).

13. Perkins, R.C., Jr., Lionel, T., Robinson, B.H., Dalton, L.A., and Dalton, L.R. *Chem. Phys. 16*, 393 (1976).

14. Onishi, S., and McConnell, H.M. *J. Am. Chem. Soc. 87*, 2293 (1965).

15. Hong, S.J., and Piette, L.H. *Cancer Res. 36*, 1159 (1976).

16. Maniatis, T.P., Venable, J.H., Jr., and Lerman, L.S. *J. Mol. Biol. 83*, 28 (1974).

17. Goldman, S.A., Bruno, G.V., and Freed, J.H. *J. Phys. Chem. 76*, 1858 (1972).

18. Mason, R., and Freed, J.H. *J. Phys. Chem. 78*, 1321 (1974).

19. Gordon, R.G., and Messenger, T. in *Electron Spin Relaxation in Liquids* (L.T. Muus and B.W. Atkins, eds.). Plenum, New York, p. 341 (1972).

20. Freed, J.H. in *Spin Labeling: Theory and Applications* (L.J. Berliner, ed.). Academic Press, New York, p. 86 (1976).

21. Robinson, B.H., Lerman, L.S., Beth, A.H., Frisch, H.L., Dalton, L.R., and Auer, C. (submitted for publication).

22. Edwardes, D. *Quart. Journ. Math. 26*, 70 (1892).

23. Robinson, B.H., Forgacs, G., Frisch, H.L., and Dalton, L.R. (in preparation).

24. Shore, J.E., and Zwanzig, R.J. *J. Chem. Phys. 63*, 5445 (1975).

25. Barkley, M.D., and Zimm, B.H. *J. Chem. Phys. 70*, 2291 (1979).

26. Robinson, B.H., Hurley, I., Scholes, C.P., and Lerman, L.S. in *Stereodynamics of Molecular Systems*, Ed., Sarma, R.H., Pergamon Press, New York, p. 283 (1979).

27. Pulleyblank, D.E., Shure, M., Tang, D., Vinograd, J., and Vosberg, H. *Proc. Natl. Acad. Sci. 72*, 4280 (1975).

28. Depew, R.E., and Wang, J.C. *Proc. Natl. Acad. Sci. 72*, 4275 (1975).

29. Hays, J.B., Mogar, M.E., and Zimm, B.H. *Biopolymers 8*, 531 (1969).

30. Jolly, D., and Eisenberg, H. *Biopolymers 15*, 61 (1976).

31. Levitt, M. *Proc. Natl. Acad. Sci. 75*, 640 (1978).

32. Miller, K.J., *Biopolymers 18*, 959 (1979).

33. Marsh, D. unpublished results quoted by J.S. Hyde and L.R. Dalton in *Spin Labeling II: Theory and Applications*, Academic Press, New York, p. 54 (1979).

34. Lerman, L.S., Wilkerson, L.S., Venable, J.H., Jr., and Robinson, B.H. *J. Mol. Biol. 108*, 271 (1976).

35. Crothers, D.M., Dattagupta, N., and Hogan, M., in *Nucleic Acid Geometry and Dynamics*, Ed. Sarma, R.H., Pergamon Press, New York, Oxford, p. 341 (1980).

36. Flexure includes all motions that introduce curvature into the helix axis.

Conformational Flexibility of the Polynucleotide Chain

Alexander Rich, Gary J. Quigley, and Andrew H.-J. Wang
Department of Biology
Massachusetts Institute of Technology
Cambridge, Massachusetts 02139

One of the widely appreciated features of protein structures is the fact that the polypeptide chain can fold into different conformations. This includes tightly folded conformations such as found in the α helix or more extended conformations as is found in the β sheet. It is less widely appreciated that the polynucleotide chain also has conformational flexibility which results in significant modifications of the chain extension. Many of these differences are associated with different puckers of the five-membered furanose ring which is a component of both the DNA and RNA chains. Here we illustrate changes in the sugar ring conformation in a variety of structures, taking examples in oligonucleotides as well as in polynucleotides. The property of polynucleotide sugar rings to alter their pucker should be looked upon as a degree of conformational freedom. It is important to have this kind of flexibility in order to form complex structures beyond the simple double helix.

The Double Helix and Ring Pucker

Double helices can be formed with both DNA and RNA molecules. These double helices have similar features in that the bases are in the center of the molecule with the familiar Watson-Crick pairing, either adenine-thymine pairs in DNA or adenine-uracil in RNA. The sugar phosphate chains are on the outside of the double helix. Although the two types of double helices are similar, they differ in one fundamental feature; the pucker of the five-membered furanose ring is different in ribose and deoxyribose.[1] The differences in ring pucker are illustrated in Figure 1. The furanose ring is viewed edge on in a plane which is defined by the ring oxygen O1' and carbons C1' and C4'; C1' has attached to it the ring nitrogen (N) of the purine or pyrimidine; C4'has attached to it the additional carbon atom C5' as indicated in the diagram. Both of these constituents, the base (N) and C5', are located on the upper side of the ring in Figure 1. It is clear that the principal difference in the conformation in the sugar ring is whether the carbon atom C2' or C3' is above the plane of the ring. Atoms which are above the plane of the ring in Figure 1 are considered to be in the endo conformation.[2] Thus the conformation C2' endo has C2' on the upper side of the ring and this correspondingly forces C3' to be on the lower side of the ring. In the other conformation C3' is endo and C2' is on the lower side of the ring. The major conformation of a deoxyribose chain is one in which the furanose ring has a C2'

273

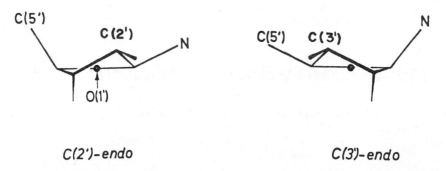

$$C(2')-endo \qquad\qquad C(3')-endo$$

Figure 1. Diagram illustrating two different conformations of the ribofuranose ring. The plane of the ring is defined by three atoms: C1' to which is attached the glycosidic bond indicated by N, O1' and C4' which is attached to the atom C5'. In this diagram, we are looking edge on at the plane defind by these three atoms. The remaining two atoms in the ring, C2' and C3' are located either above or below the plane of the ring. Atoms located above the plane of the ring are in the *endo* position. On the left C2' is in the *endo* position and on the right C3' is in the *endo* position. Although two examples of ring pucker are shown in this diagram, there are actually a number of intermediate states in which the displacement of C2' or C3' is not as great as that illustrated here. The detailed nomenclature for furanose ring pucker is complex[2]; here we elect to use only the simplified *endo* conformations.

endo conformation. In contrast, the normal polyribose chain conformation is C3' *endo*. The principal reason for the difference is associated with the additional space occupied by the oxygen atom attached to C2'in ribose which is absent in the deoxy series where only a hydrogen atom is attached to C2'. It should be emphasized that even though these are the normal conformations of these sugars in their respective double helices, the energy barriers involved in changing sugar conformation are not very great.[3]

Because of the C2' *endo* conformation of deoxyribose, the form of the DNA double helix is such that the base pairs are found on the helix axis in the familiar B form of the DNA. The bases occupy the axis of the molecule so that the familiar double helical DNA has a solid, rod-like appearance with bases in the center and two helical grooves running down the molecule. These are the major and minor grooves which occupy the space between the deoxyribose-phosphate chains (see Color Plates 2 and 6).

In contrast, the C3' *endo* conformation of ribose makes significant modifications in the RNA double helix. The base pairs are no longer perpendicular to the helix axis, but are tilted at about 15-18°. Furthermore, they are set back away from the helix axis. In fact, there is a clear space in the center of the molecule of approximately 3 Angstroms in diameter. The RNA double helix thus looks more like a molecular ribbon wrapped around an imaginary cylinder 3 Angstroms in diameter in which there is now a very deep major groove and a comparatively shallow minor groove on the outside of the helix.

These forms of the familiar double helices are not invariant. Reducing the water content of the medium by adding alcohol readily converts the familiar B form of

DNA into the A form in which deoxyribose adopts the C3′ *endo* conformation which is normally found in RNA double helices. The double helix then changes its conformation and looks more like double helical RNA (see Color Plates 3 and 7). It is noteworthy that in the conversion from the B to the A form of DNA, there is over an Angstrom difference in the phosphate-phosphate distance along each polynucleotide chain.

The reason for the change in phosphorus-phosphorus distance can be seen schematically in Figure 2 which shows the conformation of two forms of an adenosine nucleotide in a ribose polynucleotide chain. In the upper figure, the ribose is in the normal C3′ *endo* conformation and the phosphorus-phosphorus distance is near 5.9 Angstroms. In the lower part of Figure 2, the ribose is in a C2′ *endo* con-

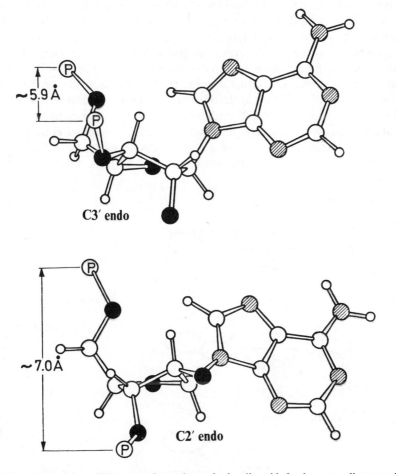

Figure 2. Illustration of two different conformations of adenylic acid. In the upper diagram with the C3′ *endo* conformation, the two phosphate groups are both above the plane of the ribose ring and are approximately 5.9 Angstroms apart. In the lower diagram, the C2′ *endo* conformation, the phosphate attached to O3′ is located below the plane of the ribose ring and the phosphates are now approximately 7 Angstroms apart. These two conformations are associated with considerable differences in the extension of the sugar phosphate chain.

formation (normally found in DNA) and the phosphorus-phosphorus distance is near 7 Angstroms. The change in distance is largely associated with a change in the pucker of the ring, although there are also small changes in the other dihedral angles in the backbone. The polynucleotide chain thus has a degree of conformational freedom which allows it to change the degree of extension in a significant way. This flexibility is used in a variety of ways in polynucleotide structures.

Polynucleotide Chain May Turn Corners With Changes in Ring Pucker

There are many examples in which the conformation of the sugar ring is changed. In some cases, the changes in pucker are associated with a change in the direction of a polynucleotide chain. An example of this is seen in Figure 3 which shows the crystal structure of adenyly-(3′,5′)-uridine (ApU) which formed a crystalline complex with 9-amino acridine.[4] In this structure, there are stacked columns of planar molecules in which the 9-amino acridine molecule alternates with an adenine-uracil base pair hydrogen bonding through the ring nitrogen of N7 of adenine. Uracil is at one end of the chain and the backbone has an extended conformation in which the molecule actually turns a corner so that its adenine residue is then hydrogen bonded to still another uracil residue (Figure 3). In the course of making this turn, one of the sugar residues adopts the unusual C2′ *endo* conformation as shown in the diagram. As seen at the right of the diagram, the uridine ribose O1′ is found at the bottom while it is found at the top of the adenosine ribose. This illustrates the extent to which the chain has radically changed direction in this structure. A somewhat similar con-

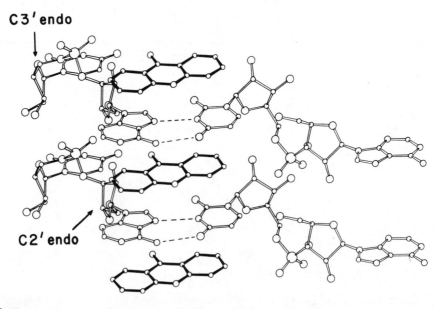

C3′ endo

C2′ endo

Figure 3. The crystal structure of the complex of ApU and 9 amino acridine. The polynucleotide chain of ApU is in an extended conformation and the chain turns a sharp corner. This can be seen by the different orientation of the two ribose residues in the oligonucleotide. The change in direction is associated with a change in ring pucker as shown.

formation is seen in the structure of adenylyl-(3′,5′)-adenosine (ApA) which crystallizes with proflavin in a similar manner.[5] However, it is not obligatory for the ribose in RNA to change pucker in coiled regions of the molecule. The examples cited above illustrate the fact that changes in pucker may be seen in some nucleotide structure determinations.

Tables I and II cite the structure of oligonucleotides which have crystallized by themselves (Table I) or together with intercalators (Table II). They are listed with description of the sugar conformation in various parts of the backbone. Although

Table I
Oligonucleotides

		5′ end	3′ end	Reference
DNA fragments	pdTpdT	C2′ endo	C2′ endo	15
	d-(pApTpApT)	C3′ endo (A)	C2′ endo (T)	16
RNA fragments	ApU	C3′ endo	C3′ endo	17
	GpC	C3′ endo	C3′ endo	18
	GpC	C3′ endo	C3′ endo	19
	ApA⁺	C3′ endo	C3′ endo	6
	A⁺pA⁺	C3′ endo	C3′ endo	6
	UpA	C3′ endo	C3′ endo	20, 21

Table II
Intercalator Complexes

	Nucleotide	Intercalator	Sugar Pucker 5′ end	3′ end	Reference
DNA fragments	dCpG	TPH⁺	C3′ endo	C2′ endo	7
RNA fragments	CpG	Acridine orange⁺	C3′ endo	C2′ endo	8
	rIodo UpA	Ethidium⁺	C3′ endo	C2′ endo	9
	rIodo CpG	Ethidium⁺	C3′ endo	C2′ endo	10
	rIodo CpG	9-Amino acridine⁺	C3′ endo	C2′ endo	11
	rIodo CpG	Acridine orange⁺	C3′ endo	C2′ endo	12
	rIodo CpG	Ellipticine⁺	C3′ endo	C2′ endo	12
	CpG	Proflavine⁺	C3′ endo	C3′ endo	13
	rIodo CpG	Proflavine⁺	C3′endo	C3′ endo	12
Non-helical	ApU	9-Amino acridine⁺	C2′ endo	C3′ endo	4
	ApA	Proflavine⁺	C2′ endo	C3′ endo	5

we have cited two examples in which oligonucleotides change sugar pucker where there is a change in the direction of the polynucleotide chain, there are several oligonucleotides in which there has been no change in the pucker even though the chain is in a fairly extended conformation. An example is seen in the structure of the trinucleotide ApApA.[6]

The Double Helix Changes Conformation Upon Intercalation

One of the conformational changes frequently encountered in double helical nucleic acids is that associated with the binding of a planar intercalator molecule which lodges between the base pairs. It does this without substantial disruption of the helix. Intercalation has two important effects: it introduces a gap between adjacent base pairs which are then separated by 6.8 Angstroms instead of the normal 3.4 Angstroms. This is due to the planarity of the intercalator which usually has unsaturated rings with π electrons which are 3.4Å thick. In addition, there is an unwinding of the double helix.

A series of structures have been solved involving intercalators, mostly in the ribonucleotide series but with one in the deoxynucleotide series (Table II). The structure of a double helical fragment of DNA together with an intercalator is shown in Figure 4. Here we see the structure of deoxy CpG which accommodates a terpyridine-platinum intercalator in the midst of a double helical fragment.[7] As seen in the diagram, the 5′ end of the double helical segment adopts the unusual C3′ *endo* conformation while the 3′ end maintains the C2′ *endo* conformation which is normal for double helical DNA.

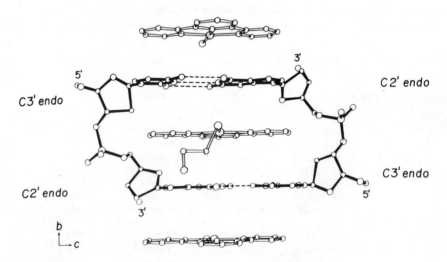

Figure 4. The structure of the deoxy CpG-platinum terpyridine intercalator complex. It can be seen that the DNA double helix has a planar intercalator lodged between the base pairs. This is associated with an unusual ring pucker on the 5′ side of the oligonucleotide segment. This conformation allows an extension of the polynucleotide chain.

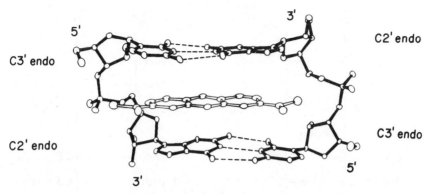

Figure 5. The structure of the CpG-acridine orange intercalator complex. It can be seen that there is an unusual conformation at the 3′ end of the oligonucleotide chain surrounding the intercalator. This mixed pucker conformation is associated with an extension of the polynucleotide chain.

Figure 5 shows the structure of the ribose dinucleoside phosphate CpG together with the intercalator acridine orange.[8] It can be seen that the 3′ end of the double helical RNA fragment has adopted the unusual C2′ *endo* conformation while the 5′ end maintains the normal C3′ *endo* conformation. A large number of intercalator structures have been solved in the ribonucleotide series as listed in Table II. Sobell and his colleagues were the first to point out that intercalation is associated with a modification of the pucker of the ribonucleotide chain on the 3′ side of the intercalator.[9-12] Although intercalation is generally associated with conformations similar to those seen in Figure 5, a number of alternative conformations are listed at the bottom of Table II. These are usually associated with the intercalators proflavin[5-12-13] and 9-amino acridine[4] both of which have the property of forming hydrogen bonds between the intercalator and the phosphate of the dinucleoside phosphate. In both cases, other modes of pucker are found. For example, in the complex of proflavin with the ribose CpG fragments, both residues have the normal C3′ *endo* conformation.[13]

For "simple" intercalators in which there is no hydrogen bonding to the phosphate residue, it is possible to make an interesting generalization about the way double helical DNA and RNA accommodate intercalator addition. This is illustrated schematically in Figure 6. At the left the DNA double helix is shown diagrammatically with the normal C2′ *endo* conformation in all residues, except for those on the 5′ side of the intercalator where the unusual C3′ *endo* conformation is adopted in a manner analogous to that which is illustrated in Figure 4. On the right the diagram shows the way in which double helical RNA accommodates an intercalator. All of the residues are in the normal C3′ *endo* conformation except for those residues on the 3′ side of the intercalator which adopt the unusual C2′ *endo* conformation. However, it can be seen that in the region immediately surrounding the intercalator (enclosed in the dashed line) the conformation of both the RNA and DNA chain are similar. Thus both molecules elongate to accommodate an intercalator by adopting a similar conformation. These conformational changes, as described by Sobell[9-12] explain the nearest neighbor exclusion. If a DNA or RNA double helix is saturated

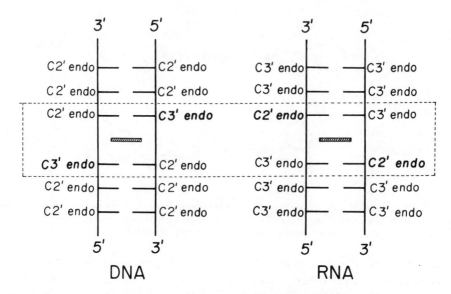

Figure 6. A schematic diagram illustrating the manner in which either DNA or RNA double helices change the pucker of the sugar residues in the region immediately surrounding a simple intercalator. Although the changes are different in the two types of double helices, the conformation in the region surrounding the intercalator enclosed by the dashed rectangle is the same in both cases.

with an intercalator, the most that can be accommodated is one intercalator for every two base pairs. The reason for this is probably associated with the necessity for mixed pucker on either side of the intercalator.

Conformational Flexibility in Nucleic Acid Macromolecules: Yeast tRNA^{Phe}

Examples of changes in the type of sugar pucker can be seen in the three-dimensional structure of yeast phenylalanine transfer RNA (tRNA[Phe]). Although most of the seventy-six nucleotides in this transfer RNA molecule adopt the C3′ *endo* conformation which is normal for a ribonucleotide chain, several residues are found to adopt the less common C2′ *endo* conformation. This is frequently associated with an interesting type of structural accommodation which is similar to those interactions described above in the oligonucleotide structures. Figure 7 is a diagram (see also Color Plate 1) which shows secondary and tertiary hydrogen bonding found in the yeast tRNA[Phe]. This schematic diagram is useful in interpreting Figures 8 through 11 which illustrate conformational changes in various parts of the molecule.[14]

Changes in Polynucleotide Direction

The principal sites using C2′ *endo* conformation in yeast tRNA[Phe] are listed in Table III. Nine examples are cited. These occur in two principal situations. In five cases (residues 7, 17, 18, 48 and 60) the C2′ *endo* conformation is adopted when the polynucleotide chain undergoes a distinct change in direction. For example,

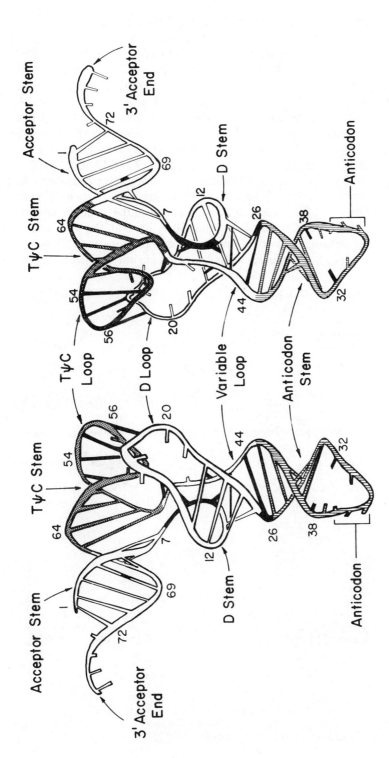

Figure 7. A schematic diagram showing two side views of yeast tRNA[Phe]. The ribose-phosphate backbone is depicted as a coiled tube, and the numbers refer to nucleotide residues in sequence. Shading is different in different parts of the molecules, with residues 8 and 9 in black. Tertiary hydrogen-bonding interactions between bases are shown as solid black rungs, which indicate either one, two or three hydrogen bonds. Those bases that are not involved in hydrogen bonding to other bases are shown as shortened rods attached to the coiled backbone.

nucleotides 1 through 7 are involved in the double helical acceptor stem but residues 8 and 9 are extended and provide a linker region which attaches one end of the acceptor stem with the beginning of the D stem. In this region, there are several conformational changes, one of which is associated with the C2′ *endo* conformation of ribose 7 where the polynucleotide chain changes direction. Other examples are found in regions where the backbone loops out away from the center where the bases are stacked. The dihydrouracil residues 16 and 17 are extended out away from the remainder of the molecule and their bases are not stacked with the other bases in tRNA. The backbone has a "bulge" or looping-out at that point (Figure 7). In order to accommodate this extended conformation, residues 17 and 18 adopt the C2′ *endo* conformation. In addition, ribose 47 in the variable loop is extended and its base thrusts away from the center of the molecule. This is associated with a C2′ *endo* conformation in ribose 48. Still another example is found in the T loop where bases 59 and 60 are excluded from the stacking of the other bases in the T stem and loop (Figure 7). This exclusion involves a C2′*endo* conformation in ribose 60. All of these conformational changes are thus associated with an abrupt change in the direction

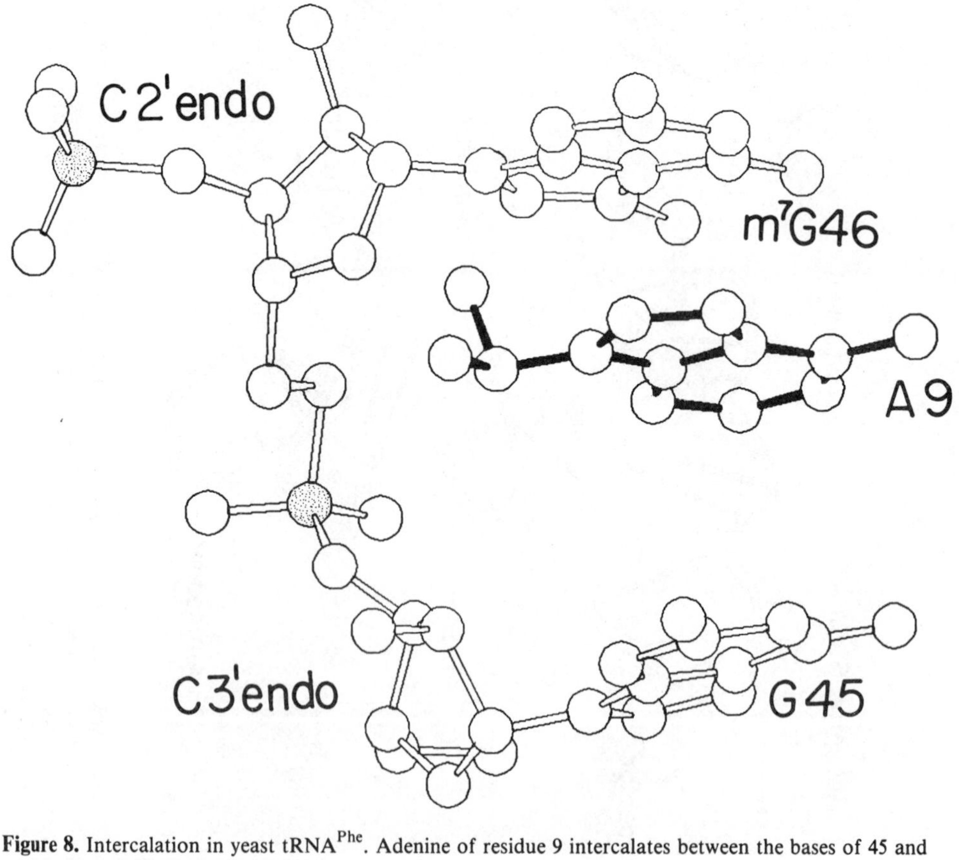

Figure 8. Intercalation in yeast tRNA[Phe]. Adenine of residue 9 intercalates between the bases of 45 and 46. This intercalation is accommodated by a change in the ribose pucker at the 3′ end of the segment. The ribose of m⁷G46 is in the unusual C2′ *endo* conformation.

of the polynucleotide chain. This directional change is accommodated by adoption of a C2′ *endo* conformation and there is a substantial change in the phosphorus-phosphorus distance along the polynucleotide chain.

It should be pointed out, however, that there are many other examples in the loop regions of tRNA[Phe] in which the polynucleotide chain changes direction but the C2′ *endo* conformation is not used. Thus, the change in pucker is a structural accommodation which may be adopted in the molecule, especially where the chain is extended, but it is not generally used when the polynucleotide chain undergoes a change in direction.

Intercalation in Yeast tRNA[Phe]

There are two parts of the molecule in which extensive intercalation is found involving pairs of nucleotides. These are in the central region with intercalation involving

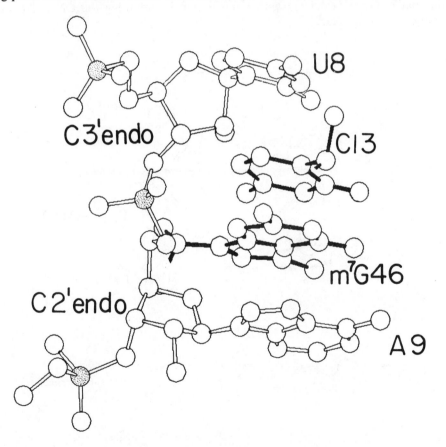

Figure 9. The nucleotides U8 and A9 have intercalated between them the bases C13 and m'G46. These bases are both hydrogen bonded to G23 in the tertiary structure of yeast tRNA[Phe]. This intercalation is accommodated by a change in the conformation of ribose of A9.

nucleotides 8 and 9 as well as in the corner of the molecule where the T and D loops interact. Figure 8 shows the conformation adopted by the sugars of residues G45 and m⁷G46 which has the adenine ring of A9 intercalated between them. Residues 45 and 46 are not involved in a double helix; nonetheless, the conformation of m⁷G46 is C2′ *endo* in a manner analogous to the conformational changes which are seen for double helical ribonucleotide fragments surrounding intercalators. The C2′ *endo* conformation is adopted at the 3′ end of the intercalator region in Figure 8. Another example is shown in Figure 9 where residues U8 and A9 are found in the extended segment which connects the acceptor stem with the D stem. It can be seen that A9 at the 3′ end of the segment has adopted the C2′ *endo* conformation even though it is not in a complementary double helical structure.

Figure 9 shows that intercalated between U8 and A9 are the bases m⁷G46 and C13 both of which are hydrogen bonded to G22. The base pair C13•G22 is part of the D stem. In Figure 9, it can be seen that the nucleotides U8 and A9 accommodate the additional distance associated with intercalation by adopting the C2′ *endo* conformation in residue A9.

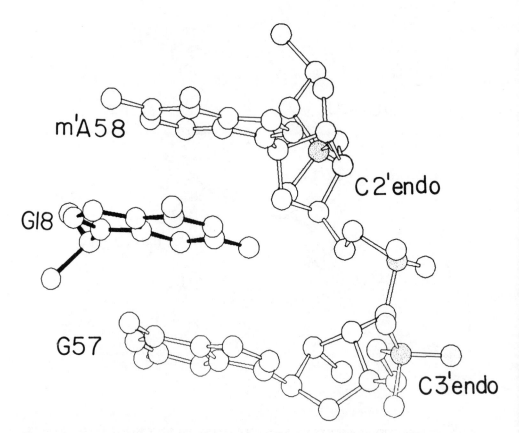

Figure 10. Diagram showing the intercalation of guanine 18 between G57 and m¹A57. The ribose of m¹A58 is in the C2′ *endo* conformation to accommodate the intercalation.

By comparing Figures 8 and 9, it can be seen that they are both portions of the same structure in which there are two interacting segments of the polynucleotide chain, both of which intercalate around each other. The polynucleotide chains are interleaved between each other so that each intercalates into the opposite member of the pair.

A similar pair of interleaved structures are found near the corner of the molecule where the bases of nucleotides 18 and 19 interact with the bases of 57 and 58. In Figure 10, it can be seen that m¹A58 and G57 are spread apart with residue G18 intercalated between them. This intercalation is associated with C2′ *endo* conformation found in m¹A58. In this case, it also adopts the C2′ *endo* conformation at the 3′ end of the oligonucleotide segment which surrounds the intercalating guanosine of G18. This is quite similar to the double helical RNA intercalator models which are described in Table II, as the bases G57 and m¹A58 are hydrogen bonded to other residues, although not in Watson-Crick pairs.

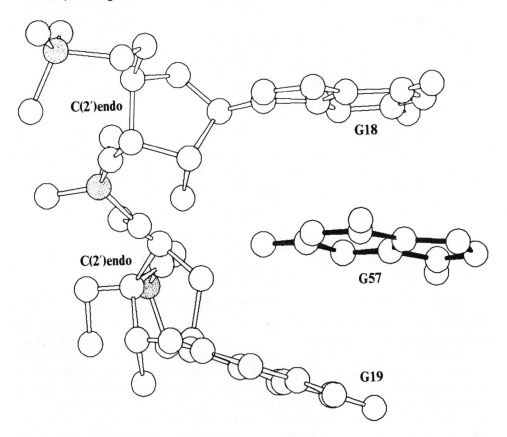

Figure 11. The intercalation of G57 between G18 and G19 is illustrated. Unlike the previous examples, the riboses of G18 and G19 are both in the unusual C2′ *endo* conformation. The C2′ conformation of G18 is probably associated with the unusual conformation of residues 16 and 17 immediately adjoining G18.

An interesting variant is seen in Figure 11 which shows the intercalation of G57 between the bases of G18 and G19. This intercalating interaction is one of the important stacking interactions which stabilize the corner of the tRNA molecule and helps maintain the interaction of the T loop and D loop. However, in this example, the less common C2′ *endo* conformation is found in both riboses of G18 and G19. By analogy with the simple intercalator structures, one would expect a C2′ *endo* conformation to be found in G19 but not in G18. Here it is likely that the unusual C2′*endo* conformation of G18 is not associated so much with the intercalation of G57 but is probably related to the fact that residues 16 and 17 are excluded from the molecule as described above so that their bases do not stack with the others. Another

Table III
C2′ *Endo* Conformations in Yeast tRNA[Phe]

Residue number	Description
Intercalation:	
9	8-9 intercalation by C13
19	18-19 intercalation by G57
46	45-46 intercalation by A9
58	57-58 intercalation by G18
Change of Direction:	
7	Extended segment at bend in chain, juncture between acceptor stem and 8-9 connection
17	Looping out of backbone at residues 16 and 17
18	Looping out of backbone
48	Looping out of variable loop backbone at U47; juncture between variable loop and T stem
60	Extended segment at bend in chain, bases 59 and 60 excluded from T stem and loop stacking

interesting feature associated with this is the fact that the bases of G18 and G19 are not only separated from each other by a distance of 6.8 Angstroms, which would be necessary for the intercalation of G57, but they are also translated relative to each other so the bases are no longer strictly on top of each other. This translation of G18 relative to G19 in the plane of the bases, can only be accommodated if both residues are in the C2′ *endo* conformation. It is likely that this feature is also related to the fact that the bases of residues 16 and 17 are excluded from stacking with the remainder of the bases in the tRNA structure. Some tRNA molecules contain only one nucleotide in this region which is excluded from the stacking, instead of the two seen in yeast tRNA[Phe]. It is possible that in these cases, with only one residue, that G18 may have the normal C3′ *endo* conformation since the translation of G18 relative to

G19 in the plane of bases may not be required.

Discussion

In the above, we have discussed the conformational flexibility found in the polynucleotide chains. This flexibility is inherent in the fact that there is a furanose ring in the chain which can adopt two different conformations which are associated with a varied extensibility of the chain. Structural studies on simple oligonucleotides reveal that conformational changes occur in regular double helical structures associated with intercalation as well as in other structures where polynucleotide chain undergo an abrupt change in direction. In the three-dimensional structure of yeast tRNA[Phe], examples are shown of both types of conformational changes. Changes in pucker due to intercalation in general follow trends which are seen in intercalation in simple double helical RNA structures even though the tRNA examples do not involve simple RNA double helices. This implies that the adoption of a C2′ *endo* conformation together with an associated extension of the polynucleotide chain is not solely limited to double helical intercalation, but may in fact be of a more general nature as shown by the examples cited above.

Nucleic acid molecules have considerable conformational flexibility. We have only a hint of this flexibility in studying the structure of the double helix itself. However, when one begins to study the interaction of double helical structures with other molecules, especially intercalators, we see that there is a method for accommodating them involving changes in extensibility. In complex globular polynucleotides such as the transfer RNA molecule, a variety of changes in pucker are observed. The molecule adopts unusual conformations associated either with chain extension, changes in direction of the polynucleotide chains or with the accommodation of other bases which intercalate into the chain even though the chains are not involved in a double helical array. As the structure of more complex polynucleotide structures are solved, it is our expectation that this will be found to be a completely general feature. This element of conformational flexibility associated with changes in ring pucker is likely to be a constant feature of polynucleotide chains when they interact with other molecules including proteins, as well as when they interact with other polynucleotide chains.

Acknowledgements

This research was supported by grants from The National Institutes of Health, The National Science Foundation, The National Aeronautics and Space Administration and the American Cancer Society. We thank J. Simpson for help in preparing the manuscript. A. H. J. W. is supported in part by the M. I. T. Center for Cancer Research (Grant No. CA-14051).

References and Footnotes

1. Arnott, S. and Hukins, D, W. L., *Biochem. Biophys. Res. Comm. 47*, 1504-1509 (1972).

2. Sundaralingam, M. in *Structure and Conformation of Nucleic Acids and Protein-Nucleic Acid Interactions*, Ed by M. Sundaralingam and S. T. Rao, University Park Press, Baltimore, Maryland, pp. 487-524 (1974).

3. Yathindra, N. and Sundaralingam, M. in *Structure and Conformation of Nucleic Acids and Protein-Nucleic Acid Interactions*, Ed by M. Sundaralingam and S. T. Rao, University Park Place, Baltimore, Maryland, pp. 649-676 (1974).

4. Seeman, N. C., Day, R. O., and Rich, A., *Nature 253*, 324-326 (1975).

5. Neidle, S., Tayler, G., Sanderson, M., Shieh, H.-S. and Berman, H. M., *Nucleic Acids Res.*, *5*, 4417-4422 (1978).

6. Suck, D., Manor, P. C., and Saenger, W., *Acta. Crysta. B32*, 1727-1737 (1976).

7. Wang, A. H.-J., Nathans, J., van der Marel, G., van Boom, J. H. and Rich, A., *Nature 276*, 471-474 (1978).

8. Wang, A. H.-J., Quigley, G. J. and Rich, A. *Nucleic Acids Res.* (in press).

9. Tsai, C. C., Jain, S. C. and Sobell, H. M., *J. Mol. Biol. 114*, 301-315 (1977).

10. Jain, S. C., Tsai, C. C. and Sobell, H. M., *J. Mol. Biol. 114*, 317-331 (1977).

11. Sakore, T. D., Jain, S. C., Tsai, C. C., and Sobell, H. M., *Proc. Natl. Acad. Sci. USA*, *74*, 188-192 (1977).

12. Sobell, H. M., in *The International Symposium on Biomoleculr Structure, Conformation, Function and Evolution.* (Ed. R. Srinivasan), Pergamon Press, Oxford, New York (in press).

13. Neidle, S., *et al.*, *Nature 269*, 304-307 (1977).

14. Quigley, G. J. and Rich, A., *Science 194*, 796-806 (1976).

15. Camerman, N., Fawcett, J. K. and Camerman, A., *Science 182*, 1142-1143 (1973).

16. Viswamitra, M. W., Kennard, O., Jones, P. G., Sheldrick, G. M., Salisbury, S., Falvello, L. and Shakked, Z., *Nature 273*, 687-688 (1978).

17. Seeman, N. C., Rosenberg, J. M., Suddath, F. L., Kim, J. J. P., and Rich, A., *J. Mol. Biol. 104*, 109-144 (1976).

18. Rosenberg, J. M., Seeman, N. C., Day, R. O., and Rich, A., *J. Mol. Biol. 104*, 145-167 (1976).

19. Hingerty, B., Subramanian, A., Stellman, S. D., Sato, T., Broyde, S. B., and Langridge, R., *Acta Crysta. B32*, 2998-3013 (1976).

20. Sussman, J. L., Seeman, N. C., Kim, S. H. and Berman, H. M., *J. Mol. Biol. 66*, 403-421 (1972).

21. Rubin, J., Brennan, T. and Sundaralingam, M., *Biochem. 11*, 3112-3129 (1972).

Structural and Dynamic Aspects of
Drug Intercalation into DNA and RNA

Henry M. Sobell
Department of Chemistry
The University of Rochester
River Campus Station
Rochester, New York 14627

and

Department of Radiation Biology
and Biophysics
The University of Rochester School
of Medicine and Dentistry
Rochester, New York 14642

Introduction

It is now over a decade since Lerman proposed his intercalation hypothesis to explain the strong binding mode of the aminoacridines to DNA.[1,2] A large body of evidence now supports this concept, and it has become increasingly apparent that—in addition to the aminoacridines—other drugs and dyes bind to DNA by intercalation. The technique of molecular cocrystallization has provided an opportunity to study the detailed structural interactions between many of these drugs with nucleic acid components. These studies have led us to propose unifying structural concepts to understand drug-DNA (and -RNA) interactions.[3-17]

This chapter begins by summarizing the results of our crystallographic studies of drug-nucleic acid crystalline complexes. It then describes detailed models of drug-DNA (and -RNA) interactions suggested by these data. Finally, it presents dynamic concepts of DNA structure that allow one to understand kinetic aspects of drug intercalation—it is possible that these concepts have broader implications to understand protein-DNA interactions. These are discussed in this chapter.

Drug-Dinucleoside Monophosphate Crystalline Complexes

Table I summarizes the pertinent crystallographic data for seven drug-dinucleoside monophosphates crystalline complexes solved in our laboratory. Figure 1 shows chemical structures of the drugs and dyes used in these studies.

Each structure has several features in common. All are 2:2 (or, in one case, 4:4), and involve two drug molecules interacting with two dinucleoside monophosphates. This reflects intercalation by one drug molecule and stacking by the other drug molecule with Watson-Crick base-pairs formed by the dinucleoside monophosphate miniature double helices. In addition, internal salt linkages connect negatively charged phosphate groups with positively charged drug molecules. This plus-minus nature of these complexes provides an important additional electrostatic component to the forces that stabilize these structures and explains the absence of counter balancing anions in the crystal lattice.

Each complex we have analyzed contains an RNA-like (rather than a DNA-like) dinucleoside monophosphate. In addition, we have synthesized 5-iodo pyrimidine containing derivatives. These heavy atom substituted ribodinucleoside monophosphates form excellent crystals with a variety of drug molecules (notice, for example, that six out of seven of these complexes involve the compound, 5-iodocytidylyl(3′-5′) guanosine). Although the presence of a heavy atom limits the accuracy of these analyses, it does not interfere with observing the important features of these structures and, from this, extracting pertinent information concerning the stereochemistry of drug intercalation.

Two different intercalative geometries have been observed in these studies (see

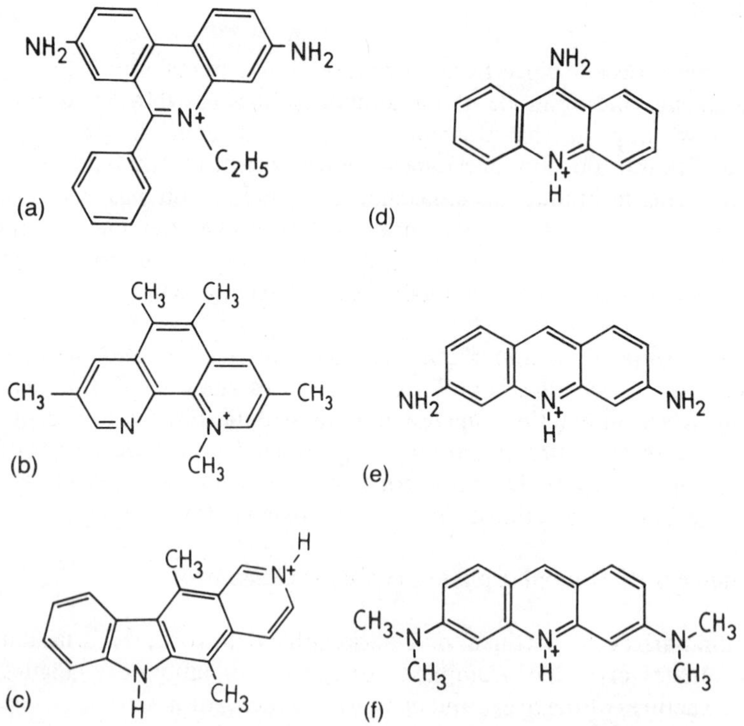

Figure 1. Chemical structures of drugs and dyes that form crystalline complexes with ribonucleoside monophosphates that have been studied by X-ray crystallography. (a) ethidium, (b) 3,5,6,8-tetramethyl-N-methyl phenanthrolinium (c) ellipticine (d) 9-aminoacridine (e) proflavine and (f) acridine orange.

Table I
Unit Cell Constants, Space Groups and Structural Data
for Drug-Nucleic Acid Crystalline Complexes

Complex	St*	Cell Constants	Space Group	Sugar Puckering	Twist Angle	References
ethidium: 5-iodo-uridylyl(3'-5')adenosine	2:2	a = 28.45Å b = 13.54Å c = 34.13Å β = 98.6°	C2	C3' *endo* (3'-5') C2' *endo*	8 ± 1°	(7)
ethidium: 5-iodo cytidylyl(3'-5')guanosine	2:2	a = 14.06Å b = 32.34Å c = 16.53Å β = 117.8°	P2₁	C3' *endo* (3'-5') C2' *endo*	8 ± 1°	(8)
3,5,6,8-tetramethyl-N-methyl phenanthro-linium: 5-iodocytidyl-(3'-5')guanosine	2:2	a = 13.99Å b = 19.12Å c = 21.31Å β = 104.9	P2₁	C3' *endo* (3'-5') C2' *endo*	11 ± 1°	(12)
ellipticine: 5-iodocytidylyl-(3'-5')guanosine	2:2	a = 13.88Å b = 19.11Å c = 21.42Å β = 105.4°	P2₁	C3' *endo* (3'-5') C2' *endo*	11 ± 1°	(12)
acridine orange: 5-iodocytidylyl-(3'-5')guanosine	2:2	a = 14.33Å b = 19.68Å c = 20.67Å β = 102.1°	P2₁	C3' *endo* (3'-5') C2' *endo*	10 ± 1°	(11)
9-aminoacridine: 5-iodocytidylyl-(3'-5')guanosine	4:4	a = 13.98Å b = 30.58Å c = 22.47Å β = 113.9°	P2₁	C3' *endo* (3'-5') C2' *endo* (A) C3' *endo* (3'-5') C2' *endo* (S)	8 ± 2° 10 ± 2°	(10)
proflavine: 5-iodocytidylyl-(3'-5')guanosine	2:2	a = 32.11Å b = 22.23Å c = 18.45Å β = 123.2°	C2	C3' *endo* (3'-5') C3' *endo*	36 ± 2°	(11)

*Stoichiometry

Figures 2-9 and Table II).

The first—observed in seven out of eight structures we have analyzed—demonstrates the following pattern of sugar puckering: C3' *endo* (3'-5') C2' *endo*. Both C4'-C5' bonds are *gauche-gauche* and glycosidic torsional angles (denoted χ) are in the low *anti* range for iodocytidine residues and in the high *anti* range for guanosine residues. The twist angle relating base-pairs above and below the intercalative drug molecule varies from 8 to 12°—this stereochemistry

Table II
The Torsion Angles For The Various
Drug-Dinucleoside Monophosphate Complexes

Torsional Angle*	Greek symbol	ethidium-iodoUpA		ethidium-iodoCpG		TMP-iodoCpG		ellipticine-iodoCpG	
		I-UpA(1)	I-UpA(2)	I-CpG(1)	I-CpG(2)	I-CpG(1)	I-CpG(2)	I-CpG(1)	I-CpG(2)
O4'C-C1'C-N1C-C6C	χ	26°	14°	29°	24°	22°	13°	25°	14°
O4'G-C1'G-N9G-C8G	χ	99°	100°	101°	109°	86°	69°	80°	74°
O5'C-C5'C-C4'C-C3'C	ϵ	56°	171°	51°	90°	82°	35°	36°	34°
C5'C-C4'C-C3'C-O3'C	ζ	98°	95°	87°	84°	93°	95°	91°	104°
C4'C-C3'C-O3'C-P	α	207°	218°	226°	225°	203°	212°	199°	234°
C3'C-O3'C-P-O5'G	β	286°	302°	281°	291°	278°	285°	285°	258°
O3'C-P-O5'G-C5'G	γ	291°	276°	286°	291°	283°	303°	281°	315°
P-O5'G-C5'G-C4'G	δ	236°	230°	210°	224°	219°	206°	212°	194°
O5'G-C5'G-C4'G-C3'G	ϵ	52°	70°	72°	55°	71°	55°	65°	47°
C5'G-C4'G-C3'G-O3'G	ζ	133°	118°	131°	134°	141°	143°	145°	141°
C4'C-O4'C-C1'C-C2'C	τ_0	-5°	16°	3°	4°	4°	3°	7°	14°
O4'C-C1'C-C2'C-C3'C	τ_1	-26°	-28°	-23°	-24°	-27°	-27°	-30°	-36°
C1'C-C2'C-C3'C-C4'C	τ_2	48°	29°	32°	33°	25°	26°	31°	30°
C2'C-C3'C-C4'C-O4'C	τ_3	-52°	-21°	-31°	-31°	-36°	-36°	-41°	-37°
C3'C-C4'C-O4'C-C1'C	τ_4	33°	3°	18°	17°	14°	16°	13°	10°
C4'G-O4'G-C1'G-C2'G	τ_0	-32°	-25°	-31°	-22°	-24°	-31°	-26°	-23°
O4'G-C1'G-C2'G-C3'G	τ_1	40°	37°	41°	33°	31°	33°	31°	29°
C1'G-C2'G-C3'G-C4'G	τ_2	-32°	-33°	-34°	-30°	-42°	-45°	-39°	-42°
C2'G-C3'G-C4'G-O4'G	τ_3	15°	20°	19°	19°	21°	21°	18°	21°
C3'G-C4'G-O1'G-C1'G	τ_4	9°	3°	6°	1°	4°	1°	3°	0°

Table II
The Torsion Angles For The Various
Drug-Dinucleoside Monophosphate Complexes

Torsional Angle*	Greek symbol	acridine orange-iodoCpG		9-aminoacridine-iodoCpG				proflavine-iodoCpG	
		I-CpG(1)	I-CpG(2)	I-CpG(1)	I-CpG(2)	I-CpG(3)	I-CpG(4)	I-CpG(1)	I-CpG(2)
O4'C-C1'C-N1C-C6C	χ	19°	15°	17°	38°	17°	20°	4°	19°
O4'G-C1'G-N9G-C8G	χ	95°	90°	109°	100°	83°	102°	103°	85°
O5'C-C5'C-C4'C-C3'C	ϵ	52°	69°	61°	71°	99°	42°	106°	82°
C5'C-C4'C-C3'C-O3'C	ζ	84°	81°	72°	85°	99°	105°	102°	79°
C4'C-C3'C-O3'C-P	α	217°	226°	236°	216°	216°	209°	224°	273°
C3'C-O3'C-P-O5'G	β	294°	284°	311°	296°	310°	287°	294°	273°
O3'C-P-O5'G-C5'G	γ	291°	299°	280°	300°	295°	294°	305°	323°
P-O5'G-C5'G-C4'G	δ	235°	228°	220°	229°	208°	222°	273°	206°
O5'G-C5'G-C4'G-C3'G	ϵ	58°	60°	67°	38°	58°	45°	3°	53°
C5'G-C4'G-C3'G-O3'G	ζ	145°	139°	110°	130°	106°	156°	122°	119°
C4'C-O4'C-C1'C-C2'C	τ_0	-3°	-4°	4°	2°	10°	-4°	6°	-4°
O4'C-C1'C-C2'C-C3'C	τ_1	-26°	-8°	-17°	-25°	-31°	-26°	-23°	-17°
C1'C-C2'C-C3'C-C4'C	τ_2	43°	16°	23°	37°	39°	44°	32°	33°
C2'C-C3'C-C4'C-O4'C	τ_3	-43°	-19°	-21°	-37°	-33°	-50°	-31°	-36°
C3'C-C4'C-O4'C-C1'C	τ_4	28°	14°	11°	21°	13°	32°	16°	24°
C4'G-O4'G-C1'G-C2'G	τ_0	-26°	-25°	-34°	-26°	-30°	-24°	3°	-1°
O4'G-C1'G-C2'G-C3'G	τ_1	28°	40°	42°	22°	46°	30°	-17°	-8°
C1'G-C2'G-C3'G-C4'G	τ_2	-20°	-42°	-33°	-12°	-42°	-21°	23°	13°
C2'G-C3'G-C4'G-O4'G	τ_3	5°	26°	13°	-1°	23°	7°	-22°	-15°
C3'G-C4'G-O1'G-C1'G	τ_4	14°	-1°	11°	14°	5°	11°	13°	10°

*The torsional angle is defined in terms of 4 consecutive atoms, ABCD; the positive sense of rotation is clockwise from A to D while looking down the BC bond.

would, therefore, give rise to significant double helix unwinding at the immediate site of drug intercalation.

The second—observed only in the proflavine-iodoCpG structure—contains all sugar

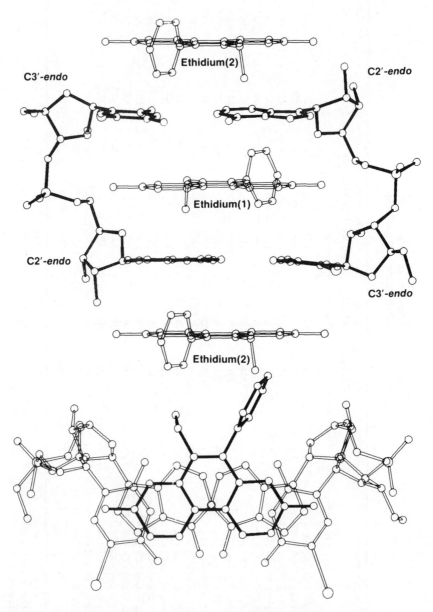

Figure 2. Structural information from the ethidium: iodouridylyl(3'-5')adenosine crystalline complex. Each sugar-phosphate chain has a mixed sugar puckering of the type: C3' *endo* (3'-5') C2' *endo*. The twist angle between adjacent base-pairs is $8 \pm 1°$. See text for additional discussion.

residues in the C3′ *endo* conformation (i.e., C3′ *endo* (3′-5′) C3′ *endo*). Again, the conformation around the C4′-C5′ bond is *gauche-gauche* and glycosidic torsional angles fall in the low *anti* range for iodocytidine residues and in the high *anti* range for guanosine residues. However, unlike the first intercalative stereochemistry, the

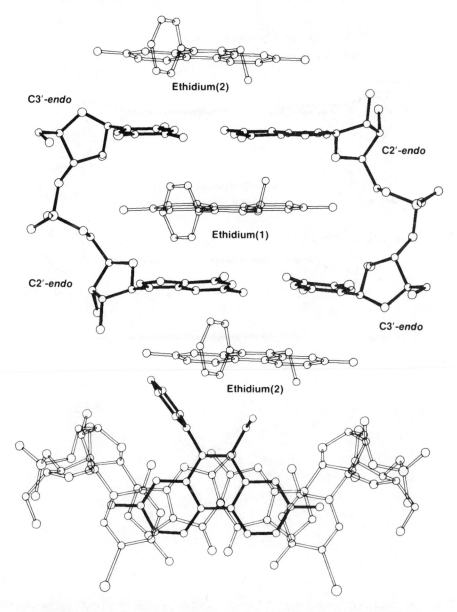

Figure 3. Structural information from the ethidium: iodocytidylyl(3′-5′)guanosine crystalline complex. Each sugar-phosphate chain has a mixed sugar puckering of the type: C3′ *endo* (3′-5′) C2′ *endo*. The twist angle between adjacent base pairs is $8 \pm 1°$. See text for additional discussion.

twist angle relating base-pairs above and below the intercalative proflavine molecule is about 36°—this stereochemistry, therefore, would *not* lead to significant unwinding in double helical DNA (or RNA) at the immediate drug intercalation site.

Neidle *et al.*[18] have described a 3:2 proflavine-CpG crystalline complex that has very

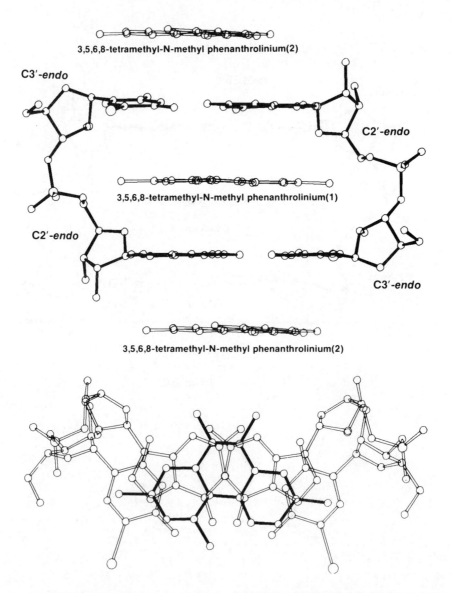

Figure 4. Structural information from the 3,5,6,8-tetramethyl-N-methyl phenanthrolinium: iodocytidylyl(3′-5′)guanosine crystalline complex. Each sugar-phosphate chain has a mixed sugar puckering of the type: C3′ *endo* (3′-5′) C2′ *endo*. The twist angle between adjacent base-pairs is 11 ± 1°. See text for additional discussion.

similar structural information. Some of the details of this structure are reported by Berman and Neidle later in this volume.[66] Their study provides independent evidence that proflavine has its own characteristic intercalative geometry in these dinucleoside monophosphate model systems.

Why does proflavine demonstrate this second type of intercalative geometry in these

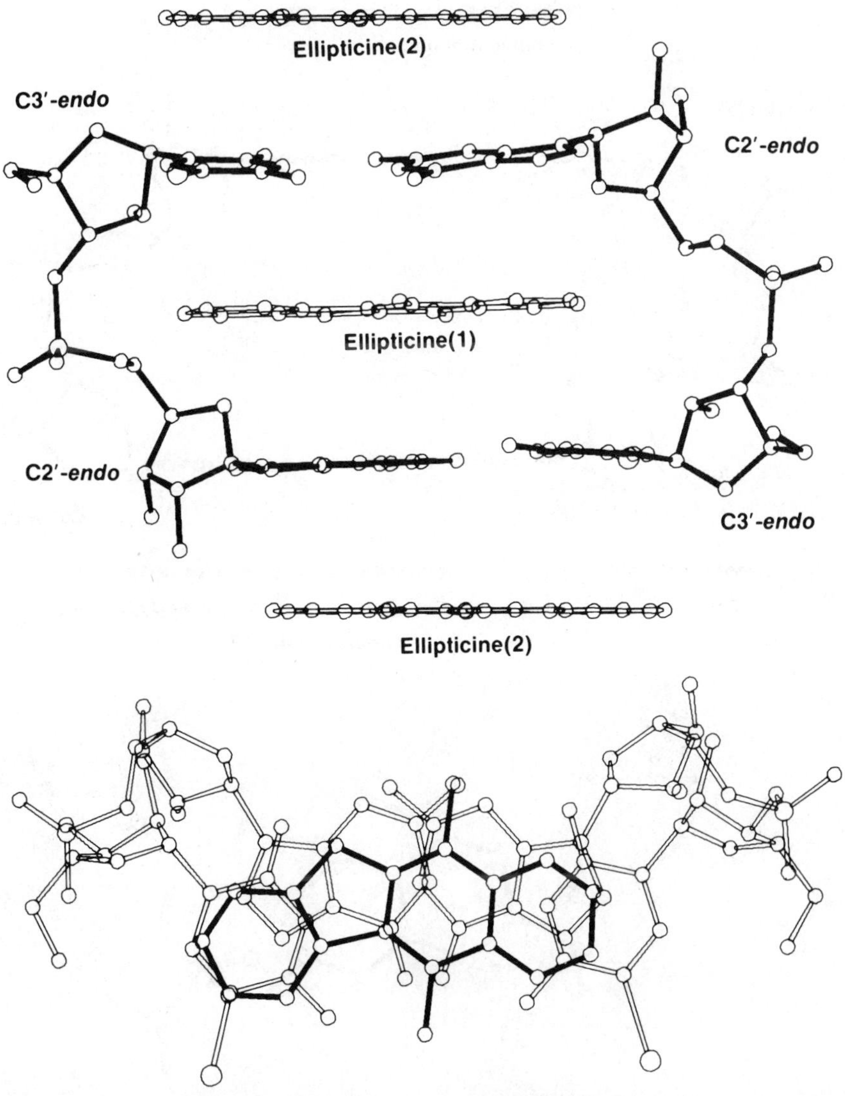

Figure 5. Structural information from the ellipticine: iodocytidylyl(3′-5′)guanosine crystalline complex. Each sugar-phosphate chain has a mixed sugar puckering of the type: C3′ endo (3′-5′) C2′ endo. The twist angle between adjacent base pairs is 11 ± 1°. See text for additional discussion.

model studies? The answer to this is provided by comparing the acridine orange- and proflavine- iodoCpG structures in detail. Although both demonstrate intercalative binding, they differ in the following way:

1) Proflavine intercalates symmetrically, forming hydrogen bonds between its amino- groups and phosphate oxygens on the dinucleotide, whereas acridine orange intercalates asymmetrically and—being a methylated proflavine derivative—is

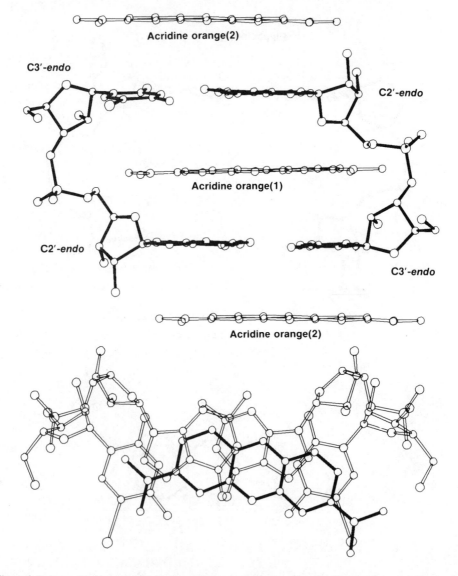

Figure 6. Structural information from the acridine orange: iodocytidylyl(3'-5')guanosine crystalline complex. Each sugar-phosphate chain has a mixed sugar puckering of the type: C3' *endo* (3'-5') C2' *endo*. The twist angle between adjacent base-pairs is 10 ± 1°. See text for additional discussion.

unable to hydrogen bond to these phosphate oxygens.

2) Base-pairs above and below the intercalated proflavine molecule are twisted about 36°, whereas, with acridine orange this value is 10°. The geometric difference between these structures mainly reflects the sugar puckering patterns that are observed—these are C3′ *endo* (3′-5′) C3′ *endo* in the proflavine-iodoCpG complex and C3′ *endo* (3′-5′) C2′ *endo* in the acridine orange-iodoCpG structure.

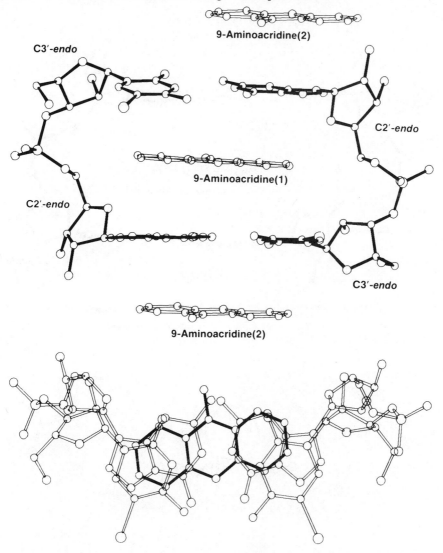

Figure 7. Structural information from the 9-aminoacridine: iodocytidylyl(3′-5′)guanosine crystalline complex, symmetric intercalative unit. Each sugar-phosphate chain has a mixed sugar puckering of the type: C3′ *endo* (3′-5′) C2′ *endo*. The twist angle between adjacent base-pairs is 10± 2°. See text for additional information.

3) Associated with these two different sugar-phosphate intercalative geometries are different distances that connect phosphate oxygen atoms on opposing chains. In the proflavine complex, this distance is 15.3Å—a distance that allows proflavine to simultaneously hydrogen bond (or span) both phosphate oxygens. In the acridine orange complex, however, this distance is 17.3Å—a distance too large for this to happen.

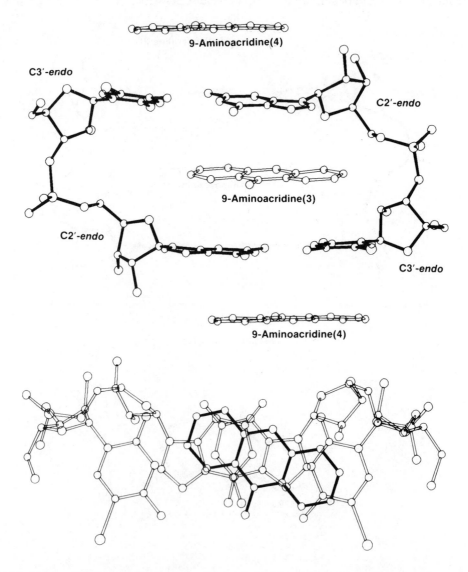

Figure 8. Structural information from the 9-aminoacridine: iodocytidylyl(3'-5')guanosine crystalline complex, asymmetric intercalative unit. Each sugar-phosphate chain has a mixed sugar puckering of the type: C3' endo (3'-5') C2' endo. The twist angle between adjacent base-pairs is 8 ± 2°. See text for additional discussion.

Hydrogen bonding, therefore, plays a key role in determining the proflavine inter-calative geometry that is observed in these model studies.

There are, however, several reasons to suspect that these proflavine studies may not be completely relevant to understanding the detailed nature of proflavine-DNA (or -RNA) binding. In the first place, proflavine has been shown to unwind DNA in much the same manner as acridine orange, unwinding DNA about -20° on the

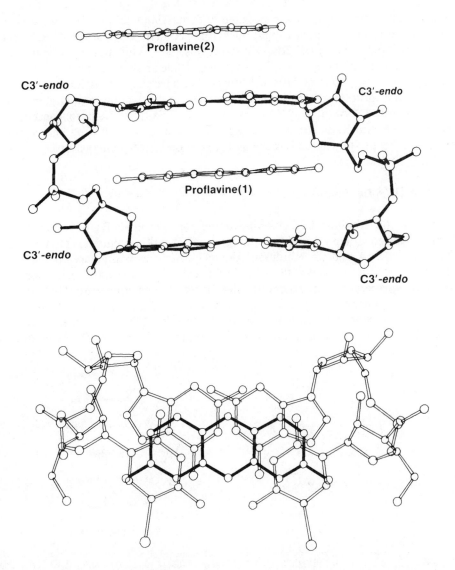

Figure 9. Structural information from the proflavine: iodocytidylyl(3'-5')guanosine crystalline complex. Each sugar-phosphate chain has a sugar puckering of the type: C3' *endo* (3'-5') C3' *endo*. The twist angle between adjacent base-pairs is 36 ± 2°. See text for discussion.

ethidium scale.[19/20] Secondly, proflavine (as well as all other intercalative drugs and dyes) demonstrates an upper binding limit of one drug molecule bound for every four nucleotides, an observation most easily understood in terms of a neighbor exclusion model in which proflavine molecules intercalate between every other basepair in the double helix. Structural evidence for neighbor exclusion has been provided by Bond *et al.*[21] in a fiber X-ray diffraction study of a terpyridine platinum compound complexed to DNA.

Although alternative models are possible,[66/67] the simplest model to understand these data postulates the mixed sugar puckering geometry C3' *endo* (3'-5') C2' *endo* to occur in drug-DNA (and -RNA) interactions.[9/13-15] This model—or, more accurately, this class of models—explains the magnitude of angular unwinding and the phenomenon of neighbor-exclusion, features characteristic of drug intercalation into DNA, and, more generally, provides a unified understanding of a large number of related drug-DNA (and -RNA) associations. Finally, it leads to concepts of dynamic DNA and RNA structure to explain kinetic features of drug intercalation and relates these to understanding the nature of protein-DNA interactions.[9/13-17] We now describe these in detail.

Ethidium-DNA Binding

To construct the ethidium-DNA binding model, we have added B DNA to either side of an idealized (deoxyribose-containing) configuration common to both the ethidium-iodoUpA and the ethidium-iodoCpG structures (see Figure 10 and Plate 17). This is done easily and without steric difficulty. An important realization that immediately emerges is the concept that drug intercalation into DNA is accompanied by a helical screw axis displacement (or dislocation) in its structure (for ethidium intercalation, we estimate that helical axes for B DNA on either side of the phenanthridinium ring system are displaced approximately +1.0Å). Base-pairs in

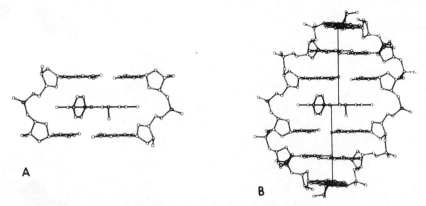

A

B

Figure 10. Steps in assembly of the ethidium-DNA binding model. (A) Idealized (deoxyribose-containing) configuration common to both the ethidium-iodoUpA and ethidium-iodoCpG crystalline complexes. (B) B DNA added above and below the intercalated deoxyribodinucleoside monophosphate to demonstrate the helical screw axis dislocation in B DNA accompanying ethidium intercalation.

the immediate region are twisted by 10° (this value has been estimated by projecting the interglycosidic carbon vectors on a plane passing midway between base-pairs and then measuring the angle between them). This gives rise to an angular unwinding of -26° at the immediate site of drug intercalation. We have also observed that inter-calated base-pairs are tilted relative to one another by about 8° in both ethidium crystal structures. This results in a small residual kink of 8° at the intercalation site, and it has been included in our ethidium-DNA binding model.

The magnitude of angular unwinding predicted by our ethidium-DNA binding model is in good agreement with Wang's estimate of ethidium-DNA angular un-winding based on alkaline titration studies of superhelical DNA in cesium chloride density gradients.[20] Moreover, the C3' *endo* (3'-5') C2' *endo* mixed sugar puckering necessarily predicts that intercalation be limited to every *other* base-pair at maximal drug-nucleic acid binding ratios (i.e., a neighbor exclusion model). We have examin-ed the stereochemistry of this model carefully. The effect of having a helical screw axis displacement every other base-pair combined with an 8° kink is to give rise to a maximally unwound and elongated DNA structure that possesses a slow right hand-ed superhelical writhe. This structure is shown in the bottom half of Plate 20.

Although ethidium and acridine orange differ somewhat in their intercalative geometries with iodoCpG (ethidium intercalates *symmetrically* from the *minor* groove, while acridine orange intercalates asymmetrically from the *major* groove), the detailed stereochemistries of the sugar-phosphate chains in both structures are remarkably similar. Our models to understand acridine orange- and proflavine-DNA binding use this information: we postulate proflavine and acridine orange to bind asymmetrically to the intercalation site—characterized by a C3' *endo* (3'-5') C2' *endo* mixed sugar puckering pattern—proflavine, however, forming only *one* hydrogen bond to the neighboring phosphate oxygen atom. We have already pointed out the possible relationship between asymmetric binding modes of these drugs and dyes to the mechanism of frameshift mutagenesis.[10]

Actinomycin-DNA Binding

Actinomycin (Figure 11) forms a 1:2 stoichiometric complex with deoxyguanosine, and the stereochemical information provided by this structure has led us to propose a model for actinomycin-DNA binding.[3-6] The model involves intercalation of the phenoxazone ring system between base-pairs in the DNA double helix and the utilization of specific hydrogen bonds, van der Waals forces, the hydrophobic in-teractions between the pentapeptide chains on actinomycin and chemical groups in the minor groove of the DNA helix (see Plate 18). Important elements in the recogni-tion of actinomycin for DNA are the guanine specificity and the use of symmetry in the interaction. These predict a base-sequence binding preference of the type d-GpC in actinomycin-DNA binding, which has been verified in model dinucleoside monophosphate solution binding studies and in synthetic polymer studies.[22/23]

Although we believe that the general features of our actinomycin-DNA binding model are correct, we have presented a slightly modified version of the model that utilizes many of the insights afforded by the drug-dinucleoside monophosphate crystal studies just described. Major features of our revised actinomycin-DNA model are as follows:

1) Intercalation of the phenoxazone ring system between base-paired d-GpC sequences accompanied by a helical screw axis dislocation of about -0.4Å (this value was about -1.5Å in our previous model because we attempted to use the precise configuration observed in the actinomycin-deoxyguanosine crystalline complex; subsequent study has shown, however, that the polymer backbone imposes stereochemical constraints which limit this dislocation to about 0.4Å). The pattern of sugar puckering at the immediate intercalation site is: C3′ *endo* (3′-5′) C2′ *endo*, as previously postulated.

2) A departure from exact twofold symmetry in the complex. This reflects the asymmetry of the phenoxazone ring system and the inability of both pentapeptide chains to interact simultaneously with (or span) guanine residues on opposite chains due to

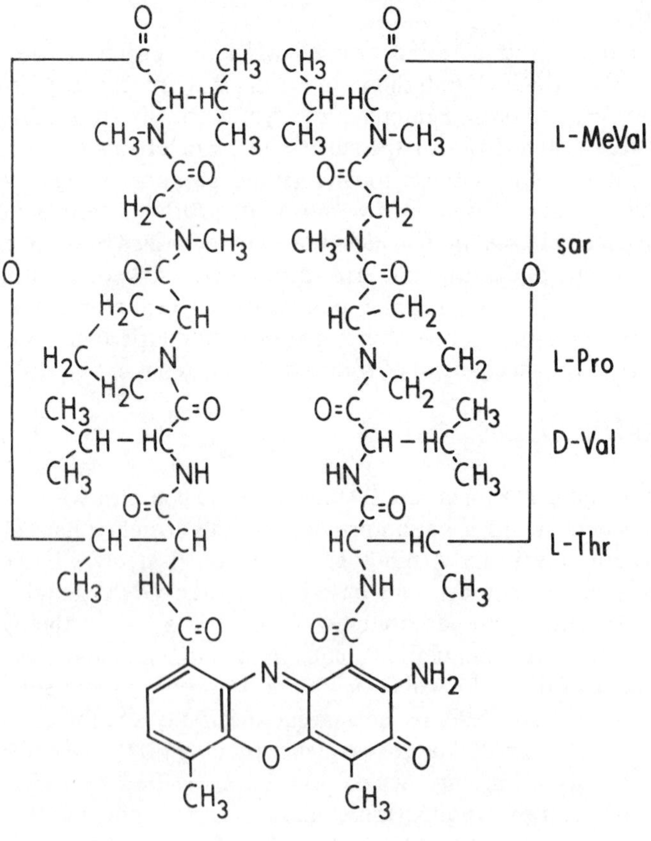

Figure 11. Chemical structure of actinomycin D.

the smaller magnitude of this helical screw axis dislocation.

3) An additional bending distortion in DNA on both sides of the intercalation site to accommodate the steric bulk of the pentapeptide chains on actinomycin. This gives rise to a kinked-type DNA structure (helical axes for B DNA sections on either side of this bend form an angle of about 40° and are displaced by about -1.6Å) which could partly explain the anomalous viscosity effects that accompany actinomycin-DNA binding.[24/25]

The ability of actinomycin to interfere specifically with DNA dependent RNA polymerase activity almost certainly reflects intercalation by the phenoxazone ring system into DNA and the presence of the two pentapeptide side chains in the minor groove of the DNA helix. Equally important are the slow on and off rate constants observed for actinomycin-DNA binding.[25/26] A more precise understanding of the pharmacological activity for actinomycin, however, must await a detailed understanding of the three-dimensional structure of the polymerase enzyme and the clarification of its processive catalytic mechanism. (See Krugh *et al*[68] and Patel[69] for further discussion of the interaction of actinomycin with DNA).

Irehdiamine-DNA Binding

Irehdiamine A (Figure 12) is a steroidal diamine that binds tightly to DNA.[27] Although the precise interaction is unknown, many aspects of the irehdiamine-DNA binding reaction suggest an intercalative-type binding mode. Thus, for example, irehdiamine A unwinds superhelical DNA as evidenced by its ability to produce a fall and rise in the sedimentation coefficient of covalently circular supercoiled DNA molecules.[19/28/29] These studies have also indicated that, on a molar basis, irehdiamine unwinds DNA roughly half as much as ethidium. Moreover, at maximal drug-nucleic acid binding ratios, the molecule demonstrates an upper binding limit of one irehdiamine to every four nucleotides.[27] Finally, although irehdiamine produces hydrodynamic effects in DNA solutions different from most intercalative drugs or dyes (i.e., there are no clear viscosity or sedimentation changes associated

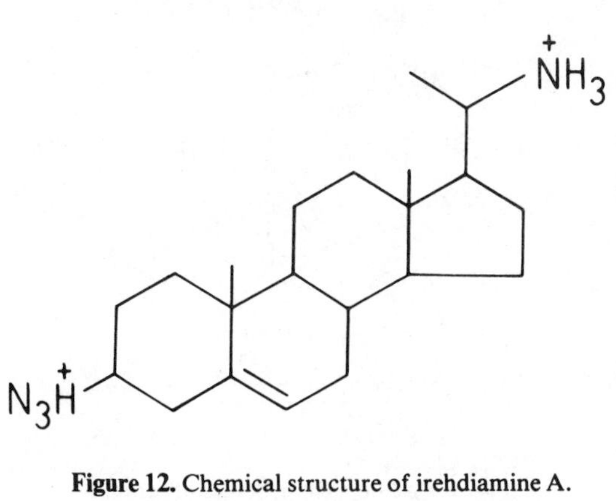

Figure 12. Chemical structure of irehdiamine A.

with irehdiamine-DNA binding), it is possible that irehdiamine mimics intercalative drug binding by binding to the kink in DNA,[9/13] a concept also known as partial intercalation.[30/31]

This is shown in Plate 19. The van der Waals surface of the irehdiamine molecule resembles a triangular wedge whose overall shape allows it to fit into the kink in DNA (as discussed later, the kink possesses the C3' *endo* (3'-5') C2' *endo* sugar puckering pattern and has glycosidic torsional angles that differ somewhat from those in the intercalative geometry—as a result, base-pairs above and below irehdiamine are only partially unstacked). Forces stabilizing the complex include electrostatic, van der Waals, and hydrophobic forces. Our model predicts an effective unwinding of -10° associated with irehdiamine-DNA binding. It also predicts a neighbor-exclusion structure, at saturating concentrations of irehdiamine, whose axial repeat length is similar to that of B DNA. This structure (denoted β kinked DNA) is shown in the top half of Plate 20.

Bis Intercalation—A Probe for β Kinked Structural Distortions in DNA

An important class of intercalative drugs and dyes that has received considerable attention in recent years are the *bis* intercalators—bifunctional intercalating agents,[32-36] that have two chromophore groups separated by about 10.2Å. These molecules intercalate into DNA in a neighbor exclusion mode (i.e., the intercalative chromophore groups are separated by two base-pairs when binding to DNA). One example is echinomycin, a polypeptide containing antibiotic that binds to DNA and inhibits RNA synthesis. This molecule has two quinoxaline ring systems connected to an octapeptide through amide linkages (Figure 13), and probably possesses a pseudo-dyad axis of symmetry that plays an important role in its structural interac-

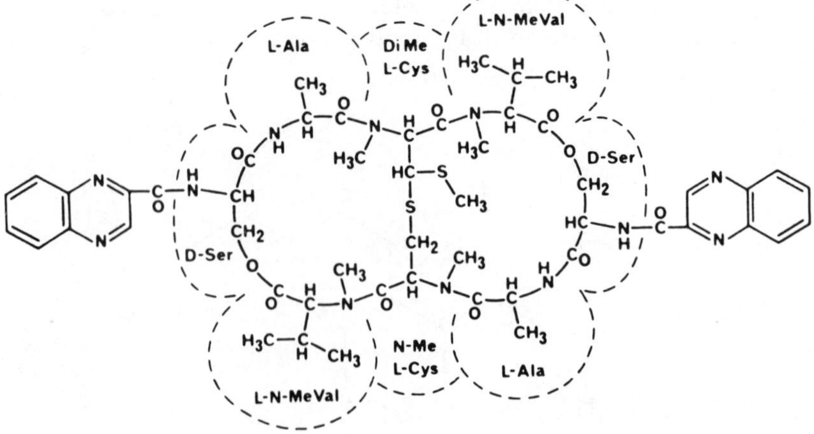

Figure 13. Chemical structure of echinomycin.

tions with DNA. The presence of these two intercalative chromophore groups held 10.2Å apart by the rigid octapeptide chains suggests that echinomycin recognizes a naturally occurring structural intermediate in RNA synthesis. We have postulated this intermediate to be β kinked DNA, a key intermediate in DNA unwinding.[14-16] It is of interest that echinomycin is some 4-5 times more potent than actinomycin in its ability to inhibit RNA synthesis, a feature that makes echinomycin an extremely potent antitumor agent. Unfortunately, however, its enhanced cytotoxicity limits its clinical usefulness.

Drugs containing two intercalative chromophore groups have been synthesized in recent years.[35/36] These include *bis* acridine and *bis* ethidium (as well as mixed ethidium-acridine) polyintercalating agents. The molecules bind DNA very tightly (i.e., binding constants approaching 10^9 liters/mole have been observed for the *bis* acridine intercalators). Although there are no data as yet concerning the kinetic rate constants for these associations, the postulate that elements of β kinked DNA are often common structural intermediates in mono- and bifunctional drug intercalation—these arise due to the fluctuations that exist at equilibrium (see below)—predicts the on rate constants for these processes to be very similar.

Dynamic Aspects of DNA (and RNA) Structure

What is the dynamic nature of DNA (and double-helical RNA) structure that gives rise to drug intercalation? How does this same dynamic structure give rise to DNA breathing—the transient disruption of hydrogen bonding between base-pairs at temperatures well below the melting temperature (as evidenced, for example, by tritium exchange,[37/38] formaldehyde reactivity,[39/40] and polarographic studies[41/42]—and how are these premelting changes related to melting at elevated temperatures?

More generally, how is the subject of DNA premelting related to understanding protein-DNA interactions? For example, what is the nature of the promoter and how does the RNA polymerase recognize and bind to the promoter? What about the single-strand specific DNA binding proteins—do their interactions with DNA involve intercalation and, if so, why? What do nucleases "see" when they cleave DNA and, related to this, why—in eukaryotic chromatin—are active genes some five times more nuclease sensitive than inactive genes? Is DNA in active genes—ribosomal genes, in particular—held by the histones in a conformation *different* from the Watson-Crick structure. Does this give rise to the differential nuclease sensitivity and to the selective inhibition of ribosomal RNA synthesis by actinomycin and echinomycin?

Here, we describe our theory to understand the dynamic nature of DNA and RNA structure in solution. Although primarily a structural theory designed to provide insight into the mechanism of drug intercalation and DNA breathing—phenomena collectively known as premelting phenomena—the theory makes important additional predictions concerning the nonuniformity in the magnitude of energy fluctuations in different regions of DNA—this reflects the presence of acoustic phonons in its structure and the heterogeneity in DNA flexibility associated with different stacking energies that stabilize various combinations of nucleotide sequences. We will review the physical nature of this effect and its biological implications here.

Introduction

We begin by envisioning DNA in solution to be continuously bombarded by solvent molecules along its length. Although the vast majority of these solvent collisions transfer energy into the polymer to give rise to anharmonic motions in its structure that eventually dissipate energy back into solution due to viscous damping, occasionally solvent molecules with the appropriate momentum (i.e., having their direction oriented along the dyad axes in DNA) collide with DNA and this gives rise to harmonic motion in the polymer.

By harmonic motion, we mean specific normal mode oscillations in DNA structure that either remain localized to a site (to produce a conformational change) or travel along the helix at the speed of sound in the form of normal mode waves. At low amplitudes, these harmonic motions cause DNA to oscillate between B DNA and a right-handed superhelical variant of B DNA that contains approximately 10 base-pairs per turn and has a pitch of about 34Å. Such structures balance the unwinding in the helix with right-handed superhelical writhe to keep the linkage invariant, a feature that creates minimal perturbation in DNA structure. Higher amplitude harmonic motions give rise to similar structures except that these contain somewhat more than 10 base-pairs per turn and have base-pairs inclined more acutely to the helix axis. Their presence could contribute to the apparent net unwinding of DNA and the altered tilt of base-pairs relative to the helix axis-features observed for DNA in solution.[43/44] At small amplitudes of oscillation, DNA behaves as an elastic body that accumulates strain energy in its structure through small changes in torsional angles that define the geometry of the sugar-phosphate backbone. These changes are localized primarily in the furanose rings of alternate deoxyribose sugar residues (normally, the puckering of the furanose ring in B DNA is C2′ *endo*; however, the effect of introducing strain energy into the helix is to alter the magnitude and direction of this puckering). At larger amplitudes of oscillation, the enhanced strain energy in the sugar-phosphate chains begins to flatten out the furanose ring. Finally, at some critical oscillation amplitude, alternate sugars "snap into" a C3′ *endo* sugar conformation with a concomitant partial unstacking of base-pairs. This structure (denoted β kinked DNA) corresponds to an inelastic distortion in DNA structure and arises from a transient high energy normal mode oscillation in the helix localized at a specific site. Such a structure is hyperflexible and would, therefore, experience enhanced energy fluctuations during its lifetime—this reflects the presence of acoustic phonons that are associated with travelling normal mode waves in DNA (see discussion below). The energy from these fluctuations can be used to further separate base-pairs (so as to allow drugs and dyes to intercalate into DNA) and to rupture hydrogen-bonds connecting base-pairs (as occurs in DNA breathing). We have qualitatively described structural aspects of these processes elsewhere.[15-17]

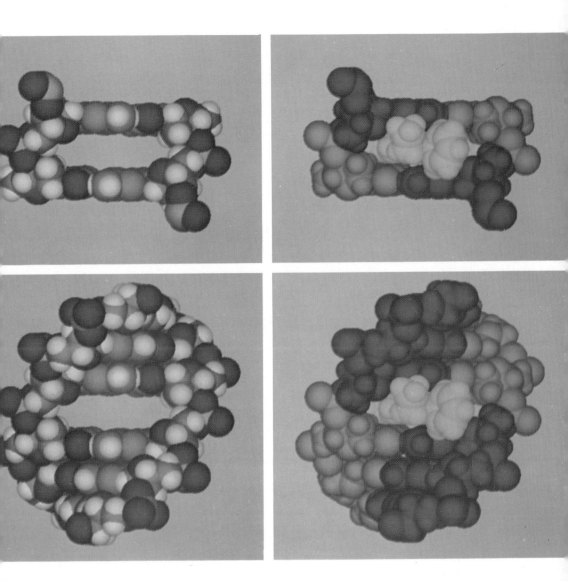

Plate 17. Space-filling representation of the steps in assembly of the ethidium-DNA binding model. *Top:* Idealized (deoxyribose-containing) configuration common to both the ethidium-iodoUpA and ethidium-iodoCpG crystalline complexes. *Bottom:* B DNA added above and below the intercalated deoxyribodinucleoside monophosphate to demonstrate ethidium-DNA binding.

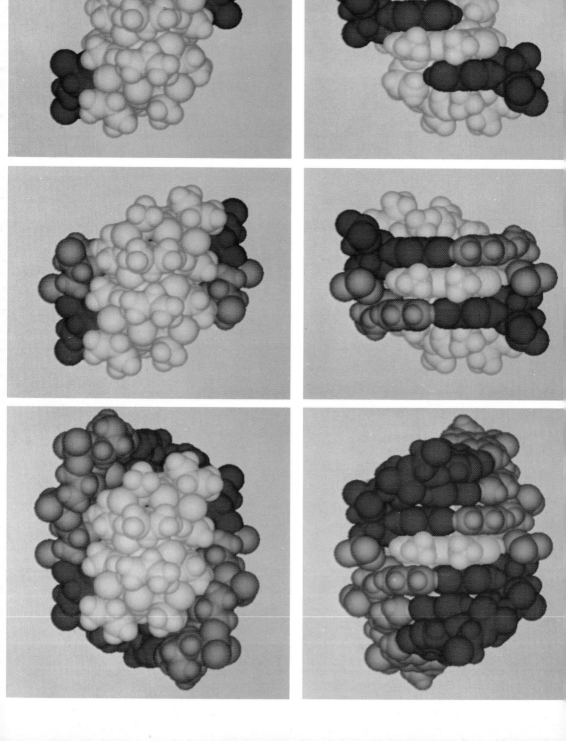

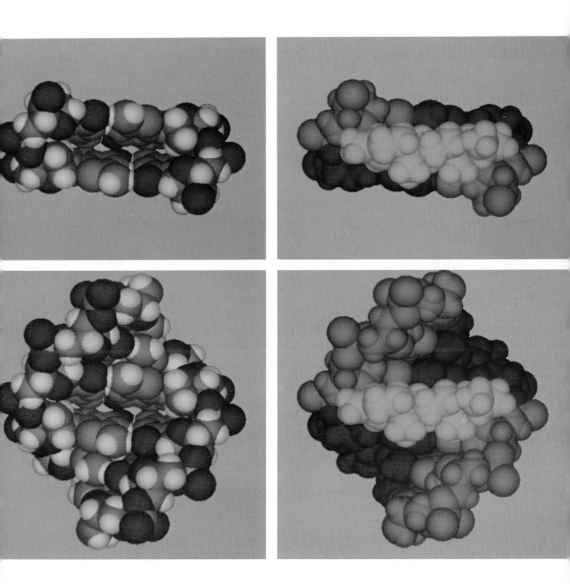

Plate 19. Space-filling representation of the steps in assembly of the irehdiamine-DNA binding model. *Top:* Irehdiamine-deoxyguanylyl(3′-5′)cytidine (dGpC) model complex. *Bottom:* B DNA added above and below the partially intercalated deoxyribodinucleoside monophosphate to demonstrate irehdiamine-DNA binding.

Plate 18. *(Opposite page)* Space-filling representation of the steps in assembly of the actinomycin-DNA binding model. *Top:* Configuration observed in the actinomycin-deoxyguanosine crystalline complex. *Middle:* actinomycin-deoxyguanylyl(3′-5′)cytidine (dGpC) model complex. *Bottom:* actinomycin-DNA binding model. See text for discussion.

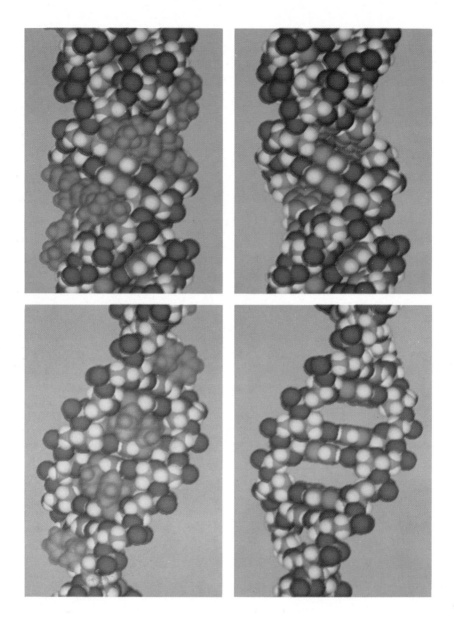

Plate 20. Space-filling representation of the irehdiamine-DNA and ethidium-DNA neighbor-exclusion models. *Top:* irehdiamine-DNA neighbor-exclusion model. *Bottom:* ethidium-DNA neighbor-exclusion model. See text for discussion.

Here, we document these conformational changes in greater detail. Using a linked-atom least squares approach for building and refining nucleic acid structures with defined stereochemistry,[45/46] we are able to follow the structural intermediates involved in drug intercalation and DNA breathing. An important related question concerns the structural nature of the B ⇌ A transition in DNA—we will now describe this.

The B ⇌ A Transition in DNA

It has been known for many years that a variety of naturally occurring and synthetic DNA polymers can exist in two major families of double-helical structures, the A and B families.[47/48] (See Plates 2-5.) Although considerable variation exists within each family, the B family of helices all contain C2' *endo* sugar residues in the high *anti* conformation, while the A family all contain C3' *endo* sugar residues in the low *anti* conformation. The B ⇌ A transition occurs in all naturally occurring DNA preparations and in most synthetic DNA polymers. First detected in oriented fibers of Na DNA at 75% relative humidity by X-ray diffraction,[49] the transition has also been observed in 70-80% ethanol-water mixtures with a variety of other physical techniques.[48/50-53] The transition occurs over a narrow range of humidity and ethanol-water concentrations and is highly cooperative. Cross-linking DNA by photodimerization inhibits the B ⇌ A transition as do a variety of intercalative drugs and dyes.

What is the structural nature of the B ⇌ A transition? Here, we show that it is possible to continuously deform B DNA to A DNA along a minimum energy pathway by changing the dihedral angles that define the geometry of the sugar-phosphate backbone. These changes primarily occur in the furanose rings of *alternate* deoxyribose sugar residues. Levitt and Warshel[54] have shown that both ribose and deoxyribose sugar rings are extremely flexible—having stable energy minima at C2' *endo* and C3' *endo* sugar conformations. The barrier separating these conformations is about 0.5 kcal/mol, a value well below the normal C-C single bond energy barrier (i.e., 2-3 kcal/mol, in most cases) and, for this reason, conformational changes in both DNA and RNA are expected to involve alterations in sugar puckering as well as in torsional angles describing the phosphodiester region of the sugar-phosphate backbone.

Figure 14 shows the conformational intermediates calculated for the B ⇌ A transition using linked-atom least squares energy minimization methods. A key structural intermediate in the B ⇌ A transition is β kinked DNA—this structure has a dinucleotide with the mixed C3' *endo* (3'-5') C2' *endo* sugar puckering as an asymmetric unit and has helical parameters that lie midway between B and A DNA. Preliminary calculations suggest its energy to be intermediate between B and A structures and to fall in a local minimum—β kinked DNA, therefore, could be a metastable intermediate in the B ⇌ A transition pathway (it is important to state, however, that these calculations involve the poly (dA-dT)$_n$ polymer sequence kinked at pTpA—we are expanding our calculations to include other DNA se-

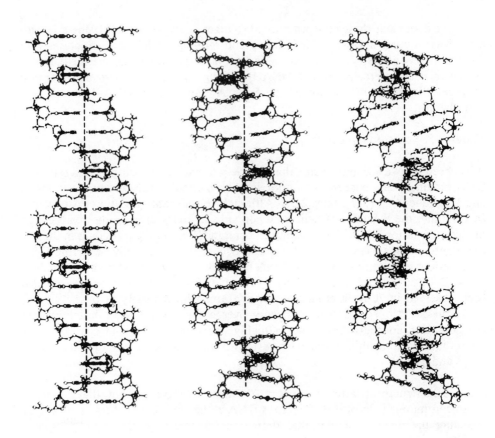

quences and will report on these at a later time). Although the B ⇌ A transition has
been calculated as a uniform transition simultaneously involving the entire polymer
length (twenty base-pairs of which are shown in Figure 14), we envision a gradation
of structures connecting B DNA with A DNA to serve as dynamic interfaces connecting the two. A moving structure such as this could leave behind either B or A DNA,
depending on the relative stabilities of these forms at given conditions.

The B DNA to β kinked DNA (and the A DNA to β kinked DNA) pathway for conformational change shown in Figure 14 reflects a specific low-frequency normal
mode oscillation in B DNA (and A DNA) structure that is excited by thermal fluctuations. Double-stranded RNA may possess a similar low-frequency normal mode
oscillation in its structure that gives rise to a conformational change from A' RNA
to β kinked RNA—this would subsequently allow double-helical RNA to undergo
drug intercalation into its structure (see below). The conversion of β kinked RNA to
the B structure may not be possible for steric reasons. We are therefore computing
the A' ⇌ B transition for double-helical RNA as well, and will describe the results of
these calculations elsewhere.

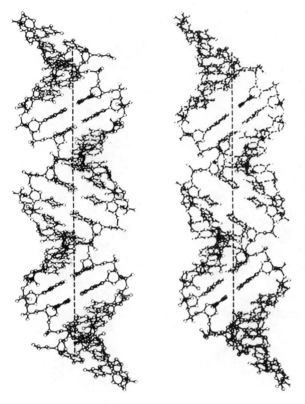

Figure 14. Conformational intermediates in the B ⇌ A transition as computed by linked-atom least squares energy minimization methods. The key structural intermediate in the B ⇌ A transition is β kinked DNA—this structure (shown on the far right) has a dinucleotide with a mixed C3′ *endo* (3′-5′) C2′ *endo* sugar puckering as the asymmetric unit and has helical parameters that lie midway between B and A DNA. The B DNA to β kinked DNA transition shown here corresponds to a low-frequency normal mode oscillation in B DNA structure. See text for further discussion.

β Kinked DNA ⇌ β Intercalated DNA
Conformational Changes

A key property of the β kinked DNA structure is its flexibility—this readily allows it to undergo further lengthening and unwinding to form the β intercalated structure (see Figure 15). These changes primarily involve alterations in the glycosidic torsional angles of the nucleotides, although a series of other minor changes occur in the torsional angles defining the phosphodiester region of the sugar-phosphate backbone. The β kinked DNA ⇌ β intercalated DNA transition is accompanied by a monotonic rise in energy without any intervening energy minimum—and corresponds to a second low-frequency (i.e., acoustic) normal mode oscillation in DNA structure that can be excited by thermal energies.

Compound Normal Mode Oscillations in DNA Structure
That Give Rise to Mono- and Bifunctional
Drug Intercalation

The information described above has allowed us to compute the compound normal

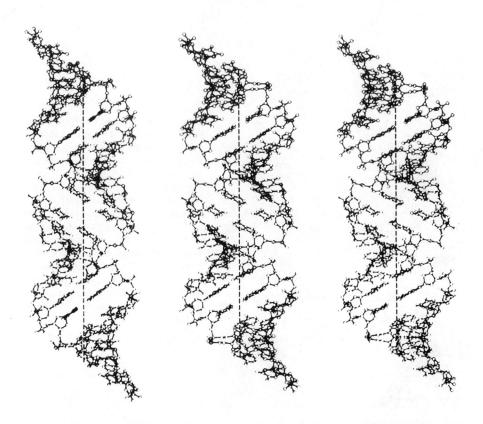

mode oscillations in DNA structure that give rise to the stereochemistries necessary for drug intercalation (both mono- and bifunctional) and DNA breathing. These calculations are in progress and we cannot yet describe them in detail. However, one can obtain considerable insight into this by considering an analogous problem in mechanics in the macroscopic domain.

Let us consider the example of an aluminum rod that is hit by a hammer perpendicular to its long axis at some initial time. If the momentum transferred by the hammer to the rod does not exceed its elastic limit, the rod undergoes (elastic) deformation at the point of impact and this gives rise to a series of elastic travelling waves having different frequencies. If the rod is finite in length, these elastic travelling waves rebound from the ends and undergo subsequent constructive and destructive interference events. As time proceeds, a standing wave having a defined frequency emerges (i.e., the rod "rings") and this vibration (along with its overtones) corresponds to a specific normal mode excited in this system.

If, however, the rod in this example were infinite in length no such normal modes will appear—the elastic travelling waves that are created by the impact of the hammer on the rod will simply continue to travel along the rod to infinity. This second example is more relevant to DNA. Here, elastic travelling waves take the form of

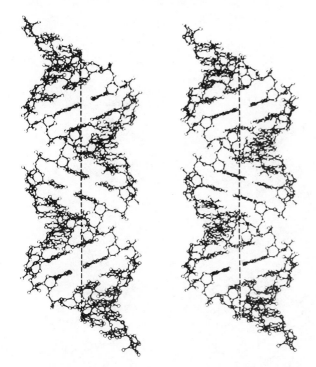

Figure 14. (continued) β kinked DNA to A DNA conformational transition as computed by linked atom least squares enery minimization methods. This series of conformational changes forms as the result of a low-frequency normal mode oscillation in A DNA structure. Double stranded RNA may possess a similar low frequency normal mode oscillation in its structure that gives rise to a conformational change from A' RNA to β kinked RNA—this would subsequently allow double helical RNA to undergo drug intercalation into its structure. See text for discussion.

normal mode waves that continuously appear due to the fluctuations (as explained earlier, these appear due to a specific class of solvent collisions with DNA—those collisions that strike DNA down its dyad axis). These normal mode waves are associated with the presence of acoustic phonons that travel along the polymer at the speed of sound, eventually undergoing viscous damping. For this reason, DNA is not expected to undergo normal modes (as defined in the classic sense) in aqueous solution.

What happens if the momentum transferred by the hammer to the rod exceeds some elastic limit? The rod bends—that is, undergoes a structural change at the point of impact—an event known as inelastic deformation. If the structural defect associated with this bend creates a hyperflexible joint (i.e., crack), then elastic energy that exists on either side of the defect will migrate into this hyperflexible region and "snap" the rod—an event known as a recoil phenomenon. With DNA, an occasional "hot" solvent collision could excite a normal mode oscillation in the B helix whose amplitude exceeds the elastic limit of the polymer —(in the section below, however, we describe how multiple fluctuations can transiently concentrate energy at flexible regions of DNA due to the presence of acoustic phonons to effect a conformational change—and we consider this to be the more probable explanation)— this gives rise to either the single- or double-kinked (that is, β kinked) DNA structures. Both struc-

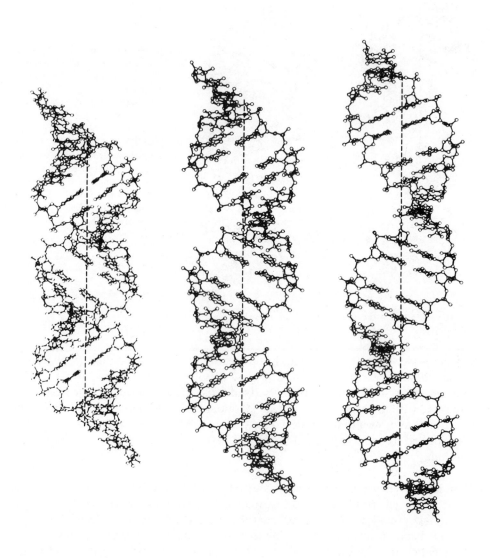

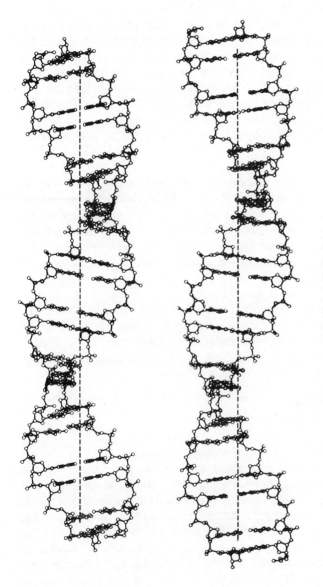

Figure 15. Conformational intermediates in the β kinked DNA ⇌ β intercalated DNA transition as computed by linked atom least squares energy minimization methods. These intermediate structures correspond to a specific low frequency normal mode oscillation in the β kinked DNA structure. See text for discussion.

tures contain hyperflexible joints (i.e., kinks) into which elastic energy can migrate—this elastic energy partially arises from the initial "hot" solvent collision with DNA (i.e., the recoil phenomenon) and from energies from other fluctuations that occur along the polymer at different times and arrive in the form of acoustic phonons. This additional (elastic) energy can be used to further separate base-pairs (to allow drugs and dyes to intercalate into DNA) and to rupture hydrogen-bonds connecting base-pairs (as occurs in DNA breathing)—motions that characterize the normal mode oscillation in the β kinked DNA structure (see previous section). Both types of normal mode oscillations (i.e., B DNA $\rightleftharpoons$ β kinked DNA, β kinked DNA $\rightleftharpoons$ β intercalated DNA) act to produce the series of conformational changes necessary for mono- and bifunctional drug intercalation—we therefore refer to these as compound normal mode oscillations in DNA structure that give rise to drug intercalation and DNA breathing.

In the rod analogy, the term "inelastic deformation" has been used to describe a permanent structural deformation associated with net work done on the system. Of course, for DNA in solution, the fluctuations produce only transient work on the polymer—DNA-kinking is a reversible process having forward and backward rate constants. The reversibility of this process reflects the small energies involved in this conformational change and—related to this—maintenance of the covalent integrity of the sugar-phosphate backbone.

Computations to describe the compound normal mode oscillations in DNA structure involve combining a gradation of structural information from the B DNA $\rightleftharpoons$ β kinked DNA and the β kinked DNA $\rightleftharpoons$ β intercalated DNA transitions. Similar calculations are being carried out to understand the dynamic nature of double-helical RNA structure. We will describe the results of these computations elsewhere.

The Influence of Acoustic Phonons on the
Magnitude of Energy Fluctuations in DNA

An important consequence of the presence of acoustic phonons in DNA structure is the altered nature of energy fluctuations along the helix at equilibrium. This is because—when calculating the magnitude of energy fluctuations at specific places on DNA—one has to take into account not only fluctuations that arise locally at this site but, in addition, fluctuations that occur at a distance and then travel to this site in the form of acoustic phonons. Since DNA is a heterogeneous polymer—having variable flexibility due to different stacking energies between base-pairs in different sequences—it follows (see below), that the probability a specific region experiences a fluctuation with a given energy depends on the elasticity of this region and of neighboring regions. Generally, more flexible DNA regions experience larger energy fluctuations—the magnitude of these enhanced fluctuations increases the probability that transient DNA conformational change can occur (i.e., B DNA $\rightarrow$ β kinked DNA) and that, once formed, this conformationally altered DNA structure experiences still larger energy fluctuations (i.e., β kinked DNA $\rightleftharpoons$ β intercalated DNA).

We have already described the physical nature of this effect and its biological importance in several previous publications.[16/17] We describe one of our more recent treatments of this effect here.[55]

The general expression for fluctuations in condensed matter is

$$\overline{\Delta w^2} = - [T(\partial P/\partial T)_V - P]^2 T(\partial V/\partial P)_T + C_v T^2 \qquad (1)$$

where $\overline{\Delta w^2}$ is the mean squared energy fluctuation, T is the absolute temperature, P is the pressure, V is the volume and C_v is the heat capacity.[56]

To calculate $\overline{\Delta w^2}$, we will need to calculate the free energy, which is given by the following expression

$$F = N\epsilon_0 + T \sum_\alpha \ln (1 - e^{-\hbar\omega_\alpha/T}) \qquad (2)$$

where N is the number of base pairs, ϵ_0 is the interaction energy per base pair at equilibrium, ω_α are characteristic frequences of longitudinal lattice oscillations.

For a linear polymer, the number of eigenvalue oscillations in the interval of wave vector dk is equal to

$$Na (dk/2\pi) = Na (d\omega/2\pi u)$$

where a is the lattice constant (i.e., for DNA, this corresponds to 3.4Å) and u is the velocity of longitudinal phonons ($u = \omega/k$).

For temperatures well below the Debye temperature, low frequencies (i.e., acoustic phonons) play a predominant role in determining the free energy. In this approximation, we obtain

$$F = N\epsilon_0 + T (Na/2\pi u) \int_0^\infty \ln(1-e^{-\hbar\omega/T})d\omega = N\epsilon_0 - Na (\pi T^2/12\hbar u) \qquad (3)$$

and, therefore

$$C_v = \partial/\partial T[F-T (\partial F/\partial T)] = Na (\pi T/6\hbar u) \qquad (4)$$

For qualitative estimates, we can consider the volume of DNA to be constant. Then

$$\overline{\Delta w^2} = Na \, \pi T^3/6\hbar u \qquad (5)$$

In DNA, the velocity of longitudinal phonons is given by $u = \sqrt{E/\varrho}$ where E is Young's modulus and ϱ is the linear density of DNA. Although ϱ is approximately constant along the polymer, E is not, due to different stacking energies between A-T and G-C base-pairs.[57] For this reason, phonons travel along the DNA molecule with nonuniform velocities. This gives rise to larger magnitude energy fluctuations in those regions of the polymer having greatest flexibility.

Does DNA Have Two Structures That Coexist at Equilibrium?

What is the structure of DNA in solution at equilibrium? Is it only B DNA? We have proposed that DNA could have *two* discrete structures that coexist at equilibrium at a given temperature—B DNA and β kinked DNA.[16/17] Here, β kinked DNA corresponds to a second order phase transition in the polymer—different regions of DNA undergoing this transition at different temperatures. Thus, DNA in solution at lower temperatures could exist only in the B form. At higher temperatures, however, permanently premelted β kinked DNA regions could appear. Due to their enhanced flexibility, these regions could be particularly prone to undergo DNA breathing motions (i.e., further base unstacking and hydrogen-bond breakage). The energies for these breathing motions arise from both local fluctuations and from fluctuations originating in neighboring regions that propagate energy in the form of acoustic phonons along DNA.

Regions of second order phase transition such as these could serve to enucleate DNA melting at higher temperatures. They may also have an important biological function—they may serve as binding sites for a variety of protein-DNA interactions. We will now describe these and other related concepts.

Protein-DNA Interactions

RNA Polymerase-Promoter Recognition

What is the nature of the promoter and how does the RNA polymerase recognize and bind to the promoter?

We propose that the promoter is, in fact, a region of DNA that has undergone a second order phase transition in its structure (i.e., β kinked DNA, as described above), and that the recognition and attachment of RNA polymerase for promoter regions reflects this. Thus, RNA polymerase may bind to the promoter by intercalating—either partially or completely—aromatic side chains into DNA and, in addition, by interacting with specific nucleotide bases and sugar-phosphate groups through hydrogen-bonds, van der Waals interactions and electrostatic interactions as it forms the tight binding complex.

An important feature of our model is the prediction that DNA exists in an altered premelted form at the promoter. Another key prediction is that this same structure accompanies RNA polymerase as it moves along the DNA template synthesizing RNA chains (i.e., this would correspond to the "bubble" often shown in schematic illustrations of RNA polymerization). Our model predicts, therefore, that intercalative drugs and dyes interfere with RNA synthesis in at least two ways: first—by binding to the promoter to interfere with the initiation of new RNA chains, and second—by binding to the β kinked DNA structure travelling with the RNA polymerase enzyme to interfere with the elongation of growing RNA chains. Evidence that ethidium does, in fact, interfere with both the initiation and the elongation of RNA chains has already appeared.[58]

Single-strand Specific DNA Binding Proteins

How do the single-strand specific DNA binding proteins (i.e., T4 gene 32 protein, fd gene 5 protein) bind to DNA? Do their interactions with single-stranded DNA involve intercalation and, if so, why? What is the structure of single-stranded DNA when it interacts with these DNA-binding proteins?

Our model predicts that single-stranded DNA assumes a premelted structure during DNA unwinding with the following structural characteristics: a perfectly alternating pattern of sugar puckering down the polynucleotide backbone (i.e., C3′ *endo* (3′-5′) C2′ *endo* (3′-5′) C3′ *endo*, and so on), with (potential) intercalating spaces between every other nucleotide base. This being the case—single-strand specific DNA binding proteins will almost certainly intercalate (either partially or completely) aromatic amino acid side chains between the bases when binding to DNA. In addition, other interactions will utilize hydrogen bonding, van der Waals interactions and electrostatic interactions—interactions all centered on recognizing structural features of the premelted single-stranded DNA structure we have postulated.

Nuclease Specificity and the Organization of DNA in Chromatin

What do nucleases "see" when they cleave DNA and, related to this, why—in eukaryotic chromatin—are active genes some five times more nuclease sensitive than inactive genes? Is DNA in active genes—ribosomal genes, in particular—held by the histones in a conformation *different* from the Watson-Crick structure and does this give rise to the differential nuclease sensitivity and to the selective inhibition of ribosomal RNA synthesis by actinomycin and echinomycin?

Transcriptionally active chromatin has been shown to be particularly sensitive to pancreatic DNase digestion—active genes being digested between 4-5 times faster than inactive genes.[59-61] DNA in active genes appears to remain bound to histones, as evidenced by staphylococcal nuclease limit digest studies[62] and measurements of the thickness of chromatin fibers actively undergoing ribosomal RNA transcription (about 70Å wide).[63] There is no evidence for a beaded structure in (ribosomal) transcriptionally active chromatin—rather, the structure appears to be an extended one in which the DNA exists in some state particularly sensitive to pancreatic DNase.

We have wondered whether this state is β kinked DNA. If pancreatic DNase recognizes kinks in DNA (see below), then it would digest this structure approximately 5 times faster than $\varkappa$ kinked DNA (a time averaged structure we postulate to exist in inactive chromatin). Since β kinked DNA has an axial repeat length similar to B DNA, the chromatin fiber could be extended about the same length as that predicted assuming B DNA in these regions. Finally, gene activation could be achieved along the lines originally suggested by Weintraub *et al.*,[64] in which the histone-DNA complex exists as a linear structure (active chromatin) or a helical structure (inactive chromatin). The flip-flop interconversion of these forms could convert the left-handed superhelical writhe in the x helix into unwinding in the β helix, this unwinding being localized at the kinks. Such a mechanism could allow control of premelting changes in DNA necessary for RNA transcription.

Inactive chromatin is now established to consist of a beaded structure in which DNA is complexed to each of the four histones (H-2a, H-2b, H-3, H-4). Each bead is an octamer of histones complexed to 140 base-pairs, most likely, as a left-handed superhelix around the periphery of the histone core (see, for example, "Chromatin," Cold Spring Harbor Symposia on Quantitative Biology, Volume XLII, 1977). An important observation concerns the pancreatic DNase digestion patterns observed from inactive chromatin—this demonstrates periodicities of approximate integral multiples of 10 nucleotide bases (i.e., fragments 10, 20, 30 ... up to 300 have been observed), suggesting the presence of nuclease sensitive sites periodically located every 10 base- pairs. In more recent experiments, Lutter has demonstrated this repeat to be somewhat variable but centered at 10.4 base-pairs on the average.[65]

We have postulated these sites to be kinks. The kink could serve as a substrate for both pancreatic DNase and splenic acid DNase II (these nucleases could partially intercalate into DNA, positioning these enzymes along the dyad axis of the kink to effect staggered cleavage about this symmetry axis.) A kink placed every 10 base-pairs in DNA gives rise to a left-handed superhelical structure with approximately the same dimensions as the nucleosome. We have postulated this structure (denoted x kinked DNA) or, more precisely, a time averaged structure of this type, to exist in the organization of DNA in inactive chromatin.

Conclusion

The subject of drug intercalation is a multidimensional one that is concerned with understanding the dynamic structure of DNA in solution and the molecular basis of its flexibility. Our X-ray crystallographic studies of drug intercalation have allowed us to propose unifying structural models to understand a large number of drug-DNA (and -RNA) interactions—these models have led us to further formulate dynamic concepts of DNA structure that give rise to drug intercalation. The phenomenon of drug intercalation is more broadly related to DNA breathing (or, DNA premelting)—and involves a second DNA structure that arises due to the fluctuations that exist at equilibrium. This DNA structure (β kinked DNA) is a hyperflexible DNA region that undergoes enhanced energy fluctuations during its lifetime due to the presence of acoustic phonon energy in DNA. These fluctuations give rise to further base unstacking (so as to allow for drug intercalation) and hydrogen-bond breakage (to allow DNA breathing).

More generally, we have asked whether DNA could have two discrete structures that coexist at equilibrium at a given temperature—B-DNA and β kinked DNA. Here, β kinked DNA corresponds to a second order phase transition in the polymer—different regions of DNA undergoing this transition at different temperatures. Such (permanently) premelted regions could enucleate DNA melting at higher temperatures. They could also serve an important biological function—they may be binding sites for a variety of protein-DNA interactions.

Thus, promoters could be permanently β kinked DNA regions recognized in part by the RNA polymerase through partial or complete intercalation of its aromatic side chains into DNA. Such a structure could migrate along with the enzyme to form the activated complex in the polymerization reaction. Histones may have evolved to control this DNA activation process through a flip-flop mechanism in which inactive DNA (x kinked B DNA) is converted into active DNA (β kinked DNA) through the use of topological constraints. Nucleases may have evolved to see kinks in DNA by partially intercalating aromatic amino acid side chains into DNA.

We can expect major advances in future years in our understanding of how proteins bind to DNA and recognize specific structural features of its base-sequence. Almost certainly, these advances will be provided by X-ray crystallography, the most powerful tool we now have to understand the structure of biological macromolecules. It will then be possible to evaluate the relevance of dynamic DNA structure—and the phenomenon of drug intercalation—to understand protein-DNA interactions.

References and Footnotes

1. Lerman, L.S. *J. Mol. Biol. 3*, 18-30 (1961).
2. Lerman, L.S. *Proc. Nat. Acad. Sci. USA, 49*, 94-102 (1963).
3. Sobell, H.M., Jain, S.C., Sakore, T.D. and Nordman, C.E. *Nature 231*, 200-205 (1971).
4. Sobell, H.M., Jain, S.C., Sakore, T.D., Ponticello, G. and Nordman, C.E. *Cold Spring Harbor Symp. Quant. Biol. 36*, 263-265 (1972).
5. Jain, S.C. and Sobell, H.M. *J. Mol. Biol. 68*, 1-20 (1972).
6. Sobell, H.M. and Jain, S.C. *J. Mol. Biol. 68*, 21-34 (1972).
7. Tsai, C.-C., Jain, S.C. and Sobell, H.M. *J. Mol. Biol. 114*, 301-315 (1977).
8. Jain, S.C., Tsai, C.-C. and Sobell, H.M. *J. Mol. Biol. 114*, 317-331 (1977).
9. Sobell, H.M., Tsai, C.-C., Jain, S.C. and Gilbert, S.G. *J. Mol. Biol. 114*, 333-365 (1977).
10. Sakore, T.D., Reddy, B.S. and Sobell, H.M., *J. Mol. Biol.* (in press).
11. Reddy, B.S., Seshadri, T.P., Sakore, T.D. and Sobell, H.M. *J. Mol. Biol.* (in press).
12. Jain, S.C., Bhandary, K.K. and Sobell, H.M., *J. Mol. Biol.* (in press).
13. Sobell, H.M., Reddy, B.S., Bhandary, K.K., Jain, S.C., Sakore, T.D. and Seshadri, T.P. *Cold Spring Harbor Symp. Quant. Biol. 42*, 87-102 (1977).
14. Sobell, H.M. in *Biological Regulation and Development*, Volume 1, Ed. Robert F. Goldberger, Plenum Publishing Corporation, pp. 171-199 (1979).
15. Sobell, H.M. in *Effects of Drugs on the Cell Nucleus*, Bristol-Myers Symposium, Volume 1, Ed. Harris Busch, Academic Press (in press).
16. Sobell, H.M., Lozansky, E.D. and Lessen, M. *Cold Spring Harbor Symp. Quant. Biol. 43*, 11-19 (1978).
17. Lozansky, E.D., Sobell, H.M. and Lessen, M. in *Stereodynamics of Molecular Systems*, Ed. Ramaswamy H. Sarma, Pergamon Press, Inc. Oxford, New York, Frankfurt, Paris pp. 265-270 (1979).

18. Neidle, S., Achari, A., Taylor, G.L., Berman, H.M., Carrell, H.L., Glusker, J.P. and Stallings, W.C. *Nature 269*, 304-307 (1977).
19. Waring, M.J. *J. Mol. Biol. 54*, 247-279 (1970).
20. Wang, J.C. *J. Mol. Biol. 89*, 783-801 (1974).
21. Bond, P.J., Langridge, R., Jennette, K.W. and Lippard, S.J. *Proc. Nat. Acad. Sci.* USA *72*, 4825-4829 (1975).
22. Krugh, T.R. *Proc. Nat. Acad. Sci. USA 69*, 1911-1914 (1972).
23. Wells, R.D. and Larson, J.E. *J. Mol. Biol. 49*, 319-342 (1970).
24. Gellert, M., Smith, C.E., Neville, D. and Felsenfeld, G. *J. Mol. Biol. 11*, 445-457 (1965).
25. Muller, W. and Crothers, D.M. *J. Mol. Biol. 35*, 251-290 (1968).
26. Bittman, R. and Blau, L. *Biochemistry 14*, 2138-2145 (1975).
27. Mahler, H.R., Green, G., Goutarel, R. and Khuong-Huu, Q. *Biochemistry 7*, 1568-1582 (1968).
28. Waring, M.J. and Chisholm, J.W. *Biochim. Biophys. Acta 262*, 18-23 (1972).
29. Waring, M.J. and Henley, S.M. *Nucleic Acids Res. 2*, 567-586 (1975).
30. Gabbay, E.J., Scofield, E.E. and Baxter, C.S. *J. Amer. Chem. Soc. 95*, 7850-7857 (1973).
31. Gabbay, E.J. in *Bioorganic Chemistry, Volume 3, Macro- and Multimoleculr Systems*, Academic Press, Inc., New York, pp. 33-70 (1977).
32. Wakelin, L.P.G. and Waring, M.J. *Biochim. J. 157*, 721-740.
33. Ughetto, G. and Waring, M. *J. Mol. Pharm. 13*, 579-584 (1977).
34. Le Pecq, J.B., Le Bret, M., Barbet, J. and Roques, R.P. *Proc. Nat. Acad. Sci. USA 72*, 2915-2919 (1975).
35. Butour, J.L., Delain, E., Couland, D., Le Pecq, J.B., Barbet, J.B. and Roques, B.P. *Biopolymers 17*, 873-886 (1978).
36. Dervan, P. and Becker, M.M. *J. Amer. Chem. Soc. 100*, 1968-1970 (1978).
37. McGhee, J.D. and von Hippel, P.H. *Biochemistry 14*, 1281-1296; 1297-1303 (1975).
38. Teitelbaum, H. and Englander, S.W. *J. Mol. Biol. 92*, 55-78; 79-82 (1975).
39. McGhee, J.D. and von Hippel, P.H. *Biochemistry 16*, 3267-3275 (1977).
40. McGhee, J.D. and von Hippel, P.H. *Biochemistry 16*, 3276-3293 (1977).
41. Palecek, E. in *Progress in Nucleic Acid Research and Molecular Biology, 18*, Eds. Davidson, J.N. and Cohn, W.E., Academic Press, Inc., New York, pp. 151-213 (1976).
42. Palecek, E. in *Methods in Enzymology, 21*, Eds. Grossman, L. and Moldave, K., Academic Press, Inc., New York, pp. 3-24 (1971).
43. Hogan, M., Dattagupta, N. and Crothers, D.M. *Proc. Nat. Acad. Sci. USA, 75*, 195-199 (1978).
44. Wang, J.C. *Cold Spring Harbor Symp. Quant. Biol. 43*, 29-33 (1978).
45. Smith, P. J.C. and Arnott, S. *Acta Cryst. A34*, 3-11 (1978).
46. Banerjee, A., Ramani, R., Lozansky, E. and Sobell, H.M., to be published.
47. Ivanov, V.I., Minchenkova, L.E., Schyolkina, A.K. and Poletayev, A.I. *Biopolymers 12*, 89-110 (1973).
48. Ivanov, V.I., Minchenkova, L.E., Minyat, E.E., Frank-Kamenetskii, M.D. and Schyolkina, A.K. *J. Mol. Biol. 87*, 817-833 (1974).
49. Franklin, R.E. and Gosling, R.G. *Acta Cryst. 6*, 673-678 (1953).
50. Brahms, J. and Mommaerts, W.F.H.M. *J. Mol. Biol. 10*, 73-88 (1964).
51. Girod, J.C., Johnson, W.C., Jr., Huntington, S.K. and Maestre, M.F. *Biochemistry 12*, 5092-5095 (1973).
52. Herbeck, R., Yu, T.-J. and Peticolas, W.L. *Biochemistry 15*, 2656-2660 (1976).
53. Erfurth, S.C. and Peticolas, W.L. *Biopolymers 14*, 247-264 (1975).
54. Levitt, M. and Warshel, A. *J. Amer. Chem. Soc. 100*, 2607-2612 (1978).
55. Lozansky, E.D. and Sobell, H.M. *Phys. Rev. Letters* (in press).
56. Landau, L.D. and Lifshitz, E.M. *Statistical Physics*, Addison-Wesley Publishing Company, p. 353 (1969).
57. Bloomfield, V.A., Crothers, D.M. and Tinoco, Jr., I. *Physical Chemistry of Nucleic Acids*, Harper and Row, Publishers, Inc. (1974).
58. Richardson, J.P. *J. Mol. Biol. 78*, 703-714 (1973).
59. Weintraub, H. and Groudine, M. *Science 193*, 848-856 (1976).
60. Garel, A. and Axel, R. *Cold Spring Harbor Symp. Quant. Biol. 42*, 701-708 (1977).

61. Flint, S. J. and Weintraub, H.M. *Cell 12*, 783-794 (1977).
62. Camerini-Otero, R.D., Sollner-Webb, B., Simon, R.H., Williamson, P., Zasloff, M. and Felsenfeld, G. *Cold Spring Harbor Symp. Quant. Biol. 42*, 43-56 (1977).
63. Foe, V.E., Wilkinson, L.E. and Laird, C.D. *Cell 9*, 131-146 (1976).
64. Weintraub, H., Worcel, A. and Alberts, B. *Cell, 9*, 409-417 (1976).
65. Lutter, L. *Nucleic Acids Research 6*, 41-56 (1979).
66. Berman, H.M., and Neidle, S. in *Nucleic Acid Geometry and Dynamics*, Ed. Sarma, R.H., Pergamon Press, New York, Oxford pp. 325-340 (1980).
67. Crothers, D.M., Dattagupta, N., and Hogan, M. in *Nucleic Acid Geometry and Dynamics*, Ed. Sarma, R.H., Pergamon Press, New York, Oxford pp. 341-349 (1980).
68. Krugh, T.R., Hook III, J.W., Balalcrishnan, M.S., and Chen, F-M., in *Nucleic Acid Geometry and Dynamics*, ed. Sarma, R.H., Pergamon Press, New York, Oxford, pp. 351-366 (1980).
69. Patel, D.J., in *Nucleic Acid Geometry and Dynamics*, ed. Sarma, R.H., Pergamon Press, New York, Oxford, pp. 185-232 (1980).

Modelling of Drug-Nucleic Acid Interactions Intercalation Geometry of Oligonucleotides

Helen M. Berman
The Institute for Cancer Research
The Fox Chase Cancer Center
7701 Burholme Avenue
Philadelphia, Pennsylvania 19111

and

Stephen Neidle
Department of Biophysics
University of London Kings College
London, WC2B 5RL
England

Introduction

The interaction of drugs with nucleic acids can involve a number of distinct processes,[1] ranging from covalent bonding to various types of non-bonded interaction. Probably the most extensively explored category is concerned with drug molecules possessing planar aromatic chromophores; the intercalating model of Lerman[2] provided the overall conceptual framework and major impetus for much of this work. The Lerman hypothesis states that these planar groups can be bound in a "sandwich" manner in between adjacent base pairs of double-stranded nucleic acids. The past few years have seen the beginnings of attempts to extend the Lerman model, so as to provide atomic-level structural information on intercalation. These attempts have brought into focus the major problems to be solved by such approaches:

- •what is the geometry of intercalation, in terms of changes in nucleic acid conformation?
- •is there indeed a singular geometry for intercalation?
- •what effect does the structure of the planar chromophore have on the geometry?
- •what are the structural differences between RNA and DNA intercalation?
- •what is the structural basis of unwinding and of neighbor exclusion?
- •what is the structural basis for sequence-preference in binding?

In the case of single-stranded nucleic acids, the geometry of drug binding is even less well defined than for a duplex situation, because of lack of the constraints imposed by Watson-Crick base pairing.

Several physico-chemical techniques have been used to attack these problems; this contribution discusses an X-ray crystallographic approach. The pioneering studies in this area of Sobell and his associates[3-5] employed planar drugs with ribonucleosides as model systems. (The crystal structure of a deoxydinucleoside complex[15] had recently been reported; a discussion of its conformational implications must await a detailed account of the structure.) Studies in our laboratories have used similar models; accordingly here we will concentrate on what these analyses have revealed about the ways in which drugs and mutagens interact with single- and double-stranded RNA. It is perhaps timely to define the limitations and potential pitfalls of this "model" nucleic acid approach, in terms of the structures evaluated to date.

The Crystallographic Data

There is now a substantial body of crystallographic data available for complexes between a variety of planar chromophores and self-complementary ribonucleosides, some of which are given in Table I. As a group these structures have shown themselves to be difficult to analyze crystallographically because:

(1) they contain many atoms in the asymmetric unit.
(2) they contain a high percentage of planar groups. Thus the "random" arrangement of atoms that is necessarily assumed for the application of current direct methods procedures is no longer valid.
(3) the planar chromophores often display large thermal motions, which limit the resolution of the data obtainable.
(4) the crystals contain large amounts of ordered and disordered water molecules.

One solution to some of these crystallographic problems is to chemically insert a heavy atom into the structure; however, in doing so one does introduce even greater limitations on the ultimate precision of the atomic coordinates. For this reason, in our laboratories we have chosen to study complexes without any heavy atoms. However, even when one does select this route and consequently obtains a result with high resolution, low R factor and relatively high precision the implications of these crystallographic results with respect to macromolecular structures are inherently limited by the fact that the crystals contain *dinucleosides* which may not be the minimal unit for the description of drug-intercalated *polynucleotides*. The

Table I

Structure	Contents of Asymmetric Unit	Observed Data	Resolution	R Factor	Reference
i[5] UpA-Ethidium Br	156 atoms	2017	1.34Å	.20	3
i[5] CpG-Ethidium Br	150	3180	1.14	.16	4
CpG-Proflavine	82	4115	.85	.11	6
ApU-9 amino acridine	75	2874	.85	.07	8
i[5] CpG-9 amino acridine	248	2251	1.34	.19	9
ApA-Proflavine	94.5	3431	.85	.11	7

reasoning behind this statement will become apparent as we examine in detail the conformational features of these complexes.

Models for Binding to RNA Duplexes

Conformational Features

The crystal structure of CpG-proflavine[6] is at first glance in a different conformational class from that of iodo CpG-ethidium Br[4] and iodo UpA-ethidium Br[3] (Figure 1). But is this really true? Examination of Table II reveals that indeed the ethidium structures differ from the proflavine one in that in the former, the pucker of ribose sugar rings are C3' *endo* at the 5' end and C2' *endo* at the 3' end while in the latter, both ribose sugars are C3' *endo*. However, when one compares all the other conformation angles in the structures with one another they are remarkably similar. Furthermore, they resemble those in A-RNA[10] more than B-DNA a fact which is not too surprising considering that all these structures are ribodinucleosides. If one can have an acceptable intercalation geometry (with base pair separations of at least 6.8Å and with or without mixed sugar puckering) which conformational features are

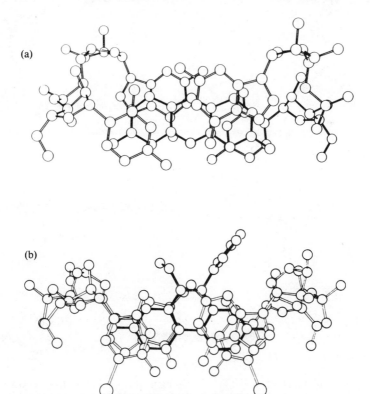

Figure 1. Views perpendicular to the plane of the base pairs of (a) CpG-proflavine and (b) iCpG-ethidium bromide.

(a)

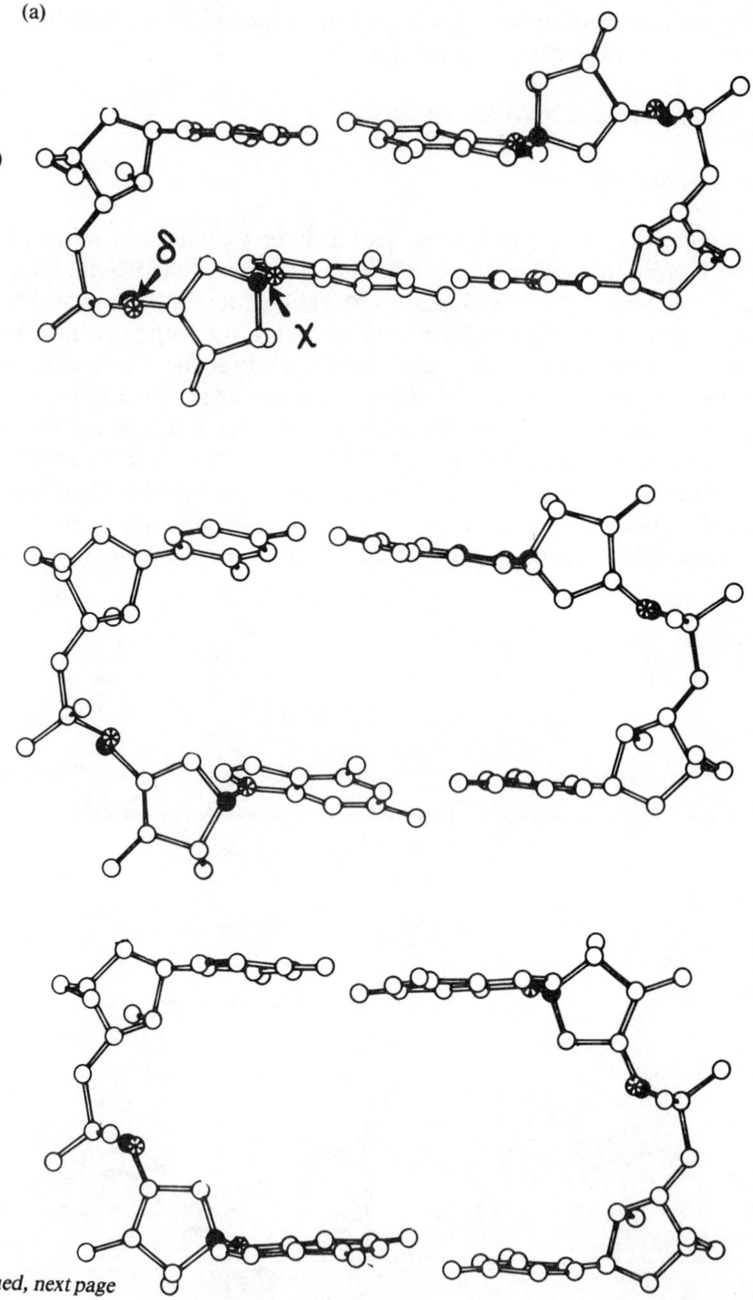

Figure continued, next page

responsible for the stretching open of the base pairs? From the crystallographic results, shown in Table II, it appears that the torsion angles around C5'-O5' (δ) and the 3' glycosidic angle (χ) are substantially increased from the values found in A-RNA. In order to verify that these two angles alone can effect the necessary base

(b)

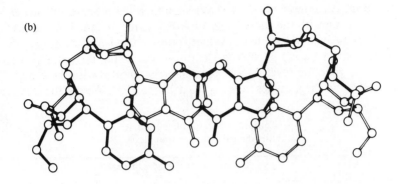

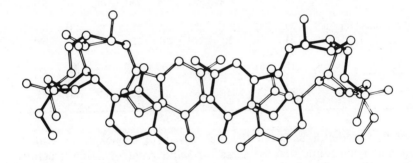

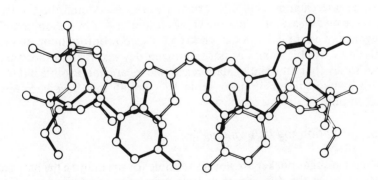

Figure 2. (a) The opening up of a model ribodinucleoside from an A-RNA conformation to that observed in CpG-proflavine. The views are parallel to the base pairs.

1) the dinucleotide in A RNA conformation

$\delta = 175°$, $\chi(3') = 14°$ 3) $\delta = 225°$, $\chi(3') = 80°$

2) $\delta = 200°$, $\chi(3') = 45°$ (b) Same as (a) but viewed perpendicular to base pairs.

pair separations for intercalation, we have simulated the process utilizing computer graphics.[11] Starting with the coordinates of A-RNA, the two torsion angles (δ and χ) were simultaneously increased to the values found in CpG-proflavine (Figure 2) and an acceptable intercalation geometry was produced. Using this same procedure, it was possible to generate several more reasonable structures with the δ angles in the range 225-235° and the χ angles ranging from 75° to 110°. We did the same calcula-

Table II
Conformational Features of Some
Ribodinucleotide-Dye Complexes

	Conformational Angles (in degrees)									Sugar Pucker
	χ	ζ	α	β	γ	δ	ϵ	ζ	χ	
CpG-proflavine	18	75	204	292	287	234	53	79	87	C3′ endo → C3′ endo
i⁵UpA ethidium bromide	26	98	207	286	291	236	52	133	99	C3′ endo →
	14	95	218	302	276	230	70	118	100	C2′ endo
i⁵CpG ethidium bromide	29	87	226	281	286	210	72	131	101	C3′ endo →
	24	84	225	291	291	224	55	134	109	C2′ endo
A RNA	13	83	213	281	300	175	50	83	13	
B DNA	85	157	159	261	321	209	31	157	85	

tion with a model dinucleoside that has mixed sugar pucker, C3′ *endo* (3′, 5′) C2′ *endo*, and again found that by increasing δ and χ, intercalation structures could be produced. We conclude from these simulations that for the ribodinucleosides studied crystallographically there are two essential conformational changes that occur when the intercalation complex forms: χ at the 3′ end and the C5′-05′ torsion angle increase by at least 50° from the values found in A-RNA. Additionally, the 3′ ribose sugar may adopt either a C2′ *endo* or C3′ *endo* conformation and other torsion angles may exhibit small but not significant deviations from those in A-RNA. This is not to say that other intercalation geometries are not possible but to date this genus with its two subsets of sugar geometries are all that have been observed at atomic resolution.

Base Pair Orientation and "Unwinding"

If the changes in sugar pucker are not responsible for opening up the base pairs, then perhaps they cause the differences in base pair orientations that are apparent in Figure 1. In that figure we see that for the ethidium complexes, which have mixed sugar pucker, the base pairs are almost completely overlapped such that the base turn angle is essentially zero. (The base turn angle is the angle between the vectors connecting the C1′ atoms of each base pair when projected on the average base plane viewed from a point perpendicular to this plane.) On the other hand, in the pro-flavine complex where all the ribose sugars are in the C3′ *endo* conformation, this angle is approximately 34°. In order to determine whether or not the sugar pucker is related to the base turn angle we have used computer graphics to manipulate model intercalation geometries with either the same or mixed sugar puckers. The backbone conformation angles in the duplex were varied slightly and each strand was moved with respect to the other with contraints imposed so that they maintained their two-fold symmetry and acceptable Watson-Crick hydrogen bonding geometry. (As has been pointed out by Levitt[12] there is considerable flexibility in the base pair geometry.) We found that for the models with both mixed and the same ribose sugar puckers it was possible to produce models with base turn angles that varied from 0 to 34° (Table III). It would appear then that this parameter is *not* dependent on the conformation of the sugar but rather on some combination of small variations in the base pair geometry and/or the backbone conformations. The value it adopts is most likely dependent on the nature of the planar chromophore.

Table III
Structural Details of Some Model Intercalation Geometries

	Conformational Angles (in degrees)									
Model	$\chi(5')$	α	β	γ	δ	ϵ	$\chi(3')$	Sugar Pucker	Base Twise	Base Turn
1	13	213	281	300	225	50	80	C3′ → C3′	10	15
2	13	213	281	300	225	50	80	C3′ → C3′	4	27
3	13	220	280	300	230	70	95	C3′ → C2′	16	11
4	13	210	300	280	235	70	95	C3′ → C2′	15	25

Why is the base turn angle such an important parameter? In a polynucleotide with a single helix axis this angle is a measure of the nucleotide turn angle and can thus serve as a measure of the helical unwinding. Whether or not these dinucleoside models imply unwinding can only be determined by building models of larger oligonucleotide units. Therefore, in models of DNA-drug intercalation in which the helix axis is deliberately constrained[13/14], it is possible to relate the base turn angle directly to the unwinding angle. However, since dinucleosides are not double helices it may be only fortuitous when this turn angle is related to the helical unwinding.

Oligonucleotide Model Building

In order to determine how these dinucleosides relate to RNA intercalation we have built tri and tetranucleotides based on the dinucleoside crystal structure results. The ground rules we applied for this exercise are that (1) the conformation at the inter-calation site must adhere to the basic geometry observed in the crystal structures of the complexes, i.e., all the torsion angles are held to the values found in A-RNA and only δ and χ were increased. (2) The two-fold relationship of the strands is maintained. (3) Watson-Crick base pairing is maintained within the limits observed in crystal structures and/or as suggested by Levitt.[12] Thus, twisting and tilting within each base pair is permissible.

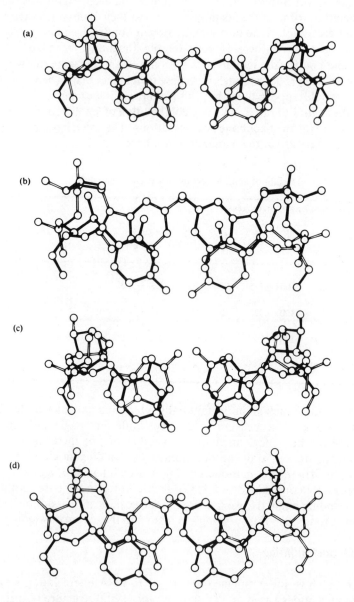

Figure 3. Four intercalation geometries
(a) C3′ endo C3′ endo small base turn angle
(b) C3′ endo C3′ endo larger base turn angle
(c) C3′ endo C2′ endo small base turn angle
(d) C3′ endo C2′ endo larger base turn angle

In the first of two models we built, the objective was to fit the intercalation geometry as closely as possible into a "normal" RNA fragment. Operationally, this means that we are attempting to superimpose a normal RNA dinucleoside on the intercalated dinucleoside so that the resultant structure is:

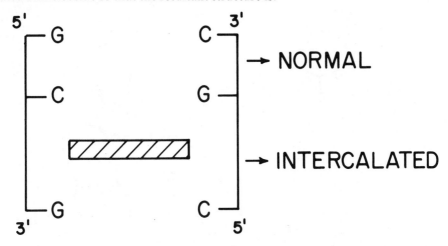

Examination of Table II shows that the χ (5') angle of the cytosine in, for example, CpG-proflavine is 17° or very close to the A-RNA value, whereas the same angle is 85° for the guanosine. *The base-paired nucleosides are thus asymmetric* in intercalated structures. The consequences of this asymmetry on model building are shown in Figures 4a, b. Figure 4a shows that if we overlap the guanosine ribose rings it is impossible to build a duplex oligomer. Figure 4b illustrates an attempt to superimpose the cytosine groups. While the guanine bases approximately overlap, the ribose sugars do not. It is obvious from the figure that some drastic conformational alterations of the A-RNA structure are necessary at the site adjacent to guanosine. Thus we conclude that it is not possible to have a normal RNA conformation adjacent to the experimentally determined intercalated one. Less obvious is that if the intercalation geometry is to be held fixed, and Watson-Crick geometry maintained, then alterations to the site adjacent to the cytosine are also needed. This was done using both manual and computer model building and the results are shown in Figure 5 and tabulated in Table IV. Residue 1, the one adjacent to the 5' cytosine, is in the trans, gauche⁻ conformation for β,γ with an unusually low δ value. Residue 3 is in the gauche⁻, trans conformation for β,γ with a trans value for ϵ. Both are unusual but energetically feasible conformers. The adjacent nucleosides are symmetric with respect to their χ angles which have values closer to the A-RNA ones. Full relaxation of the conformation to normal RNA values is possible at the next site. Other geometrical features of the model are that (1) the adjacent base pairs are about 4Å apart rather than the more usual 3.4Å, (2) there is a close contact (possibly a hydrogen bond) between the guanosine 2' hydroxyl and the ribosyl oxygen (O1') of the cytosine and (3) the phosphate oxygen at the intercalation site and the one at the guanosine adjacent site are 4Å apart. We note here that this latter phosphate geometry is ideal for Mg⁺⁺ ion coordination. In summary, this model which naturally results in extended exclusion intercalation into RNA demonstrates that the "basic" intercalation geometry must also include the adjacent sites that have significant conformational changes.

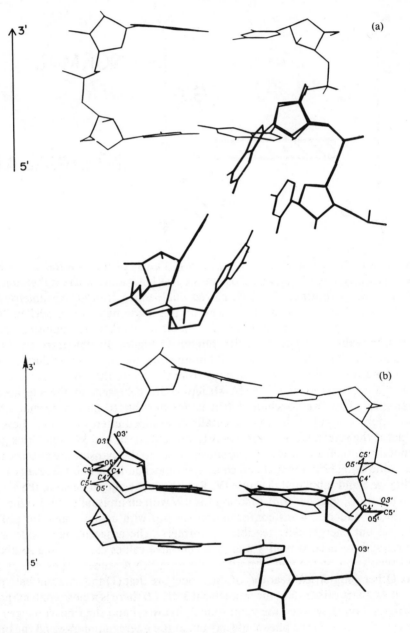

Figure 4. Attempts to fit a fragment of RNA to the intercalated site
(a) exact superposition of ribose of guanosine residues of CpG int. and GpC normal.
(b) best fit of cytosine residues of CpG int. and GpC normal.

Table IV
The Conformation Angles in Degrees
of the Intercalated Tetranucleotide

(a) The extended exclusion model (model 1)

	$\chi(5')$	α	β	γ	δ	ϵ	$\chi(3')$	Sugar Pucker	Base Twist
Residue 1	29	274	210	303	106	64	13	C3′ endo	18°
Residue 2	13	213	281	300	235	50	85	C3′ endo	4°
Residue 3	85	178	275	148	210	175	30	C3′ endo	18°

(b) The neighbor exclusion model (model 2)

Residue 1	85	171	286	143	196	176	13	C3′ endo	31°
Residue 2	13	213	281	300	235	50	85	C3′ endo	23°
Residue 3	85	172	288	143	196	176	13	C3′ endo	31°

A second model was built that would allow for neighbor exclusion binding; i.e., intercalation at every other site. The results are shown in Figure 6 and Table IV. Unlike the extended exclusion model, the adjacent site conformations are symmetrical; both are in the gauche⁻, trans conformation similar to residue 3 of model 1. In this model, the χ angles necessarily alternate and the phosphate oxygen atoms at residues 2 and 3 are 4Å apart. However, in this model this feature would also appear at every other site. So the unequal χ *angles* at the intercalation site give rise to a

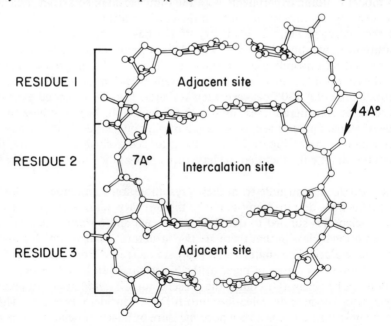

Figure 5. An extended exclusion model for GpCpGpC.

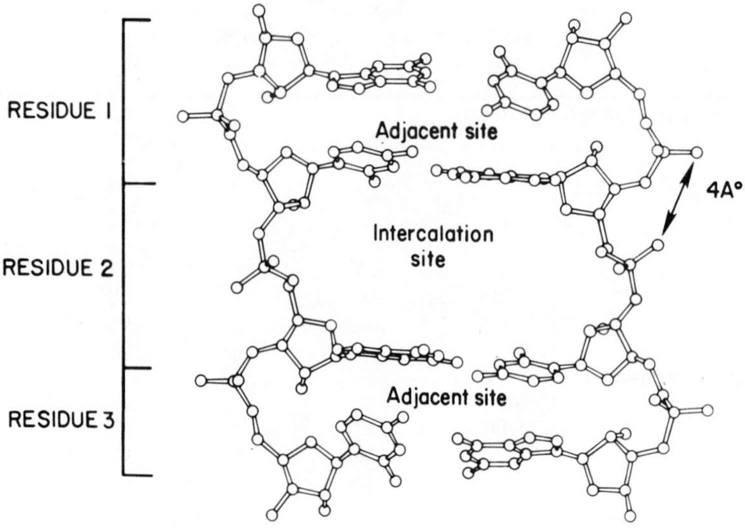

Figure 6. A neighbor exclusion model for GpCpGpC.

very distinctive (and presumably recognizable) phosphodiester geometry. Another feature of the model is that the base pairs are extremely twisted both at the intercalation site and at the adjacent site. Whether or not this type of structure would actually exist can only be answered experimentally.

Our model building experiments with the dinucleoside geometries indicate that the base turn angle is dependent on both the conformation of the nucleoside backbone and the hydrogen bonding geometry of the base pair; it is probably not a good measure of unwinding. However, with these tetranucleoside models it is possible to observe helical unwinding as we show in Figure 7. The internucleoside turn angle over the four residues in a single strand of GpCpGpC in the A-RNA conformation is approximately 90° (30° between each residue). For the extended exclusion model this angle is considerably less as shown in Figure 7b. We note here that this turn angle is related only to the conformation of oligonucleotide. The obvious conclusion from examination of Figure 7 is that the extended exclusion model of GpCpGpC is unwound, despite the fact that the base turn angle at the *intercalation site* is large.

What then do we conclude from these crystallographic and model building studies? The first is that, despite many claims to the contrary, all the crystal structures of the ribodinucleosides studied to date by both Sobell and ourselves belong to the *same* conformational class in that there are two structural changes which allow intercalation to take place. The angles δ and χ (3') are increased from the values found in A-RNA. Within this class the sugar at the 3' terminal can be C3' *endo*, C2' *endo*, or anything in between. (Of course the presence of C2' *endo* puckering makes the incorporation of these dinucleotides into B-DNA considerably easier.) The second is that the base turn angle is a very poor measure of helical unwinding, especially since none of these dinucleosides are "mini double helices." However, model building in

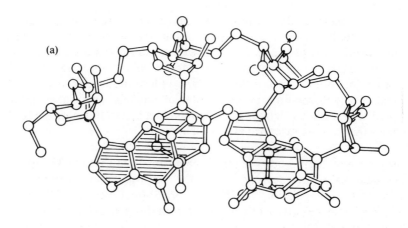

(a)

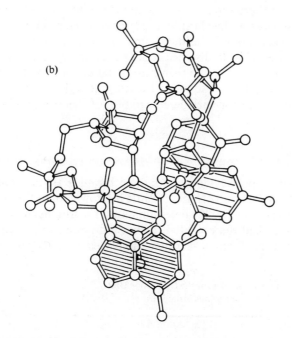

(b)

Figure 7. (a) A tetranucleotide in the A RNA conformation.
(b) The "extended exclusion" conformation for the GpCpGpC.

which the intercalation geometry is held rigorously to that which is observed experimentally shows that these models do indeed imply quite substantial unwinding.

Recognition Properties of the Model Tetranucleotides

Some important properties emerge with respect to the potential interactions of these model tetranucleotides with drugs, small ions and proteins. It is clear from these studies that because of the asymmetry of the χ angles at the intercalation site it is not possible to "polymerise" the geometry so that intercalation can occur at every site. *In other words, it is not necessary to invoke mixed sugar puckering in order to produce either intercalation or neighbor exclusion.* That is not to say that mixed sugar puckering does not occur in RNA or DNA. Indeed the small energy difference between the two conformations makes it not at all unlikely. Our model building studies also predict that while neighbor exclusion binding is certainly possible for RNA, it is less likely than a more extended exclusion binding.

In the course of these studies it also became apparent that the 02′ hydroxyl of the ribose sugar places severe restrictions on the conformational possibilities of the sites adjacent to the intercalation site. The conformational features of these neighboring sites in RNA may be sufficiently different from those in DNA as to explain why certain bifunctional drugs such as echinomycin that must necessarily span across the nucleotide backbone bind to DNA and not RNA.[16-18]

Another consequence of the alternation of the χ angles at the intercalation site is the effect it produces on the spacing of the phosphate groups between residues; the spacing between phosphate oxygen atoms in residues 1 and 2 is 6Å, whereas the spacing is 4Å between residues 2 and 3. This 4Å distance creates a geometry that is ideal for direct Mg^{2+} coordination. It is easy to see then how metal ions in competition for such sites could inhibit electrostatic binding of the drug at higher salt concentrations. Additionally, the potential for enzyme recognition of this site should not be overlooked.

Binding to Single Strands

Planar chromophores such as proflavine and ethidium bromide are also known to bind to single stranded RNA such as tRNA.[19-22] Little is known about the geometry of such binding. The first crystallographic determination of a drug-nucleoside complex[8] was of a structure that had the potential for but did not exhibit intercalative binding into a duplex. Instead, 9-amino acridine molecules were stacked between

Table V
Torsion Angles for the Nucleotide Unit
in Various Structures

	$\chi(5')$	α	β	γ	δ	ϵ	$\chi(3')$
Prof-ApA	-119	272	290	293	175	168	71
Prof-CpG	17	201	290	289	231	52	85
RNA-11	13	213	281	300	175	50	13
UpA 1	12	206	81	82	203	55	37
UpA 2	19	224	164	271	192	54	44
ApA⁺	8	223	283	298	161	53	28
A⁺pA⁺	28	209	77	93	188	56	26
ApU-9 amino-acridine	76	222	100	86	202	63	72

Hoogsteen base paired adenine and uracil residues. This structure therefore provided one kind of model of the type of drug interactions that might be operative in, for example, tRNA. More recently in our laboratories, we have examined the structure of a complex between a dinucleotide, ApA,[7] (which of course has no potential for Watson-Crick pairing) and proflavine. The adenine residues are paired as in acidic polyA[+23] and the proflavine molecules are stacked above and below these pairs in the sequence base pair, proflavine, proflavine, base pair. The striking feature of this particular structure, however, is its conformation. As is obvious from examination of Table V, the dinucleoside phosphate in this structure has little in common with the other dinucleoside phosphates studied to date and does not appear to conform with the rigid nucleotide concept.[24] Indeed, the only conformation angles that resemble the nucleotide geometry in RNA are β, γ, and δ. The explanation for this must lie in the association with the proflavine which means that our task of understanding the conformations of nucleic acids when they are in the environment of other molecules has become considerably more complicated.

Conclusions

It is apparent that, as yet, the crystallographic analyses of drug-dinucleoside complexes have not enabled the problems, posed at the beginning of this article to be unequivocally and definitively answered. Nonetheless, the structural results from a variety of these complexes do show a marked uniformity of conformational behavior, which leads one to suspect that similar features, may in part at least, be displayed at the polynucleotide level. This belief has prompted several model-building exercises. Such extrapolations from the dinucleoside systems must be performed with great care and their limitations clearly understood if they are to have any validity at all.

Acknowledgement

This research was supported in part by grants from NIH GM-21589, CA06927, RR05539, CA22780, the Cancer Research Campaign, a grant from NATO and an appropriation from the Commonwealth of Pennsylvania.

We thank Suse Broyde for her help in evaluating the oligonucleotide models and W. Stallings for discussions of various parts of this manuscript.

References and Footnotes

1. Peacocke, A. R., in "The Acridines" (ed. Acheson, R. M.) Wiley, New York (1973).
2. Lerman, L. S., *J. Mol. Biol. 3*, 18-30 (1961).
3. Tsai, C. C., Jain, S, C., and Sobell, H. M., *J. Mol. Biol. 114*, 301-315 (1977).
4. Jain, S. C., Tsai, C. C., and Sobell, H. M., *J. Mol. Biol. 114*, 317-331 (1977).
5. Sobell, H. M., Tsai, C. C., Jain, S. C. and Gilbert, S. G., *J. Mol. Biol. 114*, 333-365 (1977).
6. a. Neidle, S., Achari, A., Taylor, G. L., Berman, H. M., Carrell, H. L., Glusker, J. P., and Stallings, W. C., *Nature 269*, 304-307 (1977).
   b. Berman, H. M., Stallings, W., Carrell, H. L., Glusker, J. P., Neidle, S., Achari, A., and Taylor, G. *Biopolymers* (in press).

7. Neidle, S., Taylor, G., Sanderson, M., Shieh, H. S., and Berman, H. M., *Nucleic Acids Res. 5*, 4417-4422 (1978).

8. Seeman, N. C., Day, R. O., Rich, A., *Nature 253*, 324-326 (1975).

9. Sakore, T. D., Jain, S. C., Tsai, C. C. and Sobell, H. M., *Proc. Natl. Acad. Sci. USA 74*, 188-192 (1977).

10. Arnott, S., Smith, P. J. C., Chandrasekaran, R., in "Handbook of Biochemistry and Molecular Biology" (ed. Fasman, G. D.) Chemical Rubber Co., Cleveland, Ohio, 3rd Ed., Vol. 2, Sec. B, 411-422 (1976).

11. Berman, H. M., Neidle, S., and Stodola, R. K., *Proc. Natl. Acad. Sci. USA 75*, 828-832 (1978).

12. Levitt, *Proc. Natl. Acad. Sci. USA 75*, 640-644 (1978).

13. Alden, C. S. and Arnott, S., *Nucleic Acids Res. 2*, 1701-1717 (1975).

14. Alden, C. S. and Arnott, S., *Nucleic Acids Res., 4*, 3855-3861 (1977).

15. Wang, A. H. J., Nathans, J., van der Marel, G., van Boom, J. H., and Rich, A., *Nature 276*, 471-474 (1978).

16. Waring, M. J. and Wakelin, L. P. G. *Nature 252*, 653-657 (1974).

17. Wakelin, L. P. G. and Waring, M. J. *Biochem. J. 157*, 721-740 (1976).

18. Ughetto, G. and Waring, M. J. *Mol. Pharmacology 13*, 579-584 (1977).

19. Urbanke, C., Romer, R., and Mausa, G. *Eur. J. Biochem. 33*, 511-516 (1973).

20. Liebman, M., Rubin, J., Sundaralingam, M. *Proc. Natl. Acad. Sci. USA, 74*, 4821-4825 (1977).

21. Dourlent, M. and Helene, C. *Eur. J. Biochem. 23*, 86-95 (1971).

22. Finkelstein, T. and Weinstein, I. B. *J. Biol. Chem. 242*, 3763-3768 (1967).

23. Rich, A., Davies, D. R., Crick, F. H. C., and Watson, J. D. *J. Mol. Biol. 3*, 71-86 (1961).

24. Yathindra, N. and Sundaralingam, M. *Biopolymers 12*, 297-314 (1973).

DNA Structure and its Distortion by Drugs

D. M. Crothers, N. Dattagupta and M. Hogan
Department of Chemistry
Yale University
New Haven, Connecticut 06520

Several papers in this volume attest to a strong current interest in DNA structure and its distortion by drugs. In our contribution we summarize our use of the technique of transient electric dichroism to study DNA structure and how it is altered by drug binding.

Electric Dichroism of DNA

In an electric dichroism experiment one applies an electric field to a solution of macromolecules, and observes the change in absorbance of polarized light. If the molecules in the solution are electrically anisotropic, for example because of a permanent or induced dipole moment, the electric field causes them to orient. If, in addition, the molecules are optically anisotropic, orientation produces a change in the absorbance of polarized light.

Rod-like DNA molecules provide a simple illustration of this technique. Application of an electric field causes them to orient with their long axis parallel to the field. Since the 260 nm transition is π - π^*, it is polarized in the plane of the base, and therefore is approximately perpendicular to the helix axis. Hence orientation should cause a decrease in the absorbance of light polarized parallel to the electric field.

The absorbance change is expressed quantitatively through the reduced dichroism ϱ,

$$\varrho = \frac{A_\parallel - A_\perp}{A}$$

in which $A_\parallel$ and $A_\perp$ are respectively the absorbance for light polarized parallel and perpendicular to the electric field, and A is the absorbance without a field. The reduced dichroism of a regular helical molecule depends on two quantities: The fractional orientation Φ, and the angle α between the transition moment and the orientation axis. Specifically[1,2]

$$\varrho = 3/2(3\cos^2\alpha - 1)\Phi \tag{1}$$

341

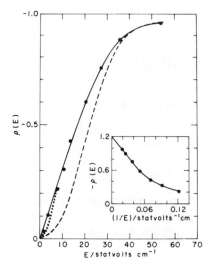

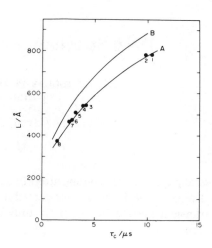

Figure 1. Dependence of the reduced dichroism ϱ on electric field strength E, compared with the calculated variation assuming a permanent (....) or induced (———) dipole moment. The solid line gives the calculated curve assuming the ion flow[3] model. The experimental points are for a 250 base pair fragment of calf thymus DNA[3][Na+] = 2.5 mM, T = 12°C.

Figure 2. Variation of the orientational relaxation time τ_c with DNA length (L). Curve A gives the theoretical curve calculated using Broersma's relationship for a cylinder[8], and curve B gives the results if an ellipsoid of revolution is assumed.

The two parameters α and Φ in equation (1) are separated by extrapolating ϱ to infinite field, and hence perfect orientation, with $\Phi = 1$. An example is shown for DNA in Figure 1. Generally, it is found that ϱ is linear in 1/E at high values of the field, so that the extrapolated intercept at 1/E = 0 is unambiguous. Linear variation of ϱ with 1/E implies that the apparent molecular dipole moment is independent of the field. However, as we have argued[3] this does not necessarily imply that the molecule contains a permanent dipole moment. Indeed, given the two-fold symmetry of the two sugar-phosphate chain in DNA, a permanent dipole moment is implausible. It has been suggested that the field-induced polarization of DNA saturates at low fields, producing a constant dipole moment[4]. Alternatively, we have proposed that orientation results from asymmetry of the counterion atmosphere induced by the flow of ions through the solution.[3] This mechanism also predicts an orienting force which is linear in the field, as observed. Involvement of the ion atmosphere is verified by the dependence of the apparent dipole moment on the ionic strength of the solution.[3]

The amplitude of the extrapolated dichroism observed for DNA in Figure 1 is about -1.2, significantly different from the expected value of -1.5 if the base transition moments were strictly perpendicular to the orientation axis. It might be argued that this results from slight bending of the DNA molecules in solution. However, it can be shown that certain drug molecules bound at low levels to these same DNA molecules exhibit an extrapolated dichroism of -1.5, which would be impossible were there significant bending. Hence we conclude that the electric field not only orients the molecules, but also straightens them against the thermal bending forces.

The observed limiting dichroism of -1.1 to -1.2 implies that the base transition moments are tipped at an angle of about 17° from the plane perpendicular to the DNA helix axis. It is believed that these transition moments lie roughly along the short axis of a DNA base pair, and thus should be roughly coincident with the C_2 symmetry axis of the helix. However, since the C_2 axis must be perpendicular to the helix axis, we conclude that an individual base transition moment must be tipped about 17° from the C_2 symmetry axis, and therefore that the short axis of each base in a base pair is not coincident with the C_2 axis. This conclusion is not compatible with the classical B form structure of DNA[5], or with any structure in which both bases in a pair lie in a plane containing the C_2 axis.

The only DNA model we know of which is consistent with our observations is that proposed by Levitt[6] on the basis of energy minimization calculations. He suggested that the base pair is not planar, but twisted. The result is a structure in which the bases in a pair have a propeller shape, with each transition moment axis tipped about 17° from the perpendicular plane, as required by our measurements.

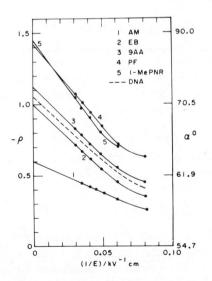

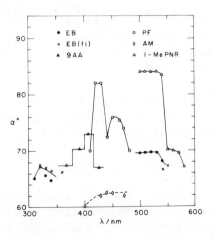

Figure 3. Representative field extrapolation for drug-DNA complexes in 2.5 mM Na⁺solution, 11°C[10] (1) AM = actinomycin, 440 nm (2) EB = ethidium, 520 nm; (3) 9AA = 9-amino acridine, 427 nm; (4) PF = proflavine, 430 nm (5) 1-MePNR = 1-methyl phenyl neutral red, 530 nm.

Figure 4. Wavelength (λ) dependence of the angle α measuring the orientation of the observed transition moment to the DNA helix axis[10]. Conditions as in Figure 6. x = ethidium dichroism detected by fluorescence.

The other main parameter measured in our electric dichroism experiments is the time required for orientation in the field. The relationship between the observed exponential time constant τ_c and the rotational diffusion constant D_r depends on whether the molecule orients in a single preferred direction (permanent dipole moment orientation) or whether the dipole moment can be induced in either direction along the helix axis. The equations are:[7]

$$\tau_c = 1/2D_r \qquad \text{(permanent moment) (2a)}$$
$$\tau_c = 1/6D_r \qquad \text{(induced moment) (2b)}$$

We verified that Equation 2b describes the field-induced orientation for DNA by comparing the measured times with values calculated from the known molecular dimensions.[3] The excellent agreement between experiment and the hydrodynamic theory of Broersma[8] is shown in Figure 2. The theory predicts that τ_c varies with L^3, so one expects to be able to detect small variations in L from measurement of τ_c.

Dichroism of Intercalated Drugs

In recent work we have extended these transient electric dichroism experiments to drug-DNA complexes.[9-11] Figure 3 shows typical dichroism values for several different drugs.[10] As shown in Figure 4, the angular orientation α of the transition moment to the helix axis depends on wavelength. These results show clearly that the intercalated chromophores are not perpendicular to the helix axis.

Figure 5 shows a schematic model that illustrates the tilt (θ_1) and twist (θ_2) found for intercalated drugs. We found that θ_1 is roughly $20 \pm 8°$, and θ_2 is $10 \pm 8°$. We propose that the tilt θ_1 is an adaptation to the unfavorable contacts that develop in a model with the DNA base pairs twisted (Figure 5A) when an adjacent base pair is flattened by the drug (Figure 5B). Tilting the drug allows a gradual variation of the angle of stacking of purines through the intercalation site (Figure 5C).

Length changes in intercalated complexes also show considerable variability. Figure 6 shows the results found compared with the simple expectation value of 3.4Å per bound drug. 9-Amino acridine is notable for its small length increment.

Kinking of DNA by Irehdiamine

Irehdiamine is a steroidal diamine which induces considerable DNA hyperchroism upon binding. We investigated the complex using the dichroism method, and found results consistent with a substantial distortion of DNA structure.[9] Small amounts of the drug were found first to decrease the length of bacterial DNA molecules, followed by a length increase when the binding approached saturation. Furthermore, as shown in Figure 7, there was a substantial reduction in the DNA dichroism with increasing degrees of binding.

A DNA length decrease followed by an increase is consistent with a kinked structure

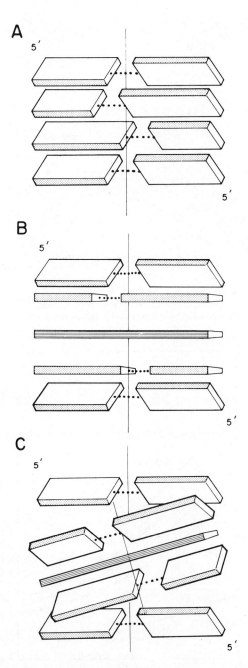

Figure 5. A model for intercalation into propeller twisted DNA. (A) Schematic representation of the twisted DNA structure suggested by Levitt[6]. (B) Hypothetical binding intermediate which depicts the unfavorable contacts between flattened base pairs and adjacent propeller-twisted base pairs. (C) Model structure consistent with the dichroism data. Unfavorable contacts have been reduced by unstacking the 5' pyrimidines. The complex is presented as being wedge-shaped as viewed along the short base pair axis to emphasize the variability seen for the twist of the complex[10].

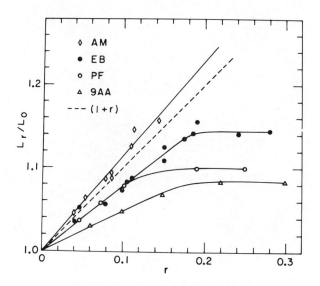

Figure 6. Dependence of the length of drug-DNA complexes on degree of binding r (drug molecules per base pair). The dotted line corresponds to the idealized 3.4Å length increase per drug.

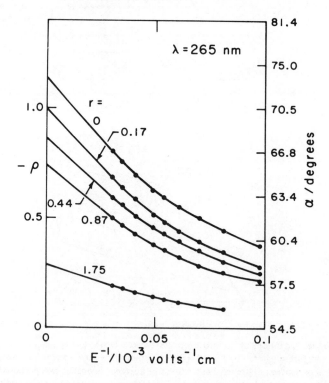

Figure 7. Field dependence of the reduced dichorism ϱ at different ratios r of added IDA per DNA base pair[9].

for the irehdiamine complex as proposed by Sobell, *et al..*[12] Furthermore, the hyper-chroism induced on binding is strongly suggestive of loss of base stacking, as are the proton NMR changes recently observed by Patel[13] for an analogous system. However, one feature of our results is not consistent with the detailed model proposed by Sobell, *et al.*: The superhelix in their β-kinked structure has a C_2 symmetry axis nearly coincident with the base pair short axis. (Actually, it lies between the two base pairs that form a section between kinks.) Hence in their model the base transition moments should be approximately perpendicular to the helix axis. Instead, we find that they are tipped by about 30° from the perpendicular plane.

Allosteric Transformation of Calf Thymus DNA by Distamycin

The final example which we present of distortion of DNA by a bound drug is an unusual case in which we observed a long range, cooperative alteration of DNA

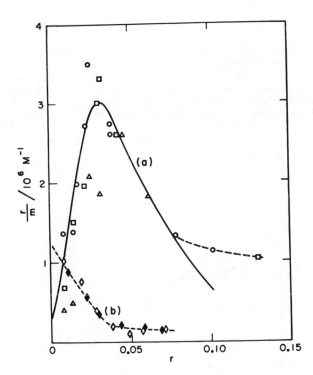

Figure 8. Scatchard plot of the binding equilibria of calf thymus (a) and *E. coli* (b) DNAs with distamycin at 11°C in 66 mM Na⁺ buffer, pH 6.5. Data were determined by a phase partition method[11]. The solid curve (a) was calculated from a statistical mechanical theory of cooperative allosteric transition of calf thymus DNA from its initial form to another structure that has higher affinity for distamycn.

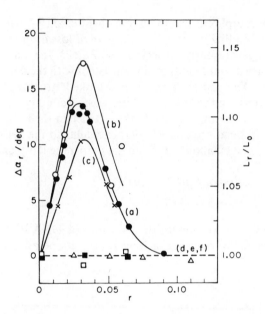

Figure 9. Apparent fractional length changes (Lr/Lo) (a, c, d, f) and ethidium orientation angle changes ($\Delta\alpha_r$) as a function of distamycin binding for calf thymus (a, b, c) and *E. coli* (d, e, f) DNA samples. T = 11°C, buffer 2.5 mM in Na$^+$, pH 7.0. (a) ●, length change for calf thymus DNA in absence of ethidium (b) O, ethidium orientation angle change with calf thymus DNA, $r_{ethidium}$ = 0.1 and variable $r_{distamycin}$ (c) x, length change for calf thymus DNA in the presence of ethidium, $r_{ethidium}$ = 0.1; (d) △, length change of *E. coli* DNA without ethidium (e) □, length change of *E coli* DNA with ethidium, $r_{ethidium}$ = 0.1; (f) ■, ethidium angle change of *E. coli* DNA, $r_{ethidium}$ = 0.1.

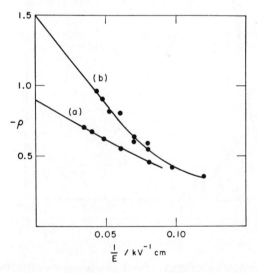

Figure 10. Extrapolation of the observed dichroism of ethidium ($r_{ethidium}$ = 0.1) to infinite field in the absence (a) and presence (b) of distamycin, $r_{distamycin}$ = 0.03. Calf thymus DNA samples, conditions as in Figure 9.

structure. We found this effect for binding of the drugs distamycin and netropsin to calf thymus DNA.[11] As shown in Figure 8, the Scatchard isotherm for drug binding (determined by a phase partition technique) is strongly cooperative (sloping upward at low r) for calf thymus DNA, but not for *E. coli* DNA.

The cooperative binding of distamycin to calf thymus DNA is accompanied by strong alteration of its structure, as summarized in Figure 9 and 10. Figure 9 shows that a striking length increase, and subsequent decrease, accompany increasing binding. Figure 10 shows that diatamycin binding, at a ratio of one drug to 30 base pairs, has a pronounced effect on the binding geometry of ethidium. The dichroism values observed for eithidium in the distamycin (r = 0.03)-DNA-ethidium (r = 0.1) complex extrapolate to -1.5. Hence the transition moments of virtually *all* the bound ethidium residues become perpendicular to the helix axis when there is only one distamycin for every 30 base pairs. This contrasts with a measured[10] tilt of 23° for the long axis of bound ethidium when no distamycin is present.

We have interpreted these results as implying a long range allosteric conversion of DNA from one structural form to another.[13] The results can be explained by a model in which form I, found in absence of drugs, is more favorable in free energy by about 20 cal per base pair. Form II has a distamycin affinity higher by about a factor 15 than form I, which accounts for the structural conversion induced by the drug. The cooperative nature of the transition is accounted for by an interfacial free energy of about 3-4 kcal/mol. As a consequence of this boundary free energy, less than 1 percent of the DNA is calculated to be in Form I in absence of the drug.

These observations of long range structural alteration of DNA by a bound drug support the possibility that DNA function can be influenced by proteins bound some distance away, as suggested by experiments on the mutual influence of two arabinose promoters.[14] However, since other DNAs did not exhibit the same structural change upon distamycin binding, we suspect that the details of base modification or sequence may be significant for the transmission of long range effects.

References and Footnotes

1. O'Konski, C. T., Yoshioka, K. and Orttung, W.H. *J. Phys. Chem. 63,* 1558-1565 (1959).
2. Allen, I. S. and Van Holde, K. E. *Biopolymers 10,* 865-881 (1971).
3. Hogan, M., Dattagupta, N. and Crothers, D. M. *Proc. Nat. Acad. Sci., USA 75,* 195-199 (1978).
4. Neumann, E. and Katchalsky, A. *Proc. Nat. Acad. Sci., USA 69,* 993-997 (1971).
5. Arnott, S. and Hukins, D. W. L. *Biochem. Biophys. Res. Commun. 47,* 1504-1509 (1972).
6. Levitt, M. *Proc. Nat. Acad. Sci., USA 75,* 640-644 (1978).
7. Tinoco, I., Jr. *J. Am. Chem. Soc. 77,* 4486-4489, (1955).
8. Broersma, S. *J. Chem. Phys. 32,* 1626-1631 (1960).
9. Dattagupta, N., Hogan, M. and Crothers, D. M. *Proc. Nat. Acad. Sci. USA 75,* 4286-4290 (1978).
10. Hogan, M., Dattagupta, N. and Crothers, D. M. *Biochemistry 18,* 280-288 (1979).
11. Hogan, M., Dattagupta, N. and Crothers, D. M. *Nature 278,* 521-524 (1979).
12. Sobell, H. M., Tsai, C., Gilbert, S. G., Jain, S. C. and Sakore, T. D. *Proc. Nat. Acad. Sci., USA 73,* 3068-3072 (1976).
13. Patel, D. and Canuel, L. L. *Proc. Nat. Acad. Sci., USA 76,* 24-28 (1979).
14. Hirsch, J. and Schleif, R. *Cell 11,* 545- 550 (1977).

Spectroscopic Studies of Actinomycin and Ethidium Complexes with Deoxyribonucleic Acids

Thomas R. Krugh, John W. Hook, III, and M.S. Balakrishnan
Department of Chemistry
University of Rochester
Rochester, New York 14627

and

Fu-Ming Chen
Department of Chemistry
Tennessee State University
Nashville, Tennessee 37203

Introduction

The structure, pharmacological activity, and the interaction of the actinomycins with double-stranded DNA has been the subject of an extensive number of studies (for example, see the review by Meienhofer and Atherton,[1] and the reviews and articles cited therein). The chemical structure of actinomycin D (just one of a large class of related molecules) is shown in Figure 1. Note that the (relatively) planar chromophore, commonly called the phenoxazone ring, has two cyclic pentapeptide rings attached to it. The combination of the intercalating chromophore and the cyclic pentapeptides is responsible for some of the unusual properties observed for the binding of actinomycin to DNA. For example, in 1968 Muller and Crothers[2] showed that the association of actinomycin to DNA is characterized by five separate rate constants. Even more interesting were Muller and Crothers' observations that the actinomycins dissociate very slowly from DNA and that the slow dissociation is apparently correlated with the pharmacological activity of the drugs (i.e. the inhibition of RNA polymerase). The dissociation relaxation curve (from calf thymus DNA) required a minimum of three separate exponential curves to fit the decay curve mathematically.[2] The actinomycins also exhibit a general requirement for the presence of guanine at the binding site (e.g., see Wells and Larson (1970)[3]; Kersten (1961)[4]; and Goldberg, et al. (1962)[5]). This unusual requirement has been frequently illustrated by citing the observation[5] that guanine-free DNA, such as poly-(dA-dT)•poly(dA-dT), does not bind actinomycin. An important exception to this observation will be presented below.

Ethidium bromide (Figure 1) is an intercalating agent with mutagenic properties. References 6-11 provide an introduction to the extensive studies on ethidium-DNA complexes as well as the important role of ethidium as a fluorescent probe. Our studies on ethidium were initiated in 1972 to determine if this relatively simple drug molecule would exhibit preferential binding to oligonucleotides. A second goal of the initial ethidium experiments was to provide another test of the use of

351

oligonucleotides as model systems for studying drug-nucleic acid complexes. In the initial experiments we showed that ethidium does exhibit a pronounced preference for binding to certain sequences with both the ribo- and deoxyribodinucleoside monophosphates.[11-13] The most striking observation was that ethidium binds more strongly to the self-complementary pyrimidine (3'-5') purine dinucleosides (or dinucleotides) than to their sequence isomers, the purine (3'-5') pyrimidine dinucleotides.[11-13] Sobell and coworkers determined the x-ray structures of cocrystalline complexes of ethidium with the self-complementary pyrimidine (3'-5') purine dinucleoside monophosphates.[14] The crystal structures of several other drug-nucleic acid complexes have also been reported during the last four years.[15-18]

Two drugs which in our studies have exhibited no marked preference between pyrimidine (3'-5') purine and purine (3'-5') pyrimidine deoxydinucleotides are daunorubicin and actinomine (Figure 1).[19] Daunorubicin (frequently called

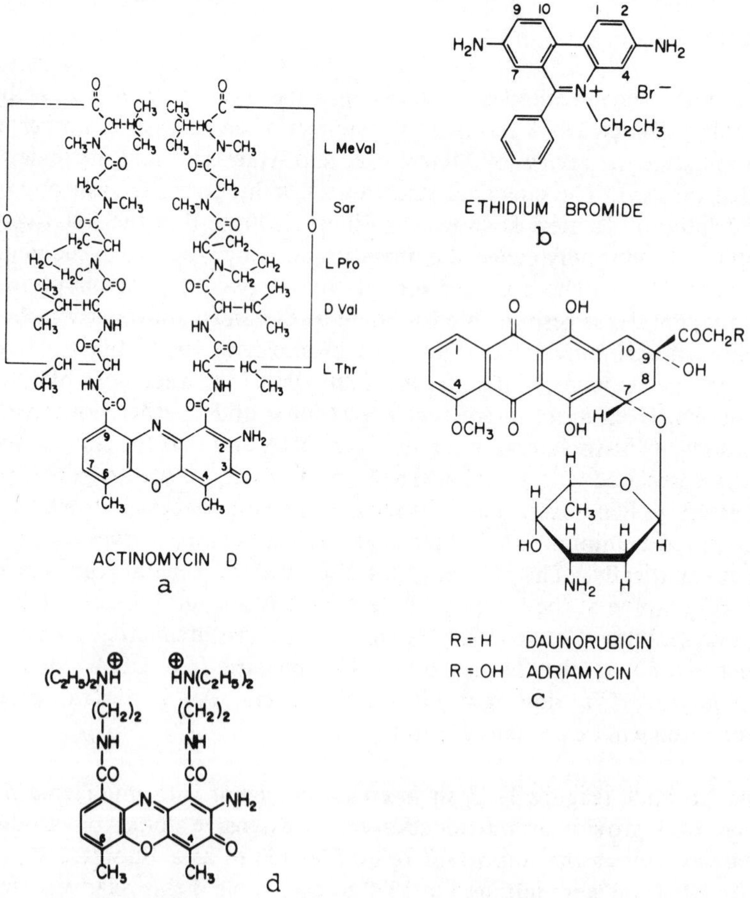

ACTINOMYCIN D
a

ETHIDIUM BROMIDE
b

R = H   DAUNORUBICIN
R = OH   ADRIAMYCIN
c

d

Figure 1. The chemical structures of (a), actinomycin D, (b) ethidium bromide, (c), daunorubicin and adriamycin, and (d), actinomine.

daunomycin) and the related antibiotic adriamycin are intercalating drugs which are important drugs for cancer chemotherapy.[20-22] Actinomine is an actinomycin related derivative which has been used to model the interaction of the actinomycin chromophore in binding to DNA.[2] It should be noted, however, that actinomine has two positively charged amino groups at neutral pH,[2] whereas the actinomycins are uncharged, and thus actinomine is as much different from actinomycin (two positively charged side chains versus two uncharged cyclic pentapeptide groups) as it is similar (identical phenoxazone rings).

Ethidium and Actinomycin Bind Non-competitively to Calf Thymus DNA at Low r Values

In the model oligonucleotide binding studies, ethidium bromide has exhibited a marked preference for binding to pyrimidine (3'-5') purine binding sites when compared to the purine (3'-5') pyrimidine binding sites;[11/12/13/23/24] the (dG-dG)•(dC-dC) site has also been shown to be a favorable binding site in the oligonucleotide binding studies.[25] On the other hand, actinomycin exhibits a general preference for binding to guanine (3'-5') pyrimidine sequences,[3/26-29] although (dG-dG)•(dC-dC) is also a favorable binding site.[2/3] In considering the simultaneous binding of actinomycin D and ethidium bromide to calf-thymus DNA, one expects that if the sequence preferences discussed above hold true when these molecules bind to DNA, then ethidium and actinomycin D should bind primarily in a non-competitive fashion, at least at low levels of total bound drug. This expectation has been confirmed by Reinhardt and Krugh,[7] as illustrated by the data in Figure 2a. The roughly parallel lines (at least for low r values) of the ethidium bromide titrations (with and without actinomycin D) are indicative of non-competitive binding.[30/31] On the other hand,

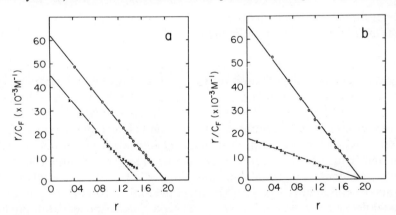

Figure 2.(a) Competition of actinomycin D with ethidium bromide for binding sites on DNA. Fluorescence Scatchard plots of EthBr (concentration 1.0-29.1 μM) in 50 mM Tris-HCl (pH 7.5), 0.2 M NaCl buffer (o) and in the presence of Act D, [DNA•P]/[Act D] = 5 (x).; (b) Competition of actinomine with ethidium bromide for binding sites on DNA. Fluorescence Scatchard plots of ethidium bromide (concentration 1.0-28.3 μM) bound to calf thymus DNA ([DNA•P] = 3.5 μM) in 50 mM Tris-HCl (pH 7.5), 0.2 M NaCl buffer (o) and in the presence of actinomine, ([DNA•P]/[actinomine]) = 0.5 (x). Reproduced with permission from Ref. 7.

the binding of ethidium bromide to calf thymus DNA in the presence of actinomine is characterized by a competition for binding sites, as illustrated by the data in Figure 2b. This competition is consistent with the oligonucleotide binding studies with actinomine, which shows no strong preference for purine (3'-5') pyrimidine or pyrimidine (3'-5') purine sequences.[7] These types of approaches, at both the oligonucleotide and polynucleotide levels, will lead to a more complete understanding of the sequence preferences that drugs exhibit, and more importantly, a molecular understanding of the origin of these sequence preferences for various classes of drugs.

A further test of the relative preferences for ethidium binding to the (dC-dG)• (dC-dG) sites compared to the (dG-dC)•(dG-dC) sites was done using the method of continuous variation to study complex formation of ethidium to the deoxyhexanucleotides (dG-dC)$_3$ and (dC-dG)$_3$. The data in Figure 3 show that the most favorable complex formed in the ethidium experiment with (dG-dC)$_3$ involves the complex formation between two ethidiums and two (dG-dC)$_3$ strands (i.e., the lines intersect at the 0.5 mole fraction point). With (dC-dG)$_3$ the stoichiometry of complex formation is 3 ethidiums: 2 strands (or 3 ethidiums per helix), since the maximal point of complex formation is at the 0.6 ethidium mole fraction point. These stoichiometries of complex formation have been independently verified by circular dichroism and fluorescence titrations [M. S. Balakrishnan and T. R. Krugh, to be published]. The 3:2 complex formation with (dC-dG)$_3$ is consistent with intercalation at the three dC-dG sequences of the double helix while the 2:2 stoichiometry for the binding of ethidium with (dG-dC)$_3$, which has two dC-dG sequences, is evidence that ethidium has a strong preference for binding at the dC-dG sequences (as opposed to the three (dG-dC) sequences).

Fluorescence Lifetime Measurements of Ethidium
Complexes in Solution and in the Solid State

The cocrystalline complexes of ethidium bromide with the dinucleoside monophosphates UpA and CpG used for the x-ray analysis had a stoichiometry of two ethidiums and two dinucleoside monophosphates.[14/15/32] From the solution experiments we have concluded that when UpA or CpG are present in large excesses compared to the ethidium concentration, the predominant complex formed is one in which essentially all the ethidium is intercalated into a miniature double helix.[6] Reinhardt and Krugh[7] have performed fluorescence lifetime measurements on two separate cocrystalline complexes of ethidium with CpG (2:2 stoichiometry) as well as on a solution of ethidium (1.3×10^{-4} M) with CpG (6.9×10^{-3} M). In solution the experimental decay curve was well represented by a single exponential decay function with a lifetime of 23 nsec, a value which is in close agreement with the previously measured value for ethidium intercalated into double-stranded RNA.[33] In the cocrystalline complexes two distinct fluorescence decays were observed with lifetimes of approximately 4.5 and 11 nsec. The two distinct lifetimes observed in the crystalline state are consistent with the two environmentally distinct ethidiums observed in the x-ray structures and thus the fluorescence experiments provide a link

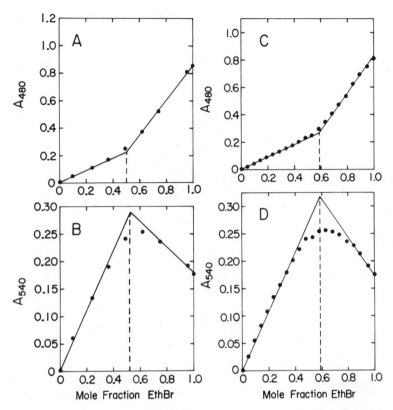

Figure 3. Absorbance vs. mole fraction EthBr for the continuous variation experiment involving EthBr plus (dG-dC)₃ or (dC-dG)₃. The concentration of EthBr plus hexanucleotide strand was kept constant at 150 μM throughout the experiment. (dG-dC)₃ absorbance monitored at 480 nm (A) and 540 nm (B). (dC-dG)₃ absorbance monitored at 480 nm (C) and 540 nm (D). Figures A and B reproduced with permission from reference 25. Figures C and D from M. S. Balakrishnan and T. R. Krugh, to be published.

between the solid state studies and the solution studies.

Daunorubicin and Adriamycin Facilitate Actinomycin Binding to poly(dA-dT)•poly(dA-dT)

As noted above, the classic example of the actinomycin requirement for guanine at the intercalation site is the observation that actinomycin D does not bind to double-stranded poly(dA-dT)•poly(dA-dT). However, the circular dichroism spectra of solutions of poly(dA-dT)•poly(dA-dT) with various combinations of either daunorubicin or adriamycin clearly show that these drugs facilitate the binding of actinomycin to this polynucleotide (Figure 4). We believe that this is the first example of cooperative binding of two intercalating drugs. Neither ethidium bromide nor acridine orange facilitates the binding of actinomycin D to poly-(dA-dT)•poly(dA-dT) (although both of these molecules intercalate into the polynucleotide) which illustrates the specificity of the cooperative interaction of ac-

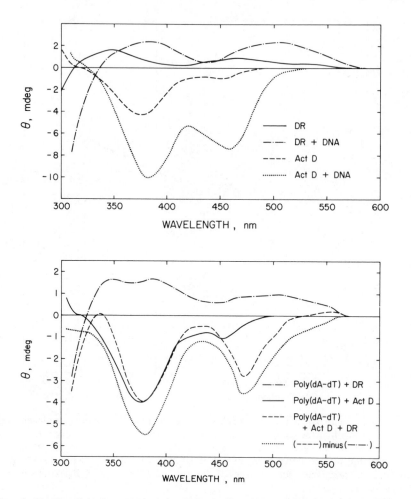

Figure 4. Circular dichroism spectra of solutions of: *a*, 8.5 × 10⁻⁶M daunorubicin (DR) and ac-tinomycin D (Act D) alone and in the presence of 8.5 × 10⁻⁵M DNA; *b*, 8.5 × 10⁻⁶ M daunorubicin plus 8.5 × 10⁻⁵M poly(dA-dT)•poly(dA-dT); 8.5 × 10⁻⁶M actinomycin D plus 8.5 × 10⁻⁵M poly(dA-dT)•poly(dA-T); 8.5 × 10⁻⁶M actinomycin D plus 8.5 ×10⁻⁶ daunorubicin plus 8.5 ×10⁻⁵M poly(dA-dT)•poly(dA-dT). The curve ...... was calculated by subtracting the poly(dA-dT)•poly(dA-dT) + DR spectrum from the poly(dA-dT)•poly(dA-dT) + DR + Act D spectrum which, to a first approxima-tion, is an estimate of the circular dichroism spectrum of actinomycin D when bound to poly(dA-dT)•poly(dA-dT). All spectra were recorded on a Jasco J-40 circular dichroism instrument in a 4-cm path length cell at 20°C. All solutions contained 10 mM potassium phosphate buffer, pH 7.0. Essentially similar results were obtained at lower drug-to-phosphate ratios, as well as in the presence of 0.1 M NaCl. The spectra in which adriamycin were used in place of daunorubicin gave qualitatively similar results. The circular dichroism spectra in which ethidium bromide is used in combination with actinomycin D and poly(dA-dT)•poly(dA-dT) do not show the appearance of the large negative band in the 440-480 nm region. Reproduced with permission from reference 22.

tinomycin and daunorubicin when both are bound to poly(dA-dT)•poly(dA-dT). It will be interesting to explore the nature of the structural changes in the confor-mation of poly(dA-dT)•poly(dA-dT) that result from the binding of either daunorubicin or adriamycin because these conformational changes in the double helical geometry of poly(dA-dT)•poly(dA-dT) are presumably responsible for the facilitation of actinomycin D binding to this polynucleotide.

Enzyme Inhibition Studies

Actinomycin D is a very effective inhibitor of RNA polymerase when native DNA is used as a template (e.g., see references 1, 2, 34). However, when poly-(dA-dT)•poly(dA-dT) is used as a template there is no inhibition of *E. coli* RNA polymerase (Figure 5), an observation which has also been taken as evidence for the non-binding of actinomycin D to this guanine-free polynucleotide. If sufficient daunorubicin is added to the solutions to facilitate actinomycin D binding to the template, one observes that actinomycin D does not enhance the inhibition of *E. coli* RNA polymerase (Figure 5, from A. H. McHale, S. S. Holcomb, R. Josephson, and T. R. Krugh, unpublished data). In other words, even though actinomycin D binds to poly(dA-dT)•poly(dA-dT) in the presence of daunorubicin, actinomycin D does not appear to interfere with RNA polymerase activity. These experiments suggested a direct test of the hypothesis that the inhibition of RNA polymerase is closely coupled to the extremely slow dissociation of actinomycin from calf thymus DNA[2] since the failure of actinomycin to inhibit RNA polymerase predicts a rapid dissociation rate for the actinomycin D-poly(dA-dT)•poly(dA-dT) daunorubicin complex. We have experimentally verified that actinomycin D dissociates much faster from poly(dA-dT)•poly(dA-dT) than from calf thymus DNA.

Actinomycin Dissociation Kinetics

Absorption, CD and fluorescence studies in our laboratory have revealed that ac-

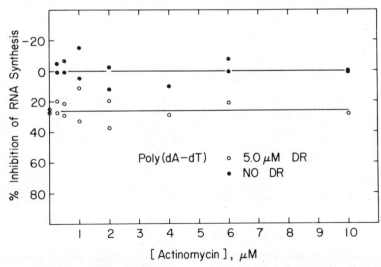

Figure 5. Drug induced inhibition of RNA synthesis using DNA dependent RNA polymerase from *E. coli* with poly(dA-dT)•poly(dA-dT) as the template as a function of actinomycin D concentration. [A. H. McHale, S. S. Holcomb, R. Josephson and T. R. Krugh, unpublished data.]

tinomycin D and its fluorescent derivative, 7-amino-actinomycin D, bind to poly(dG-dC)•poly(dG-dC) roughly an order of magnitude more strongly than to calf thymus DNA. The saturation binding appears to occur at a phosphate-to-drug ratio (P/D) of approximately 8:1. There is some uncertainty in this ratio, however, due to conflicting values for the molar extinction coefficient of poly(dG-dC)•poly(dG-dC).[35-37]

Three representative dissociation curves for actinomycin D-poly(dG-dC)•poly(dG-dC) complexes are shown in Figure 6. Two characteristics are evident in these dissociations: (1) The time dependence of the absorbance is well described by a single exponential; and (2) The apparent dissociation lifetime, τ, depends on the phosphate/drug (P/D) ratio (Figure 7). The obvious linearity of the dissociation curves throughout the whole range of observed τ's precludes explaining the P/D dependence of τ with two exponential processes in which the relative magnitudes of the individual ΔA's vary. This would be the case, for example, if there were two classes of binding sites whose relative populations vary with P/D. Any two exponential model in which the apparent τ varies over a factor of two, as is observed here, would predict that at least one of the three curves in Figure 6 should exhibit a significant concave curvature. Such curvature is not observed

It is intriguing to note that the dissociating complex's behavior is dependent on its history. For instance, when the 7.1:1 (P/D) complex is 60 percent dissociated, there is approximately one drug molecule bound for every 18 phosphates, as at the beginning of the 17.7:1 dissociation shown in the middle curve of Figure 6. However, though the phosphate/bound drug ratio is the same in these two cases, the behavior

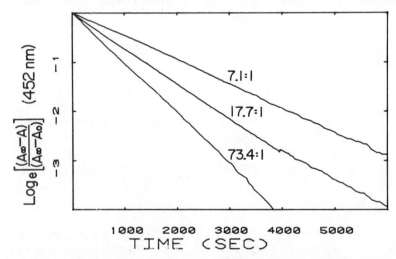

Figure 6. Absorbance changes due to dissociation of actinomycin D from poly(dG-dC)•poly(dG-dC) in 0.1 M NaCl, 10^{-4} M EDTA, 0.01 M sodium phosphate buffer at pH 7.0 and 22°C. The three curves are for phosphate-to-drug ratios of 7.1:1, 17.7:1, and 73.4:1 as marked. In these experiments the initial drug concentration ranged from 7 to 10 × 10^{-5}M. Dissociation was induced by mixing with a 5 percent solution of sodium dodecyl sulfate (SDS).[2]

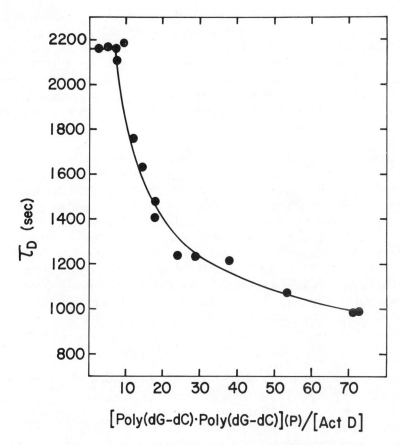

Figure 7. Dissociation lifetimes for the poly(dG-dC)•poly(dG-dC)-actinomycin D complex plotted as a function of phosphate/drug ratio. The conditions are the same as in Figure 6.

is quite different. One possible explanation for this behavior would be that the polynucleotide structure might be locally distorted by the drug molecule in some manner which slowed the dissociation of the other drug molecules bound to the polynucleotide. If relaxation from such a perturbed structure was long with respect to the dissociation time of the drug, the dissociation kinetics of the drug would appear as a single exponential with τ dependent upon the initial P/D ratio. Such a structural distortion might be expected to alter the circular dichroism (CD) spectrum of the polynucleotide. A 7.1:1 (P/D) mixture of poly(dG-dC)•poly(dG-dC) and actinomycin D was dissociated with SDS, and the 240-500 nm CD spectrum was observed periodically following the initial mixing. The change in θ at all wavelengths qualitatively followed the dissociation curve observed in the visible absorbance band at 452 nm. No significant changes in the CD spectrum were observed past 7500 seconds (the dissociation was monitored for 22,000 seconds). Experiments in which the dissociation was monitored with CD at single wavelengths of 250, 290, and 460 nm all gave the same observed lifetimes (within experimental error) as the absorbance experiments.

Preliminary experiments on the temperature dependence of the dissociation time constants for an actinomycin D - poly(dG-dC)•poly(dG-dC) complex at a 15:1 P/D initial ratio indicate that the activation energy is ~22 kcal/mole. The activation energy for the slowest dissociating component of the actinomycin D - calf thymus DNA complex was ~19 kcal/mole. It will be interesting to compare the activation enthalpies and entropies as a function of the P/D ratio in these systems, as well as to compare these values with other native and synthetic DNAs with varying lengths and compositions.

The transmission of the distortions of the nucleic acid conformation along the double helix involved in the saturation dependence of the actinomycin D poly(dG-dC)•poly(dG-dC) dissociation and in facilitation of actinomycin binding to poly(dA-dT)•poly(dA-dT) by daunorubicin may also be an important component in the selective recognition of nucleic acid sequences (as for example, in promoter-operator complexes), and in the transmission of thermal stability from the G-C region to the A-T region which has been observed in block oligonucleotides such as $d(C_{15}A_{15})•d(T_{15}G_{15})$.[38-40] Hogan et al. (1979)[41] have recently observed that the binding of distamycin induces a cooperative transition of calf thymus DNA to a new form with higher affinity for the drug and altered structural properties of the DNA. Several other DNA's tested did not show this phenomenon,[41] which further illustrates the interesting questions concerning allosteric conformational changes and sequence specificity which have yet to be answered.

The single-exponential decay observed in the dissociation of the actinomycin-poly(dG-dC)•poly(dG-dC) complex suggests that the multiple dissociation time constants for the actinomycin D-calf thymus DNA complex (Muller and Crothers, 1968;[2] and below) are most likely a result of the heterogeneity of the actinomycin D binding sites on natural DNA's. In a natural DNA containing several classes of binding sites of differing affinities for actinomycin D, the fraction of drug molecules bound to a given class of binding sites will change as one approaches saturation of all binding sites. At low D/P ratios (low saturation), the stronger sites would be expected to have a larger fraction of the bound actinomycin D. As more drug is added and saturation is approached, a larger fraction of the bound drug will be located at the weaker sites. If dissociation from each class of binding sites is a single exponential with τ dependent on the class of binding sites, then the relative contribution of each exponential will vary as a function of the relative population of drugs bound to that class of sites. Several dissociation curves for actinomycin from calf thymus DNA at different D/P ratios are shown in Figure 8. The dissociation of the complex can be mathematically described as a sum of three exponential terms.

The data in Figure 8 were analyzed using equation 1 and a non-linear least squares fitting routine. In each dissociation experiment, there is a long-lived component, τ_s, with a lifetime of about 1000 seconds. The relative magnitude of this slow component is a function of the D/P ratio as illustrated in Figure 9. At a D/P ratio equal to 1:200 the slow component is the dominant feature, accounting for almost 80 percent of the observed absorbance change. As the D/P ratio increases, the faster com-

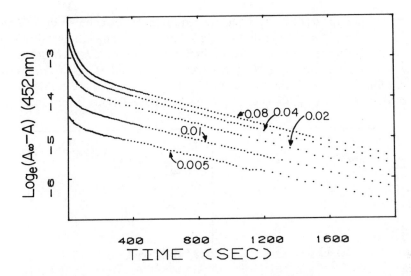

Figure 8. Absorbance changes due to dissociation of actinomycin D from calf thymus DNA in BPES buffer (pH 7.0) at 24°C. The initial concentration of DNA was 3.1×10^{-3}M in each experiment.

ponents become relatively more important. By a D/P ratio equal to 1:40, ΔA_s represents less than half of the observed change. The observed dependence of ΔA_s on D/P fits well with the site preference model discussed above.

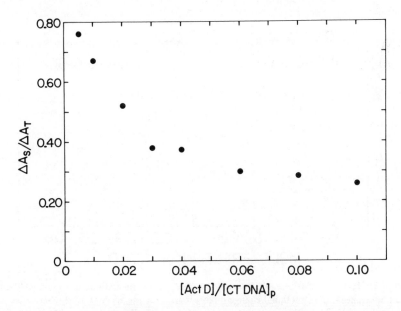

Figure 9. Dependence of the fractional absorbance change due to the slow component on the drug-to-phosphate ratio in the dissociation of the actinomycin D-calf thymus DNA complex. Conditions are as in Figure 8.

Since actinomycin apparently has a general requirement for gaunine in binding to DNA, it is tempting to speculate that the population of strong binding sites would depend on the GC content of the DNA. Preliminary results on *Cl. perfringens* DNA (30 percent G + C), calf thymus DNA (42 percent G + C), and *M. lysodeikticus* DNA (72 percent G + C) at a P/D ratio equal to 55:1 (low saturation) are consistent with this hypothesis. The observed values of τ_s in each of these DNA's is very close to the

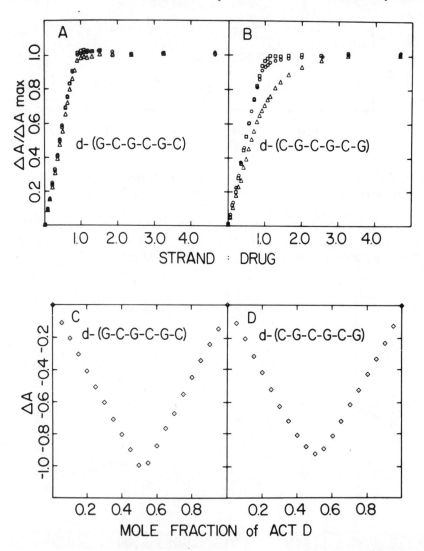

Figure 10. (A) Fractional change in absorbance ($\Delta A/\Delta A_{max}$) for the addition of d-(G-C-G-C-g-C) to 7-amino-actinomycin D; (B), fractional change in absorbance for the addition of d-(C-G-C-G-C-G) to 7-amino-actinomycin D. The values were obtained at o, 485 nm, □, 508 nm and △, 550 nm. (C). Change in absorbance (ΔA values) at 425 nm as a function of the molefraction of actinomycin D in a solution with d-(G-C-G-C-G-C). (D). Same as (C) except that d-(C-G-C-G-C-G) was used. In both (C) and (D) the total concentration of nucleotide plus actinomycin was kept constant at 2.0×10^{-4}M.

low saturation lifetime observed for poly(dG-dC)•poly(dG-dC). While it may be interesting to speculate that a (dG-dC)•(dG-dC) sequence may be a member of the strong binding class, more data are needed to make any firm assignment.

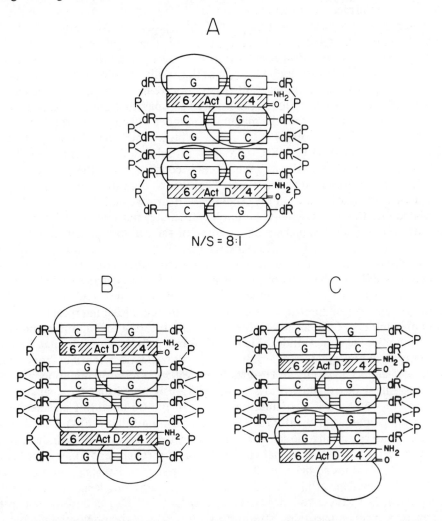

Figure 11. (A), Schematic illustration of one possible complex of actinomycin with d-(G-C-G-C-G-C) in which two actinomycins are intercalated into a hexanucleotide duplex from the minor groove, i.e., a stoichiometry of one actinomycin per hexanucleotide. The spacing of the actinomycins represents a binding of one actinomycin for every 8 nucleotides (i.e., N/S = 8:1). (B), Schematic illustration of two actinomycins intercalated into a d-(C-G-C-G-C-G) duplex, as in (A). Note, however, that the actinomycins are intercalated into (dC-dG)•(dC-dG) sequences in contrast to Figure 11A where the actinomycins are intercalated into (dG-dC)•(dG-dC) sequences. (C), An alternative complex which is consistent with the 1:1 strand:drug stoichiometry is shown in Figure 11C. This mixed type of complex retains the concept of preferential binding of actinomycin to (dG-dC)•(dG-dC) sequences, as well as retaining a maximum saturation of one drug per 8 nucleotides (i.e., there must be at least 4 base pairs between the two actinomycin chromophores if the pentapeptides from both groups are to be located in the minor groove of the helix).

The visible spectral changes observed when either actinomycin D or 7-amino actinomycin D binds to nucleotides or to DNA provide a useful means of monitoring
complex formation. Figures 10C and 10D show the change in absorbance as a function of the molefraction of the nucleotide for the binding of actinomycin to the two
self-complementary hexanucleotides d-(G-C-G-C-G-C) and d-(C-G-C-G-C-G). The
data show that actinomycin binds to both of these self-complementary hexanucleotides with an apparent stoichiometry of one actinomycin per hexanucleotide
(i.e., two actinomycins per duplex). Because this result had several implications with
respect to the sequence preferences of the actinomycins, we used visible absorption
(Figures 10A and 10B) and fluorescence spectroscopies to monitor complex formation. These data confirm that the maximum number of actinomycins that will bind
to these deoxyhexanucleotides is two per duplex. The intercalation of two actinomycins into the d-(G-C-G-C-G-C) duplex (as schematically illustrated in Figure
11A) at the two (dG-dC)•(dG-dC) sequences is consistent with the kinetic dissociation experiments presented above (Figure 7) as well as our other binding studies with
poly(dG-dC)•poly(dG-dC). On the other hand, if the two actinomycins are both intercalated into the d-(C-G-C-G-C-G) duplex, then they must either be located in the
two (dC-dG)•(dC-dG) sequences, or else one of the actinomycins must intercalate
from the major groove. Previous experiments[26-28] and model building studies[29/32]
tend to discount the likelihood of intercalation from the major groove. However,
another possibility is that only one of the actinomycins intercalates while the second
is stacked on the end of the duplex, as schematically illustrated in Figure 11C.
Although the optical spectra do not allow for any definitive interpretation of the
types of complexes formed, the visible absorption titration of 7-amino-actinomycin
D with the two hexanucleotides (Figures 10A and 10B) do illustrate that 7-amino-
actinomycin D binds differently to these two hexanucleotides. The absorption maximum of 7-amino-actinomycin D shifts from ∼505 nm to ∼550 nm upon intercalation into oligo- or polynucleotides.[42] We previously noted that the change in absorbance at ∼550 nm is particularly sensitive to the formation of an intercalated
complex.[42] Note that in the d-(G-C-G-C-G-C) titration (Figure 10A) the values for
$\Delta A/\Delta A_{max}$ are essentially the same for all three wavelengths, whereas in the d-(C-G-
C-G-C-G) titration the fractional change in the absorbance at 550 nm lags behind
the $\Delta A/\Delta A_{max}$ values at 485 and 508 nm. Thus these experiments (Figures 10A and
10B) provide evidence that different equilibria govern complexation of 7-amino-
actinomycin D to these two deoxyhexanucleotides. We also comment that in the
presence of excess nucleotides (strand:drug ratio > 2:1) the predominant complex
presumably is one in which essentially all of the actinomycins are intercalated.
Although these results are consistent with the formation of the complexes
schematically illustrated in Figures 11A and 11C (during the initial part of the titrations), we emphasize that the binding of drugs to oligonucleotides may exhibit complex behavior. A more detailed discussion of these and related data will be presented
in a subsequent paper.

In summary, the present experiments have demonstrated several new phenomena
concerning the influence of the conformational state of the nucleic acids upon their
interaction with drugs. These types of experiments may also serve as models for

understanding the concepts of protein-nucleic acid recognition and (allosteric) control, as well as for providing a molecular basis for new drug design.

Acknowledgments

This research was supported by NIH Grants CA-17865, CA-14103, as well as by a Research Career Development Award (CA-00257), and an Alfred P. Sloan Fellowship (to TRK), and a MARC Fellowship (to FMC). The authors acknowledge the collaboration of M. Petersheim and S. Lin during a portion of the kinetic studies.

References and Notes

1. Meienhofer, J., and Atherton, E. in *Structure-Activity Relationships Among the Semisynthetic Antibiotics* (Perlman, D., Ed.), pp. 427-529, Academic Press, New York (1977).
2. Muller, W., and Crothers, D. M., *J. Mol. Biol. 35*, 251-290 (1968).
3. Wells, R. D., and Larson, J. E., *J. Mol. Biol. 49*, 319-342 (1970).
4. Kersten, W., *Biochim. Biophys. Acta 47*, 610 (1961).
5. Goldberg, I. H., Rabinowitz, M. and Reich, E., *Proc. Natl. Acad. Sci. USA 48*, 2094 (1962).
6. Reinhardt, C. G. and Krugh, T. R., *Biochemistry 16*, 2890-2895 (1977).
7. Reinhardt, C. G., and Krugh, T. R., *Biochemistry 17*, 4845-4854 (1978).
8. Le Pecq, J.-B. in *Methods of Biochemical Analysis 20*, (Glick, D., Ed.) pp. 41-86, John Wiley and Sons, New York (1971).
9. Lee, C.-H. and Tinoco, I., Jr., *Nature 274*, 609-610 (1978).
10. Patel, D. J. and Shen, C., *Proc. Natl. Acad. Sci. USA 75*, 2553-2557 (1978).
11. Krugh, T. R., Wittlin, F. H., and Cramer, S. P., *Biopolymers 14*, 197-210 (1975).
12. Krugh, T. R., and Reinhardt, C. G., *J. Mol. Biol. 97*, 133-162 (1975).
13. Krugh, T. R. in *Molecular and Quantum Pharmacology* (Bergmann, E. D. an Pullman, B., Eds.), pp. 465-471, Reidel Pub. Co., Dordrecht, Holland (1974).
14. Tsai, C.-C., Jain, S. C., and Sobell, H. M., *Proc. Natl. Acad. Sci. USA 72*, 628-632 (1975).
15. Sobell, H. M., Jain, S. C., Sakore, T. D., Reddy, B. S., Bhandary, K. K., and Seshadri, T. P. in *International Symposium on Biomolecular Structure, Conformation, Function, and Evolution, Madras, India, January 4-7, 1978*, (Srinivasan, R., Ed.) Pergamon Press, Inc., New York (in press).
16. Neidle, S., Archai, A., Taylor, G. L., Berman, H. L., Carrell, H. L., Glusker, J. P., and Stallings, W. C., *Nature 269*, 304-307 (1977).
17. Seeman, N. C., Day, R. O., and Rich, A., *Nature 253*, 324-326 (1975).
18. The reader is also referred to the articles in this volume by Drs. H. M. Berman, H. M. Sobell, A. Rich, and their coworkers, and the references therein.
19. Krugh, T. R., Holcomb, S., Moehle, W. E., unpublished results.
20. Henry, D. W., in *Cancer Chemotherapy*, (Sartorelli, A. C., Ed.) Amer. Chem. Soc. Symp. *30*, Amer. Chem. Soc., Washington, D.C., pp. 15-17 (1976).
21. Arcamone, F., Cassinelli, G., Franceschi, G., Penco, S., Pol, C., Redaelli, S., and Selva, A., in *International Symposium on Adriamycin* (Eds., Carter, S. K., Di Marco, A., Ghione, M., Krakoff, I. H., and Mathe, G.) 9-22, Springer-Verlag, New York, New York (1972).
22. Krugh, T. R., and Young, M. A., *Nature 269*, 627-628 (1977).
23. Patel, D. J., *Biopolymers 15*, 533-558 (1976).
24. Patel, D. J., *Biochim. Biophys. Acta 442*, 98-108 (1976).
25. Kastrup, R. V., Young, M. A., and Krugh, T. R., *Biochemistry 17*, 4855-4865 (1978).
26. Krugh, T. R., *Proc. Natl. Acad. Sci. USA 69*, 1911-1914 (1972).
27. Patel, D. J., *Biochemistry 13*, 2396-2402 (1974).
28. Krugh, T. R., Mooberry, E. S., and Chiao, Y.-C. C., *Biochemistry 16*, 740-747 (1977).
29. Sobell, H. M., *Prog. Nuc. Acid Res. and Mol. Biol. 13*, 153-190 (1973).

30. Le Pecq, J.-B., and Paoletti, C., *J. Mol. Biol. 27*, 87-106 (1967).
31. A more rigorous analysis of the binding data is actually required to accurately characterize the complex equilibria involved, but the qualitative conclusion that these two drugs bind in a non-competitive manner at low r values appears to be justified (see the discussion in Reinhardt and Krugh, reference 7).
32. Jain, S. C., Tsai, C.-C., and Sobell, H. M., *J. Mol. Biol. 114*, 317-331 (1977).
33. Burns, V. W. F., *Arch. Biochem. Biophys. 145*, 248 (1971).
34. Wells, R. D., in *Progress in Molecular and Subcellular Biology 2*, (Hahn, F. E., Ed.) 21-32, Springer-Verlag, New York (1971).
35. Wells, R. D., Larson, J. E., Grant, R. C., Shortle, B. E., and Cantor, C. R., *J. Mol. Biol. 54*, 465-497 (1970).
36. Pohl, F. M., Jovin, T. M., Baehr, W., and Holbrook, J. J., *Proc. Natl. Acad. Sci. USA 69*, 3805-3809 (1972).
37. Lee, K.-R., this laboratory
38. Burd, J. F., Wartell, R. M., Dodgson, J. B., and Wells, R. D., *J. Biol. Chem. 250*, 5109-5113 (1975).
39. Burd, J. F., Larson, J. E., and Wells, R. D., *J. Biol. Chem. 250*, 6002-6007 (1975).
40. Early, T. E., Kearns, D. R., Burd, J. F., Larson, J. E., and Wells, R. D., *Biochemistry 16*, 541-551 (1977).
41. Hogan, M., Dattagupta, N., and Crothers, D. M., *Nature 278*, 521-524 (1979).
42. Chiao, Y-C.C., Rao, K.G., Hook, J.W., Krugh, T.R., and Sengupta, S.K. *Biopolymers 18*, 1749-1762 (1979).

The Coil Form of Poly(rU):
A Model Composed of Minimum Energy Conformers
That Matches Experimental Properties

S. Broyde
Biology Department
New York University
New York, New York 10003

and

B. Hingerty
Biology Division
Oak Ridge National Laboratory
Oak Ridge, Tennessee 37830

Introduction

In the study of the coil form of ribopolynucleotides, poly (rU) has attracted special attention. This is due to the fact that its properties are insensitive to temperature in the range 15°-45°C.[1]   In this respect it differs from other sequences, whose characteristics come to resemble those of poly (rU) once the temperature is raised. The same is true of UpU versus the other ribodinucleoside monophosphates.[2] The optical[2-4] and NMR[5-8] evidence indicates that UpU and poly (rU) are largely unstacked above 15°C.

The detailed conformational characteristics of poly (rU) in solution are thus of interest because of their uniqueness at ordinary temperatures. A series of uracils, and possibly even one subunit, in single stranded regions of RNAs is likely to possess distinctive features of shape, which could be salient to their functions. For example, the region specifying termination of transcription at the end of the tryptophan operon of *Escherichia coli* has a 3′ - terminal m-RNA transcript whose sequence is C-A-U-U-U-U$_{OH}$.[9] A series of as many as 6 - 8 uridine residues is, in fact, a common feature of the termination sites at the 3′ end of m-RNA molecules.[10] These are not necessarily all encoded in the DNA.

A key work in the study of poly (rU) is that of Inners and Felsenfeld,[1] who measured the limiting characteristic ratio C_∞, the mean square unperturbed end to end distance divided by the product of the number of bonds in the chain and the mean square bond length. A value of 17.6 was obtained, which indicated that the coil was relatively extended and restricted in conformation.

A number of workers have calculated coil models that agree with this characteristic ratio. Inners and Felsenfeld[1] calculated a polynucleotide model that matched this value, employing conformers from Sundaralingam's 1969 survey of the crystallographic nucleotide and polynucleotide literature.[11] Another such calculation, based on the same data, was made by Delisi and Crothers,[12] who took into account the interdependence of the δ, ϵ and β,γ angle pairs. (See Fig. 1 for structure, numbering scheme and conformational angle designations for UpU.) Olson and Flory have made classical potential energy calculations for the polynucleotide backbone.[13] Using the statistical weights calculated for each minimum in the δ,ϵ and β,γ energy surfaces they reproduced the experimental characteristic ratio. In a recent work, Yevich and Olson have made extensive additional calculations which further characterize coils.[14] Tewari, Nanda and Govil calculate coil models with proper characteristic ratios,[15] employing the conformations obtained from quantum mechanical calculations.[16] However, the β,γ = g-,g- conformations are given added weights of about 2 kcal/mole. These calculated coil models do not agree with each other in conformational detail. Moreover, they were calculated for the polynucleotide backbone without considering the influence of the bases on conformation. Porschke has made the important observation, employing temperature jump techniques, that coils of different sequences have different conformational states.[18] It must be mentioned, however, that coils of poly(rU) and poly(rA) have

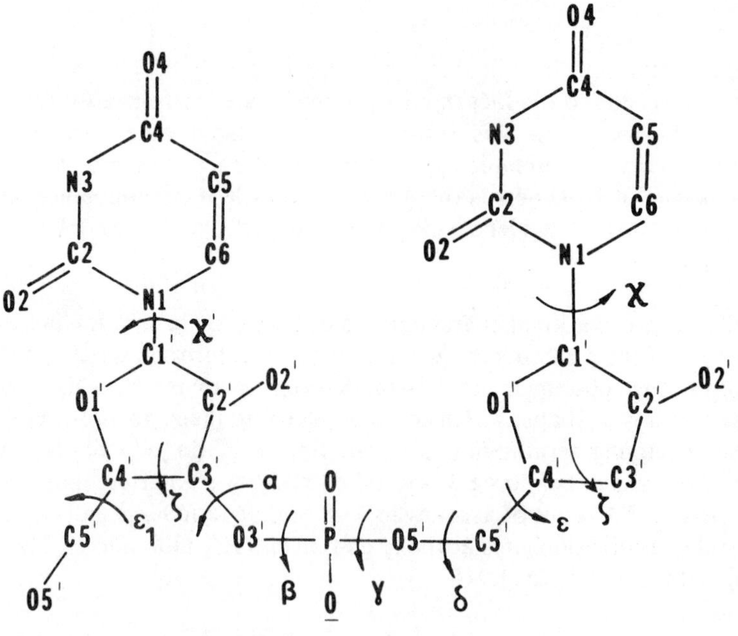

UpU

Figure 1. Structure, numbering scheme and conformational angle designations for UpU. The dihedral angles A - B - C - D are defined as follows: χ',χ: O1' - C1' - N1 - C6 α: P - O3' - C3' - C4' β: O5' - P - O3' - C3' γ: C5' - O5' - P - O3' δ: C4' - C5' - O5' - P ϵ,ϵ_1: C3' - C4' - C5' - O5'. The angle A - B - C - D is measured by a clockwise rotation of D with respect to A, looking down the B - C bond. A eclipsing D is 0°. Sugar pucker is described by the pseudorotation parameter P[25].

similar characteristic ratios.[1]

In the present study, our aim was to obtain a conformational model specific to the poly (rU) coil. This model was designed to match the experimental characteristic ratio[1] as well as conform to other criteria. The model is composed of minimum energy conformers of UpU. These were obtained by classical potential energy calculations in which all torsion angles and the ribose pucker were variable parameters. Its conformational characteristics match the NMR findings of Sarma and co-workers on UpU[7] and poly (rU).[8] In addition, its calculated persistence length is of the same order of magnitude as an experimental value observed for a perturbed denatured RNA chain,[17] while helical duplexes have persistence lengths that are very much larger.[19]

Methods

The potential energy calculations were carried out as detailed previously[20] including Van der Waals, E_{nb}, electrostatic, E_{el}, torsional, E_{tor}, and ribose strain, E_{st}, contributions to the energy, E.

$$E = E_{nb} + E_{el} + E_{tor} + E_{st} \tag{1}$$

$$E_{nb} = \sum_{i<j}\sum (a_{ij}r_{ij}^{-6} + b_{ij}r_{ij}^{-12}) \tag{2}$$

$$E_{el} = \sum_{i<j}\sum 332\, q_i q_j r_{ij}^{-1} D^{-1} \tag{3}$$

$$E_{tor} = \sum_{k=1}^{8} V_{o,k}(1 + \cos3\theta_k) \tag{4}$$

$$E_{st} = \sum_{l=1}^{5} K\tau_l (\tau_l - \tau_{o,l})^2 \tag{5}$$

r_{ij} is the distance in angstroms between atoms i and j, q_i is the partial charge assigned to atom i, $V_{o,k}$ is the barrier to internal rotation for the dihedral angle k and θ_k is the value of that angle, $K\tau_l$ is a force constant, τ_l is the strained ribose bond angle, and $\tau_{o,l}$ is the value that angle adopts at equilibrium. Also, k denotes the eight independent dihedral angles and l the five deoxyribose bond angles. Partial charges were taken from Renugopalakrishnan, *et al.*[21] The dielectric constant (D) was assigned a value of 4 and the torsional barriers as well as the parameters a_{ij} and b_{ij} were taken from Lakshminarayanan and Sasisekharan.[22/23]

The energy of ribose was calculated previously by Dr. T. Sato, some of whose results have been reported by Sasisekharan.[24] In his work, the energy was minimized as a function of the pseudorotation parameter, P, the puckering amplitude, θ_m (notation of Altona and Sundaralingam[25]), and the bond angles of O1' - C1' - C2' (α_1) and O1' - C4' - C3' (α_2). These completely define the deoxyribose coordinates. For these calculations, τ_0 was taken to be 113.5° for C-O-C, 110.0° for C-C-O and 109.5° for

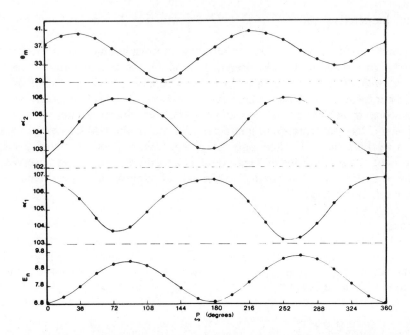

Figure 2. Energy, E_n, of ribose (kcal/mol); α_1, the bond angle O1′ - C1′ - C2′ (deg); α_2, the bond angle O1′ - C4′ - C3′ (deg); and θ_m the puckering amplitude (deg) as a function of P, the pseudorotation parameter. E_n was given previously in Stellman, *et al.*[49]

C-C-C. K_τ values employed were, in kcal/mol rad², 66.5 for C - O - C, 59.9 for C - C- 0, and 54.0 for C - C - C. These are 70 percent of the values obtained experimentally,[26] and were devaluated by Sato in order to obtain a better fit between observed and calculated conformations. Results of Sato's calculations for ribose are presented in Fig. 2. His results for deoxyribose have been given previously.[20] These energies and the other variables in Fig. 2 were incorporated in the energy calculations of the present work, and by linear interpolation permitted a continuous variation in ribose energy as a function of puckering. Bond lengths and angles were taken from Arnott, *et al.*[28] The eight backbone torsion angles and the ribose pucker were variables in a 9 parameter minimization of the energy, employing a modified version of the Powell algorithm.[29] Starting conformations included the minima obtained previously for UpU, calculated with fixed ribose pucker, as well as the low energy forms calculated for deoxydinucleoside monophosphates.[20/31]

The statistical weight w_i, of the ith minimum is given by:

$$w_i = A_i \exp(-\Delta E_i/RT)/Z \tag{6}$$

where ΔE_i is the energy of the ith minimum, A_i is the relative area associated with it, and

$$Z = \sum_i A_i \exp(-\Delta E_i/RT) \tag{7}$$

Olson has detailed procedures for calculating statistical weights from energy contour maps.[36] To evaluate the relative area, A_i associated with each minimum, the torsion angles other than β and γ were fixed at their values at the minimum. β and γ were separately varied in one degree intervals and the energy was calculated at each point until the 1 kcal/mole contour was located. The contours are approximately elliptical and have sheer walls beyond 1 kcal/mole. Indeed, the 1 kcal/mole and the 3 kcal/mole edges virtually coincide when the other torsion angles are kept fixed. The characteristic ratio, C_∞ was computed by the method of Flory,[32/33] expanded by Olson[34-37] for polynucleotides, with the matrix method of Eyring.[38]

The persistence vector calculation is done similarly, as described by Olson[39] and Yevich and Olson.[15]

Results and Discussion

UpU Minimum Energy Conformations

Table I presents low energy conformers of UpU. In these new calculations the ribose pucker as well as the backbone dihedral angles were variable parameters. Furthermore, the trans and g- domains of the C4' - C5' torsion ϵ were explored, in addition to the g+ region that had been studied previously with ribose pucker fixed.[30] With C-3'-endo type pucker, the three lowest energy forms have ϵ = g+ and the O3'-P and O5'-P torsions β,γ are respectively t, g-, (skewed), g-,t and g-, g- (A form). These are at energies of less than 1 kcal/mole. The β,γ = g+, g+, ϵ = g+ and the Watson-Crick[37] (β,γ = g-,t, ϵ = t) conformations are at somewhat higher energies. With C-2'-endo pucker, there are three conformers with almost equally low energy. In two of them ϵ is trans, together with β,γ rotations of t,g+ and g+,g+. In the third form β,γ = t, g- and ϵ is g+. Other minimum energy conformations are the Watson-Crick and the B form, although they are of higher energy. Conformers with ϵ = g- were not found below 3 kcal/mole.

Statistical weights of these conformers are given in Table II. Also shown are the ranges and the relative areas of the 1 kcal/mole contours in the β,γ plane associated with each minimum. The β,γ = t, g- conformation has the highest statistical weight in each puckering domain.

ORTEP drawings of the three lowest energy conformations in each puckering region are shown in Figure 3. The interesting point is that these conformers are predominantly not stacked. The A form has bases highly overlapping and essentially coplanar. The β,γ = g-, t conformer 2 also has considerable stacking, but the alternate g-,t conformation 4 (not shown) is unstacked. This contrasts with, for example, the low energy conformations of dApdA, which are predominantly stacked.[20]

Experimental Observations on UpU and Poly (rU)

Our goal in the present work was to mesh these and other theoretical studies with experimental observations by others on UpU and poly rU, to obtain a model for the

Table I
Minimum Energy Conformations of UpU[a]

#	χ'	ε₁	α	β	γ	δ	ε	χ	P	ΔE	Description: $\beta, \gamma; \epsilon$
						C-3'-endo Region					
1.	15	56	282	225	319	103	52	56	13	0.	t, g-; g+
2.	44	78	287	320	170	190	57	10	6	0.5	g-, t; g+
3.	3	58	210	316	278	180	51	17	8	0.9	g-, g-; g+; A
4.	45	62	204	326	144	203	50	4	12	1.9	g-, t; g+
5.	45	62	190	41	84	196	76	26	14	2.6	g+, g+; g+
6.	11	172	199	290	170	212	136	10	-5	3.6	g-, t; t
						C-2'-endo Region					
1.	31	55	289	197	48	224	141	11	180	0.0	t, g+; t
2.	21	176	279	91	84	251	176	54	185	0.04	g+, g+; t
3.	33	59	289	202	301	124	59	77	170	0.1	t, g-; g+
4.	15	178	211	271	172	192	150	78	202	2.1	g-, t; t
5.	27	59	202	255	317	170	62	81	179	3.1	g-, g-; g+; B

[a] ΔE is the energy difference in kcal/mol between the local minimum and the global minimum in each puckering domain. Dihedral angles and P are in degrees.

coil form of poly (rU). The NMR findings of Sarma and co-workers [7] reveal that the conformation of UpU changes little on elevating the temperature from 20°C to 89°C. These NMR studies indicate the following approximate conformational features for UpU in solution: ribose pucker 53 percent C-3′-endo, 47 percent C-2′-endo; $\epsilon = g+$, 83 percent. The NMR properties of poly (rU) are virtually identical with those of UpU in this temperature range[8] (although a different form of the polymer exists below 15°C[39]). Consequently, the low energy forms of UpU are similar to the conformations of poly (rU). The present poly (rU) conformational model was designed to match these data. In addition, it must match the measured characteristic ratio C_∞ of 17.6 obtained for poly (rU).[1] Another experimental quantity taken into account (although not obtained for poly (rU) coils) is the approximately 75 Å persistence length measured for a denatured perturbed RNA chain.[17] The theoretical study by Olson of the dependence of the unperturbed dimensions of polynucleotides on the orientation of the phosphodiester bonds[43] was also considered. She finds that the experimentally observed high values of C_∞ can be matched theoretically if conformers with $\beta = $ trans make an important contribution to the coil form.

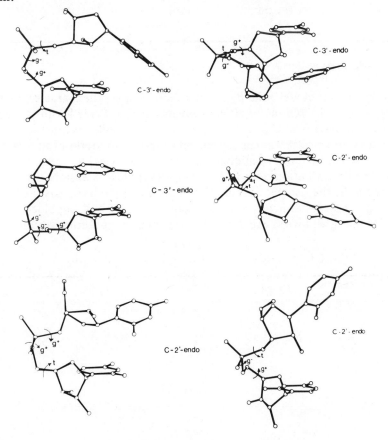

Figure 3. Lowest energy conformers of UpU.

Table II
Calculated Statistical Weights, w, of
UpU Minimum Energy Conformations

Minimum #	β Range, (°)	γ Range, (°)	Relative Area	w
		C-3′-endo Region		
1.	219-228	316-325	.27	.67
2.	317-323	167-174	.14	.15
3.	309-322	275-282	.30	.16
4.	322-329	140-147	.16	.016
5.	37-46	80-86	.12	.0037
6.	281-298	168-172	.23	.0014
		C-2′-endo Region		
1.	191-208	43-51	.45	.27
2.	83-105	82-87	.37	.21
3.	181-211	297-307	1.0	.51
4.	262-280	168-177	.54	.0096
5.	244-261	311-324	.74	.0025

Characteristic Ratios

The first step in our model building effort was to compare the characteristic ratio of an energy weighted assembly of all conformers listed in Table I with the measured value. (Conformer 5 with C-3′-endo pucker was not included because it is sterically disfavored in polymers[44/45].) The calculated characteristic ratio in this case is 4.9. Quantum mechanical conformational calculations also yield low results.[14] Our low value is due to contributions by conformers with very low extensions. The extension is reflected in h, the rise per nucleotide residue along the helix axis (assuming the construction of a regular helix from each conformational building block). Table III gives h for the three lowest energy conformers in each puckering region. Negative

Table III
Rise Per Residue of UpU Minimum Energy Conformation

Minimum #	β, γ; ε	h
	C-3′-endo Region	
1.	t,g-; g+	3.78
2.	g-,t; g+	.80
3.	g-,g-; g+	2.64
	C-2′-endo Region	
1.	t,g+; t	-3.83
2.	g+,g+; t	-0.15
3.	t,g-; g+	5.51

values of h indicate a left handed helix. It is seen that the minima β,γ = g-, t and g+, g+ are very compact forms. Therefore, these are improbable in the extended coil, although they may be important under other conditions; for example, the turn in the anticodon loop of tRNAs is negotiated via the conformation β,γ = g-, t (C-3'-endo minimum 2) and involves a U base.

In the next step, we selected from among our low energy conformers combinations that would agree with the NMR results:[7] 53 percent C-3'-endo, 47 percent C-2'-endo, 83 percent ϵ = g+. Since the g- region of ϵ is not represented among conformers below 3 kcal/mole mole, the 17 percent non g+ conformers were assigned to the trans domain. The likeliest candidate for this role on energetic grounds is conformer 1, with C-2'-endo pucker, which has β,γ = t, g+. The remaining 30 percent

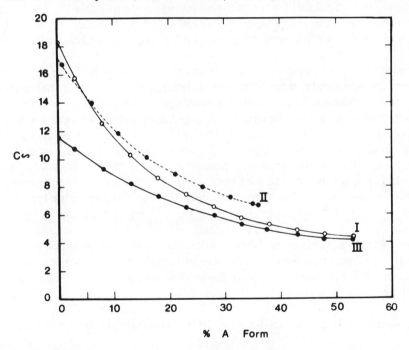

Figure 4. Characteristic ratio, C_∞, vs. percent A Form.

Curve I:
β,γ = t, g+, ϵ = t, C-2'-endo: 17 percent
β,γ = t, g-, C-2'-endo: 30 percent
balance of conformational blend:
β,γ = t, g-, C-3'-endo and A Form.

Curve II:
β,γ = g-, t, ϵ = t, C-3'-endo: 17 percent
β,γ = t, g-, C-2'-endo: 47 percent
balance of conformational blend:
β,γ = t, g-, C-3'-endo and A Form.

Curve III:
β,γ = g+, g+, ϵ = t, C-2'-endo: 17 percent
β,γ = t, g-, C-2'-endo: 30 percent
balance of conformational blend:
β,γ = t, g-, C-3'-endo and A Form.

C-2'-endo contribution is assigned to conformer 3, β,γ = t, g-. It has the highest statistical weight for C-2'-endo pucker and is very extended (h=5.51Å). For the C-3'-endo contribution, the β,γ = t, g- and the A forms are the energetically plausible choices. The characteristic ratio was then calculated for a series of polymers in which the energy weightings, ΔE_ψ, were changed so that the C-2'-endo conformers were present in the above proportions and the percent A form varied from 0 percent to 53 percent, in tandem with the C-3'-endo β,γ = t,g- conformation. Results are given in Figure 4, Curve I. We see that the experimental characteristic ratio is matched with about 1 percent A form and 52 percent β,γ =t, g-. The present coil model is thus composed of 82 percent conformers with β,γ = t, g-, 52 percent being C-3'-endo and 30 percent C-2'-endo. The remaining 18 percent has 17 percent, β,γ = t, g+, ϵ = t and 1 percent A form. The pseudo-energies, ΔE_ψ, for the conformers in this model were calculated from equation 6 and are given in Table IV. The relative areas of Table II and the above weights were employed. Comparing ΔE_ψ with ΔE of Table I, we find that relative energies are adjusted by less than ~1 kcal/mole.

Numerous other conformational combinations of those listed in Table I, all coinciding with the NMR data, were examined to determine if the correct characteristic ratio could be achieved. This involved a search employing other ϵ = trans conformers in combination with the two β,γ = t, g- forms and/or with the β,γ = g-, t forms. As expected, no polymers containing the low extension conformers of Table III could achieve the necessary high C_∞. Figure 4, Curve III shows, for example, how the characteristic ratio varies with percent A form if the ϵ = trans conformation with β,γ = t, g+ is replaced with the C-2'-endo β,γ =g+, g+ conformer 2. It is possible to calculate various two state poly (rU) coil models with the correct characteristic ratio. One example consists of the A form in combination with the C-2'-endo β,γ = t, g+ conformation, at either ~25 percent or ~92 percent A form. However, these do not match the NMR results. Our earlier two state coil model for poly (dA)[46] also does not quite match the recently obtained NMR data for dApdA.[47] We are presently calculating multi-state models for the poly dA coil that agree with the NMR findings.

Another satisfactory coil model for poly (rU) was obtained which employed the

Table IV

Conformational Properties of Poly rU Coil Models
that Match Experimental Observations

Model #	$\beta, \gamma; \epsilon$	Pucker	Percent	ΔE_ψ, kcal/mole
1	t, g-; g+	C-2'-endo	30	0.72
	t, g+; t	C-2'-endo	17	0.59
	t, g-; g+	C-3'-endo	52	-0.39
	g-,g-; g+ (A)	C-3'-endo	1	2.03
2	t, g-; g+	C-2'-endo	47	1.27
	g-, t; t	C-3'-endo	17	-0.17
	t, g-; g+	C-3'-endo	36	0.18

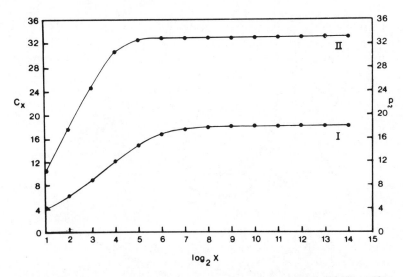

Figure 5. Characteristic ratio, C_x, (Curve I) and persistence vector magnitude $|P|$ (Curve II, as a function of degree polymerization x for a polymer composed of the following conformers (Model 1):

$\beta, \gamma = $ t, g-, C-2'-endo, 30 percent
$\beta, \gamma = $ t, g-, C-3'-endo, 52 percent
$\beta, \gamma = $ t, g+, $\epsilon = $ t, C-2'-endo, 17 percent
$\beta, \gamma = $ g-, g-, C-3'-endo, 1 percent

C-3'-endo Watson-Crick conformer 6, $\beta,\gamma = $ g-, t, $\epsilon = $ t as 17 percent of the conformational blend, together with 36 percent $\beta,\gamma = $ t, g-, C-3'-endo and 47 percent $\beta,\gamma = $ t, g- C-2'-endo. It shares the predominant characteristics of the previous model in being about 83 percent $\beta,\gamma = $ t, g-. Figure 4, Curve II, shows C_∞ vs. percent A form for a polymer in which the A form -t, g- (C-3'-endo) proportion is varied from 0 to 36% A form, the other conformers being held as stated above. The experimental characteristic ratio is achieved at 0 percent A form. This model is less likely in view of the lower statistical weight of the Watson-Crick conformation versus that of the $\beta,\gamma = $ t, g+, $\epsilon = $ t form in its puckering domain. The 3.8 kcal/mole difference between ΔE_v (Table IV) and ΔE (Table I) of the Watson-Crick conformation also shows this. A combined model consisting of both the Watson-Crick conformation (8.5 percent) and the $\beta,\gamma = $ t, g+, $\epsilon = $ t (8.5 percent) form also reproduces C_∞. Table IV summarizes the conformations in the present coil models.

The calculated characteristic ratio reaches 99 percent of its limiting value with x, the number of residues in the polymer $= 2^9$ (Figure 5, Curve I). By contrast, helices require 2^{12}–2^{13} residues.[39]

Persistence lengths

The persistence lengths of these poly (rU) coil models are consistent in order of magnitude with the 75 Å experimental determination of this quantity for denatured

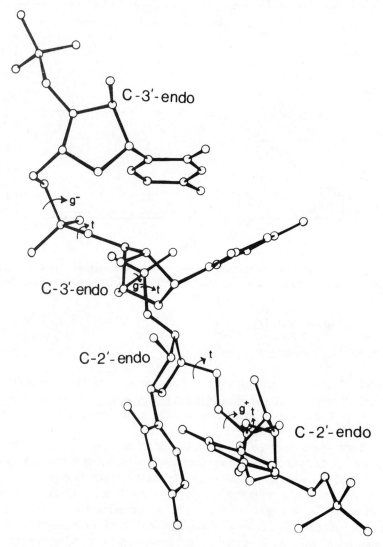

Figure 6a. Segment of poly rU coil model 1. Two subunits have $\beta,\gamma = $ t,g-, $\epsilon = $ g+, one subunit has $\beta,\gamma = $ t,g+, $\epsilon = $ t.

RNA.[17] The computed limiting persistence vector has a magnitude of 31.5 Å for the first model. For the second model, the persistence vector magnitude is 25.2 Å. The measured quantity is probably larger than the calculated value because the latter is for an unperturbed polymer. The persistence vector converges to 99 percent of its limiting value with 2^5 residues (Figure 5, Curve II) while helices require 2^{11} subunits.[39]

Models

Figure 6 shows ORTEP[41] drawings of segments of the two models, containing the

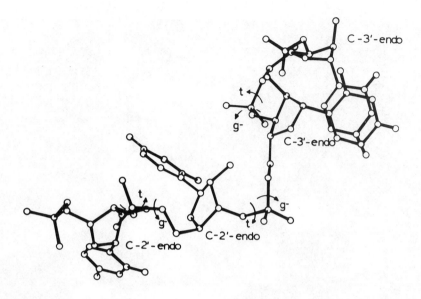

Figure 6b. Segment of poly rU coil model 2. Two subunits have $\beta,\gamma = $ t,g-, $\epsilon = $ g+, one subunit has β,γ = g-,t, $\epsilon = $ t.

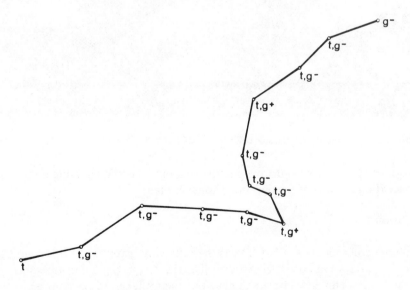

Figure 7. Phosphorus atoms in a segment of poly rU coil model 1.

three conformers present in each. The bases are unstacked in the first model (except for the 1 percent A form, which is not shown). There is stacking at the Watson-Crick conformations of the second model, constituting 17 percent of the blend. The NMR data suggest that 8 ± 5 percent of the bases are stacked, so either model or a combination of both can roughly agree with these results; however, the first model is better. The location of P atoms in a poly (rU) coil containing approximately correct

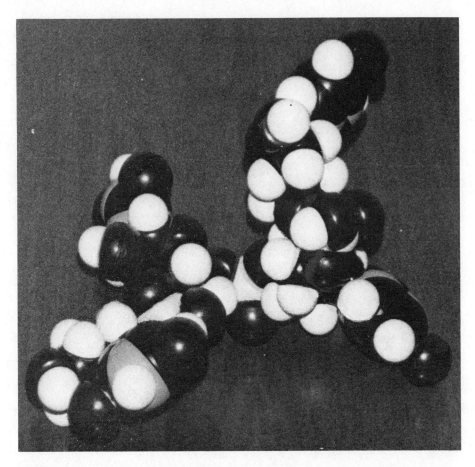

Figure 8. Corey - Pauling - Koltun space filling molecular model of conformers shown in Figure 6a.

proportions of the three conformers in the first model is shown in Figure 7. Figure 8 is a space filling model of the tetramer shown in Figure 6a.

Conclusion

The present coil model of poly (rU) has the following features: (1) It is composed of low energy conformers of UpU. Specifically, the β,γ = t, g- conformations, which have the highest statistical weights in each of their puckering domains constitute 82 percent of the building blocks. The remainder is 17 percent β,γ = t, g+, ϵ = t and 1 percent A form. The t, g+ conformation is the second most probable conformer with C-2'-endo pucker. A second, less likely model has β,γ = g-, t, ϵ = t (Watson-Crick form), C-3'-endo pucker as 17 percent of the blend. These results agree with theoretical studies by Olson[44] showing that coils must possess a large proportion of conformers with β = trans. The structure of pdTpdT in the crystal,[48] which has β,γ = t, g- is also pertinent. (2) It matches NMR data on UpU[7] and poly (rU)[8] in solution. (3) The calculated characteristic ratio agrees with the value of 17.6 measured for poly (rU) coils.[1] (4) The calculated persistence lengths are of the same order of magnitude as was measured for denatured RNA and DNA.[17] Persistence lengths of helical, double stranded DNA, on the other hand, are about 600 Å.[19] (5) Calculated characteristic ratios and persistence vectors in this model converge to their limiting values at 2^9 and 2^5 residues, respectively, while helices require many more subunits for convergence.[39]

The unique conformational features of a series of uracils are different from the A type helices found in ordered RNA single strands, double strands and RNA-DNA hybrids. Since a series of uracils is typically found at the transcription termination site of m-RNAs, the distinctive shape may be recognized by RNA polymerase as the signal to release the RNA transcript from its DNA template.

Acknowledgement

We thank Wilma Olson for much helpful advice. Research sponsored jointly by the U.S. Public Health Service under NIH Grant 5ROIGM 24482-02 and the Office of Health and Environmental Research, U.S. Department of Energy, under contract W-7405-eng-2b with the Union Carbide Corporation.

References and Footnotes

1. Inners, L. and Felsenfeld, G. *J. Mol. Biol. 50*, 373 (1970).
2. Warshaw, M. and Tinoco, I. *J. Mol. Biol. 20*, 29 (1966).
3. Richards, E., Flessel, C. and Fresco, J. *Biopol. 1*, 431 (1963).
4. Simpkins, H. and Richards, E. *Biopol. 5*, 551 (1967).
5. Ts'o, P. O. P., Kondo, N., Schweizer, M. and Hollis, D. *Biochemistry 8*, 997 (1969).
6. Alderfer, J. and Ts'o, P. O. P., *Biochemistry 16*, 2410 (1977).
7. Lee, C., Ezra, F., Kondo, N., Sarma, R. H. and Danyluk, S. *Biochemistry 15*, 3627 (1976).
8. Evans, F. and Sarma, R. H. *Nature 263*, 567 (1976).
9. Wu, A. and Platt, T. *Proc. Natl. Acad. Sci. USA 75*, 5442 (1978).
10. Gilbert, W. in *RNA Polymerase*, Losick, R. and Chamberlin, M., eds., Cold Spring Harbor Laboratory (1976), p. 193.
11. Sundaralingam, M. *Biopol. 7*, 821 (1969).
12. Delisi, C. and Crothers, D. *Biopol. 10*, 1809 (1971).
13. Olson, W. K. and Flory, P. *Biopol. 11*, 25 (1972).
14. Tewari, R., Nanda, R. and Govil, G. *Biopol. 13*, 2015 (1974).
15. Yevich, R. and Olson, W. *Biopol. 18*, 113 (1979).
16. Pullman, B., Perahia, D. and Saran, A. *Biochim. Biophys. Acta 269*, 1 (1972).
17. Mingot, F., Jorcano, J., Acuna, M. and Davila, C. *Biochim. et Biophys. Acta 418*, 315 (1976).
18. Porschke, D., *Biochemistry 15*, 1495 (1976).
19. Godfrey, J. and Eisenberg, H. *Biophys. Chem. 5*, 301 (1976).
20. Broyde, S., Wartell, R., Stellman, S. and Hingerty, B. *Biopol. 17*, 1485 (1978).
21. Renugopalakrishnan, V., Lakshminarayanan, A., and Sasisekharan, V., *Biopol. 10*, 1159 (1971).
22. Lakshminarayanan, A. and Sasisekharan, V. *Biopol. 8*, 475 (1969).
23. Lakshminarayanan, A. and Sasisekharan, V. *Biopol. 8*, 489 (1969).
24. Sasisekharan, V. *Jerusalem Symp. Quant. Chem. Biochemistry 5*, 247 (1973).
25. Altona, C. and Sundaralingam, M. *J. Am. Chem. Soc. 94*, 8205 (1972).
26. Snyder, R. G., and Zerbi, G. *Spectrochim. Acta 23*, 391 (1967).
27. Stellman, S., Hingerty, B., Broyde, S., and Langridge, R. *Biopol. 14*, 2049 (1975).

28. Arnott, S., Dover, S., and Wonacott, A. *Acta Cryst. B. 25*, 2192 (1969).
29. Powell, M. *Computer J. 7*, 155 (1964).
30. Broyde, S., Wartell, R., Stellman, S., Hingerty, B. and Langridge, R. *Biopol. 14*, 1597 (1975).
31. Hingerty, B. and Broyde, S. *Nucleic Acids Res. 9*, 3429 (1978).
32. Flory, P. J., *The Statistical Mechanics of Chain Molecules*, Interscience Publishers, N. Y., (1969), pp. 22-25, 114-117, 281-286.
33. Flory, P. *Proc. Natl. Acad. Sci. USA 70*, 1819 (1973).
34. Olson, W. and Flory, P. *Biopol. 11*, 1 (1972).
35. Olson, W. and Flory, P. *Biopol. 11*, 57 (1972).
36. Olson, W. *Biopol. 14*, 1775 (1975).
37. Olson, W. *Macromolecules 8*, 272 (1975).
38. Eyring, H., *Phys. Rev. 39*, 746 (1932).
39. Olson, W. *Biopol.* In Press.
40. Crick, F. and Watson, J. *Proc. Roy. Soc., Ser. A. 223*, 80 (1954).
41. ORTEP: A Fortran Thermal Ellipsoid Plot Program for Crystal Structure Illustrations. Caroll K. Johnson, Oak Ridge National Laboratory, Oak Ridge, Tennessee.
42. Young, P. and Kallenbach, N. *J. Mol. Biol. 126*, 467 (1978).
43. Olson, W. *Biopol. 14*, 1797 (1975).
44. Olson, W. *Nucleic Acids Res. 2*, 2055 (1975).
45. Yathindra, N. and Sundaralingam, M. *Proc. Natl. Acad. Sci. USA 71*, 3325 (1974).
46. Hingerty, B. and Broyde, S. *Nucleic Acids Res. 5*, 3249 (1978).
47. Cheng, D. and Sarma, R. H. *J. Am. Chem. Soc. 99*, 7333 (1977).
48. Camerman, N., Fawcett, J. K., and Camerman, A. *J. Mol. Biol. 107*, 601 (1976).
49. Stellman, S., Hingerty, B., Broyde, S., and Langridge, R. *Biopol. 14*, 2049 (1975).

The Flexible DNA Double Helix
Theoretical Considerations

Wilma K. Olson
Department of Chemistry
Rutgers University
New Brunswick, New Jersey 08903

Introduction

The macroscopic properties of the DNA double helix in dilute solution depend dramatically upon chain length.[1/2] In the range of low molecular weight the behavior of this molecule closely resembles that anticipated for a rigid rodlike helix found in the solid state. In contrast, extremely long DNA chains of molecular weight 10^7 or more adopt on the average very compact conformations in comparison to the extended regular helix. At this size range the molecule further qualifies as a random coil in the sense that the distribution of chain conformations is completely random or Gaussian. In between these extremes of chain lengths the DNA exhibits the limited motions of a gradually and smoothly bending wormlike model.[3-5] No single mathematical model yet proposed can adequately represent the physical properties of DNA chains in this intermediate category.

The detailed local molecular motions that give rise to the macroscopic flexibility of DNA appear on the basis of available experimental evidence to be highly restricted. According to various physical measurements (that include Raman spectroscopy,[6] CD,[7/8] NMR,[9/10] electron microscopy,[11/12] infrared dichroism,[13/14] electric birefringence,[15] wide-angle X-ray scattering,[16/17] and others) DNA in dilute aqueous salt solutions at neutral pH exhibits the characteristic B-type backbone observed directly by X-ray crystallographic analyses of low molecular weight DNA analogs[18] and deduced indirectly from X-ray fiber diffraction studies.[19] These solution measurements, however, cannot detect the minor variations of backbone conformation seen within the B- family of structures in the solid state. Fluctuations in polynucleotide conformation are undoubtedly responsible for the well known "breathing" of the double helix in solution.[20-24] At temperatures far below the normal thermal transition point some base pairs of DNA occur in non-hydrogen bonded states open to interactions with the aqueous environment. Protons normally involved in Watson-Crick base pairing exchange[20-22] rapidly with deuterium or tritium in the solvent while exocyclic amino and endocyclic imino groups on the purines and pyrimidines react easily with formaldehyde probes.[23/24] Since the bases attached to the DNA maintain strong stacking interactions[22] under conditions of breathing, these molecular motions presumably are limited in scope.[25]

The computations of polymer chain dimensions outlined below demonstrate how the rigidity of the local conformation of the polynucleotide backbone is magnified into the enormous degree of flexibility of long DNA double helices. The molecular model we adopt is consistent with limited local breathing motions of DNA. In addition, the model accounts satisfactorily for the gross dimensions of the polymer over a broad range of molecular weights. The chain bends smoothly without the occurrence of drastic turns or kinks. The flexibility of the chain as described by the spatial density distribution functions is further consistent with the macroscopic descriptions previously attributed to DNA of various chain lengths (e.g., rigid rod, wormlike coil, Gaussian). Despite the local stiffness in the polynucleotide backbone, longer molecules exhibit a tendency to bend backwards into hairpin loops and to cyclize into "circular" structures. The predicted probabilities of occurrence of such structures appear to be related to the sizes of loops and rings observed in experimental studies.

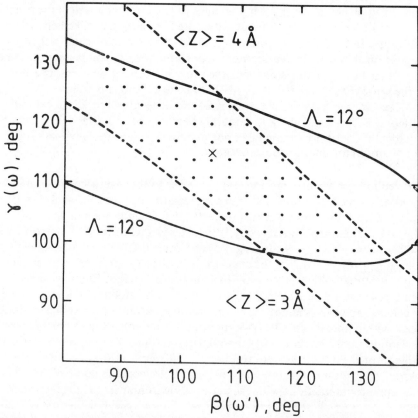

Figure 1. Contour diagram of the $\Lambda = 12°$ base stacking angle (solid line) and the $<Z> = 3\text{Å}$ and 4Å base stacking distances as a function of the phosphodiester angles for a flexible B-DNA helix. Points denoted on the diagram, where base overlap is also observed, are the 81 discrete conformational states included in the model. The × denotes the conformation repeated in the reference helix of the model. Rotation angles are defined with respect to *trans* = 0°. Rotation angle nomenclature $\omega'\omega$ used in previous publications is noted in parentheses.

Residue Flexibility

For the purpose of simplicity the molecular motions attributed in this study to the DNA double helix are limited to minor angular variations in the phosphodiester rotation angles (β and γ) of each chain repeating unit. These two angles are chosen on the basis of X-ray,[18/26] NMR,[9/27/28] and theoretical studies[29/30] that implicate them as the major source of flexibility in the polynucleotide backbone. Upon assignment of fixed values to the remaining two C-C and two C-O bonds of each nucleotide repeating unit the distance between successive phosphorus atoms is fixed and imaginary virtual bonds[31] may be drawn to span each chemical residue. As a consequence of this mathematical device, statistical mechanical treatment of polymer chain averages is greatly simplified.[31/32]

The extent of motion associated with the flexible B-DNA helix of this study is indicated by the contour surface in Figure 1. In the absence of reliable estimates of conformational energy, the β and γ angles are restricted with equal probability to the 81 points denoted on the surface. In each of these states adjacent bases may be regarded as stacked in that (1) the angle Λ between base planes is 12° or less, (2) the mean distance $<Z>$ between base planes falls in the range 3-4 Å, and (3) the bases exhibit partial overlap. Only one point on the surface (denoted in Figure 1 by ×), however, may be considered the building block of an ideal double helix with both base stacking and Watson-Crick base pairing (cf. seq.). The limited motions set by the restrictions in Λ, $<Z>$, and overlap reproduce the radius of gyration $<s^2>$ of DNA spanning a range of chain lengths between 2^8 and 2^{13} nucleotide units.[32]

The relative motions of adjacent bases in the above model are also illustrated by the adenine dinucleoside triphosphate structures of Figure 2. The dimer represented in Figure 2(a) is a reference conformation, that when repeated throughout two complementary DNA strands, produces a theoretical 13-fold double helix. This duplex (denoted in Figure 1 by ×) is characterized by structurally ideal linear hydrogen bonds of the Watson-Crick variety.[32/33] Minor (i.e., ~20°) variation of β and/or γ from the above reference conformation is sufficient to distort and in some cases to break interstrand hydrogen bonds. The dimers represented in Figures 2(b) and 2(c) are illustrative of the extrema of flexibility allowed in the present model. Upon changes in the phosphodiester conformation the bases are found to slide back and forth in a manner reminiscent of the oscillating single-stranded base stacking suggested a number of years ago by Davis and Tinoco[34] to account for the observed CD of dinucleoside monophosphates. As evident from the high base overlap in Figure 2(b), this conformation is more tightly wound (by ~20°) than the reference state. The bases in Figure 2(c), in contrast, are more loosely wound (by ~-12°) than those in the reference structure. As detailed below, the bases in the latter unwound dimer are more exposed to the outside of the helix and hence more susceptible to interactions with the chemical probes of breathing and intercalation studies.

The alternate views in Figure 3 parallel to the local helical axes of the above three dimers illustrate the orientational motions in adjacent bases in our flexible model.

The bases attached to the reference backbone in Figure 3(a) adopt the classical parallel self-alignment and horizontal tilt of B-DNA in the solid state. In contrast,

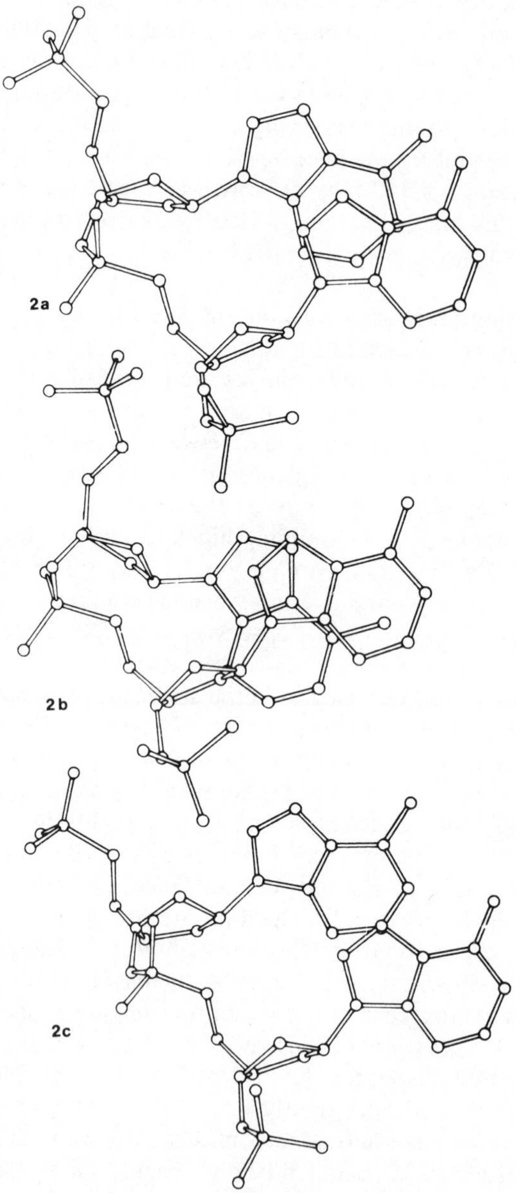

Figure 2. Detailed molecular representations in the pdApdAp dinucleoside triphosphate of the internal motions of the flexible B-DNA helix. All views are drawn perpendicular to the plane of the 5′-adenine moiety to illustrate base stacking. (a) The $\beta,\gamma = 105°$, $115°$ reference conformation. (b) The tightly wound $\beta,\gamma = 132°$, $99°$ state of extreme flexibility. (c) The loosely wound $\beta,\gamma = 87°$, $132°$ state of extreme flexibility.

the bases in the highly wound dimer in Figure 3(b) form a wedge of 11° that opens toward the core of the helix. This orientational feature appears to disfavor interactions between the base and the aqueous environment. As evident from Figure 3(b) the two bases of the tight dimer also tilt by -9° from the horizontal alignment found in the reference dimer. The bases attached to the loose dimer in Figure 3(c) describe a wedge of 12° that opens toward the outside of the helix. Furthermore, these bases upon tilting 52° from the normal horizontal arrangement move close to the outer surface of the structure. Such conformational changes appear to favor base-solvent associations.

The relative motions of the bases and the backbone about the reference conformation of our flexible model are apparent from the overlapping structures in Figures 4(a) and 4(b). Superimposed upon each of the two extreme conformations (darkened bonds) at the O(3′)-P-O(5′) linkage is the reference dimer structure (light bonds). Terminal phosphates and bases are displaced in opposing directions by the two types of motion (winding vs. unwinding). The wedges introduced between the bases are also apparent in the figure.

Polymer Flexibility

As illustrated in the perspective drawings of Figure 5, the above limitations upon local flexibility in the DNA backbone produce a smoothly bending structure that approximately resembles a regular helix of the same chain length. The flexible

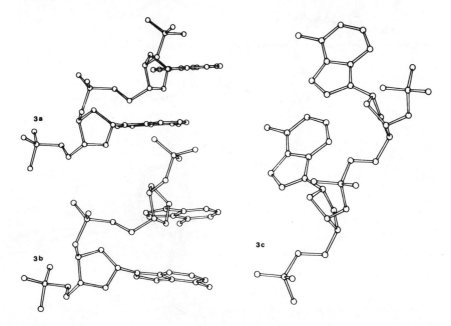

Figure 3. Additional views drawn perpendicular to the local helical axes of the three dimers described in Figure 2.

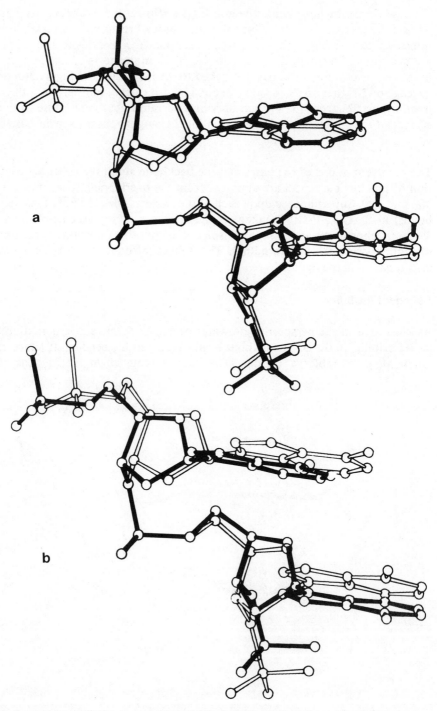

Figure 4. Comparative molecular representations of flexibility in the B-DNA model. The extreme conformations (dark bonds) are superimposed at the central phosphodiester linkage upon the reference state (light bonds). (a) $\beta,\gamma = 132°, 99°$. (b) $\beta,\gamma = 87°, 132°$.

molecule illustrated in Figure 5(a) is a 128-residue single-stranded B-DNA chain chosen at random by Monte Carlo methods.[32/35] This structure represents a single "snapshot"of the chain as it passes through the multitude of conformations (81^{128} or approximately 10^{171}) accessible in the present scheme. For comparison, the regular helix generated by 128 consecutive occurrences of the reference dimer conformation appears in Figure 5(b). For clarity, the sugar and base moieties are omitted in the two figures. Segments of the chains are represented instead by virtual bonds connecting successive phosphorus atoms. The flexible, wormlike chain possesses a pseudohelical backbone and pseudohelical axis that changes direction continually. The individual turns of the pseudohelix also vary considerably in size as a consequence of local winding and unwinding of the dimer chain units. The gradual bending of the pseudohelical backbone describes a trajectory with a large radius of curvature. The end-to-end separation of the flexible chain is shorter than that of the regular helix.

The flexible DNA backbone is best described by the complete array of conformations it can assume. A comprehensive picture of this array is provided by the three-dimensional spatial probability density function $W_0(\mathbf{r})$ of all possible end-to-end vectors.[32/36] This function represents the probability per unit volume in space that the flexible chain terminates at vector position $\mathbf{r}$ relative to the chain origin $\mathbf{0}$ as reference. For short chains $W_0(\mathbf{r})$ may be estimated with reasonable accuracy by direct Monte Carlo calculations and for long chains in terms of a three-dimensional Hermite series expansion of the Gaussian centered at the so-called persistence vector (cf. seq.).[36/37] Distribution functions obtained using the Hermite approximation

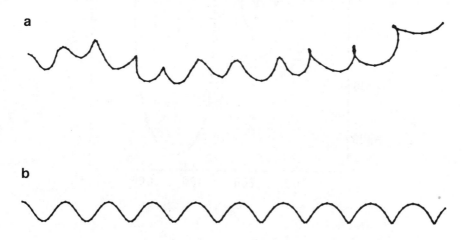

Figure 5. Computer generated perspective representations of a 128-residue single- stranded B-DNA chain. The chain backbone is represented by the sequence of virtual bonds that connect successive phosphorus atoms in the chain. (a) A representative flexible helix generated by Monte Carlo methods. (b) The reference helix of the model.

with correction terms through fourth order for flexible DNA chains of length $2^5 =$ 32, $2^6 = 64$, and $2^7 = 128$ are presented in Figure 6. The three-dimensional distribution functions are represented by three two-dimensional contour slices in the xz helical plane of the reference dimer and hence in the xz plane of the regular helix depicted in Figure 5(b). The regions enclosed by the contours account for approximately 90 percent of all spatial arrangements accessible to the chains of specified length. Superimposed upon the contours in Figure 6 are the backbone trajectories of three flexible Monte Carlo chain of length 128. This group of random molecules is seen to describe approximately the three different distribution functions. The trajectories of the three molecules are defined with respect to the coordinate system of the reference helix. The first few segmensts of the chains are roughly superimposable

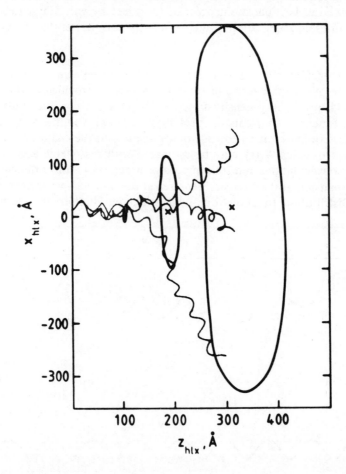

Figure 6. Two-dimensional contours that delineate from left to right the array of conformations described by a flexible B-DNA helix of lengths 2^5, 2^6, and 2^7. Contours are based upon the three-dimensional spatial probability density function $W_0(\mathbf{r})$ and are drawn in the xz helical plane of the reference structure illustrated in Figure 5b. Persistence vectors **a** for the three chain lengths are denoted by ×. Superimposed upon the contours are the xz projections of three of the 81^{128} possible random helices of length 128. The flexible helices are also drawn in the internal coordinate system of the reference helix.

upon the corresponding positions of the reference helix. Segments more removed from the chain origin, however, are found to deviate appreciably from the regular structure.

The position, size, and shape of the density distribution functions of flexible DNA depend markedly upon chain length. The three distribution functions presented in Figure 6 are described with respect to the persistence vector **a** characteristic of the chain length. These positions denoted by the ×'s in Figure 6 are the mean coordinates ($<x>$, $<y>$, $<z>$) of all the possible spatial configurations of the chains in the fixed coordinate system. Because the DNA helix is subject to structural constraints of fixed bond lengths and valence angles as well as to limitations in internal rotations, **a** is a non-null vector at all chain lengths. The limiting asymptotic coordinates of a_∞ are attained by chain length 2^{11} = 2048 after which all distribution functions are centered about position (20.0Å, 17.6Å, 553.4Å) in the reference helix coordinate system. As evident from Figure 6 at short chain lengths there is little deviation of the flexible chain from the average structure defined either by **a** or by the reference helix and the distribution is confined to a very limited domain. Since the distribution functions exhibit approximately cylindrical symmetry about the helical z-axis, the volume of space available to a chain of length 32 is best described as a small ellipsoid. With the perpetuation of smooth bending to DNA of longer chain lengths the random helices are found to deviate significantly from both **a** and the reference helix. The volume encompassing chain termini for DNA of length 128 is approximately 14,000 times the volume available to a chain of length 32 and about 50 times that found at length 64. The spatial distribution function at x = 128 is similar to the cap of a rather flat mushroom. By the time the DNA contains 2^8 or 256 residues, the chain backbone is able to bend back upon itself. The distribution functions associated with chains of length 2^7 (enclosed by dashed lines) and 2^8 (shaded area) are compared in Figure 7. The volume described by the termini of the longer chains in the figure is similar to that described by a huge bouquet of flowers. As evident from the shaded contour region in Figure 7, it is still difficult (although not impossible) for a flexible helix of 256 residues to terminate in the "stem" region of the distribution. At chain lengths beyond 2^{10} = 1024 residues the distribution function once again becomes ellipsoidal in shape. At this chain length the DNA is equally likely to describe a pseudohelix longer or shorter than the persistence vector. A chain of 1024 units is, however, somewhat more flexible in its motions along the x- and y-axis than along the z-axis of the reference helix. At length 2^{13} = 8096 the distribution volume may be described as a sphere and the probability function is an ideal Gaussian centered at **a**. Beyond this size, the distribution functions are simply larger spheres centered at a common point. At very large chain lengths the magnitude of a_∞ is small compared to the size of the distribution. At this limit the probability distribution may be represented in the classical fashion as a Gaussian centered at the origin.

Circle and Loop Formation

The long more tortuous DNA chains described above are potentially capable of

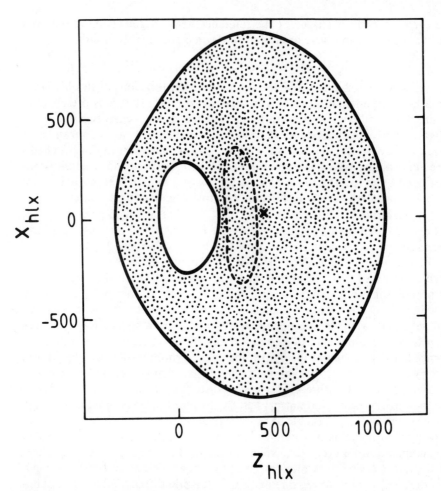

Figure 7. Two-dimensional contours of probability density for flexible B-DNA helices of lengths 2^7 (dashed line) and 2^8 (shaded area). The persistence vector at 2^8 is denoted by $\times$.

folding into the cyclic and looped structure long observed in experimental systems. Recent biochemical data indicate that the sizes of so-called foldback duplexes formed upon extremely fast renaturation of DNA are nonrandom.[38/39] These hairpin structures, whether looped (i.e., single-stranded with bases mismatched) or unlooped (i.e., double-stranded with bases paired), span a narrow distribution of lengths between 70-600 base pairs. Ring structures formed by annealing a variety of eukaryotic DNA fragments exhibit a strikingly similar size preference (300-600 residues).[40/41] As outlined below, this chain length dependence appears to reflect the relative probabilities that the termini of variously sized fragments of our flexible DNA model collide.

Cyclic or looped structures of DNA arise when the termini of a molecule or of a molecular fragment are confined to a particular three-dimensional arrangement in

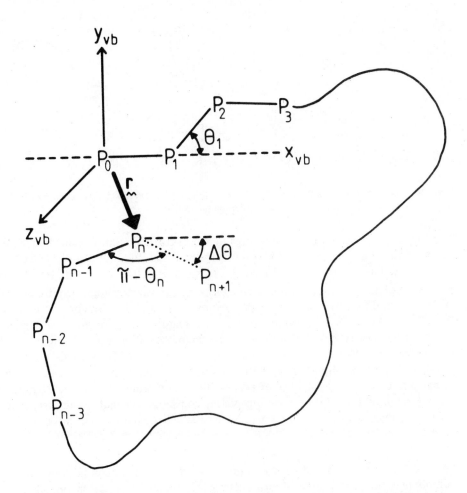

Figure 8. Polynucleotide segment of n virtual bonds connecting atoms 0 through n in a conformation approaching the requirements for cyclization or loop formation. Reference frame is defined along the first virtual bond. Valence angle supplements are denoted by θ. $\Delta\theta$ is the angle between the hypothetical n + 1 bond (dotted) and the first bond.

space. As illustrated in Figure 8, for a DNA helix comprising n virtual bonds, the end-to-end separation of the chain describes a vector $\mathbf{r}$ in the reference frame (x_{vb}, y_{vb}, z_{vb}) along the first bond (P_0 to P_1). Ring closure requires that $\mathbf{r} = 0$. This vector, however, is nonzero in cases of loop formation. In the studies outlined below to estimate the probability of foldback loops, the appropriate $\mathbf{r}$ vectors describe a cylindrical shell of diameter $16 \pm \delta d$ Å where $d^2 = 2(x^2 + y^2)$ and thickness $0 \pm \delta z$ Å in the (x,y,z) *helical* coordinate system of the initial virtual bond. This separation insures that the terminal bases of the loop may associate via hydrogen bonding. Previous estimates of polynucleotide loop closure[42/43] based upon Jacobson-Stockmeyer theory[44] assume $\mathbf{r}$ to include all points on a spherical shell of ~ 16Å diameter centered at P_0. This approximation ignores the geometrical constraints of base pairing at the ends of a loop.

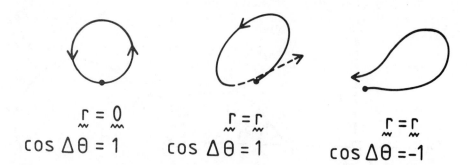

Figure 9. Schematic illustrating the distance (r) an orientational (cos $\Delta\theta$) constraints associated with DNA ring closure, *anti* loop formation, and *syn* loop formation.

The formation of DNA rings and loops of size n further requires that the terminal bonds of the chain are oriented in an arrangement that maintains the valence bond angles at atoms P_0 and P_n and also the torsion angles about bonds $P_0...P_1$ and $P_{n-1}...P_n$ within acceptable limits. Fulfillment of the valence angle requirements in cyclization and loop closure constrains the imaginary n + 1 virtual bond connecting P_n to P_{n+1} in Figure 8 to approximately parallel angular orientations $\Delta\theta$ with respect to the initial virtual bond ($P_0...P_1$) of the chain. As illustrated in Figure 9, the parameter $\tau = \cos \Delta\theta$ must fall in the range 1-$\delta\tau$ to 1 in both DNA closed circles and in so-called *anti*[45] or superhelical loops but must vary between -1 and -1 + $\delta\tau$ in *syn*[45] or hairpin loops. The assurance of acceptable torsion angles about the P_n juncture further requires that the intersecting planes containing virtual bonds n-1, n, and n + 1 describe an angle ω_n within the range of allowed local molecular motions of our model.

Upon elaboration of Jacobson-Stockmayer theory to include conditions of angle compliance, the probability Q(r) of occurrence of cyclic (**r** = **0**) or looped (**r** = **r**) DNA segments of length n is given by the product[46]

$$Q(\mathbf{r}) = [W_0(\mathbf{r})\delta\mathbf{r}] [T_r(\tau)\delta\tau] [\sum_{\eta}\Omega_{rr}(\omega_n)\delta\omega_n]. \tag{1}$$

In this expression $W_0(\mathbf{r})$ is the spatial probability density function and $\delta\mathbf{r}$ is the admissible departure of **r** from the value required for cyclization or loop formation; $T_r(\tau)d\tau$ is the probability that τ assumes a specific value (± 1) when **r** = **r** and $\Omega_{rr}(\omega_n)\delta\omega_n$ is the probability of occurrence of torsion angle ω_n for bond n when **r** = **r** and $\tau = \tau$. Included in the summation of Equation 1 are the various admissible conformations of ω_n indexed by η. As described elsewhere the $T_r(\tau)$ term may be expanded in averaged Legendre polynomials $<P_k>_{\tau = \tau}$ with argument $\tau_r = 0$.[46] The averaged polynomials used in the expansion here are evaluated in terms of the scalar chain moments $<\tau^m r^{2p}>$ with m = 0, 1, 2 and p = 0, 1, 2. Torsional correlations are assumed to be random and $\sum_{\eta}\Omega_{rr}(\omega_n) \delta\omega_n$ is taken to be unity.

In contrast to the analysis presented here, cyclization or loop closure of DNA based upon Jacobson-Stockmayer theory ignores all angular correlations and assumes that $W_0(\mathbf{r})\delta\mathbf{r}$ is a one-dimensional Gaussian distribution centered at the chain origin.[44] As a first correction in this paper, we assume $W_0(\mathbf{r})\delta\mathbf{r}$ at each chain length to be a three-dimensional Gaussian function centered about the persistence vector **a**. This ellipsoidal distribution is the basis function of the three-dimensional Hermite expansion functions presented in Figures 6 and 7. As a second correction we evaluate $T_r(\tau)\delta\tau$ in a Legendre polynomial expansion.[46]

Data describing the probability of DNA ring closure $Q(0)$ with chain length n are presented at various levels of approximation in Figure 10. As expected, the probability of loop formation based upon Jacobson-Stockmayer theory (curve 1), decreases linearly with chain length. As chain length increases in this scheme the spherical volumes associated with a one-dimensional Gaussian increase in size. Since all points within the spheres are equally likely and since the total probability of all conformations of a given sized chain is unity, the probability of occurrences at $\mathbf{r} = 0$ decreases with n. In marked contrast the $Q(0)$ of cyclization based upon a three-dimensional Gaussian centered at a (curve 3) exhibits a maximum at 2^8 repeating units. As noted in the contour surfaces above for chains less than 2^8 residues, a locally "stiff" DNA backbone cannot easily fold back toward the chain origin. At chain lengths greater than 2^8 units where the molecule is more flexible and $\mathbf{r} = 0$ is an attainable state, the spatial density distributions increase in volume. Following the argument given above for Gaussian spheres the probability of conformations where $\mathbf{r} = 0$ also decreases with n beyond 2^8 in the ellipsoidal scheme. Curves 1 and 3 coalesce at 2^{10} units where the magnitude of **a** is now small compared to the volume enclosed by the distribution function.

As evident from curve 2 in Figure 10, introduction of the valence angle correlation term $T_0(1)\delta\tau$ into the one-dimensional Gaussian expression of curve 1 depresses Q over all n up to n = 2^{12}. A local maximum in Q now occurs at 2^7 residues but the most probable cyclized chain length remains at 2^0. The angular correction factors for all $n \leq 2^{12}$ similarly decrease the probability of loop closure in curve 3 to curve 4. Below chain length 2^8 $T_0(1)$ approaches 0 in the "stiff" DNA chain and $\log_{10}Q(0)$ lies far below the range of data plotted in Figure 10. According to curve 4 cyclization occurs with greatest and approximately equal probability in DNA rings of length 2^8 and 2^9.

In the absence of angular correlations the probability of condensing a locally "stiff" DNA backbone into loops of end diameter $\sim 16\text{Å}$ parallels the behavior of curve 3 in Figure 10. Since the constraints upon $\mathbf{r}$ in loop closure are less stringent than those in cyclization, the computed loop probabilities are several orders of magnitude larger than those reported in Figure 10. The angular correlation factors $T_r(1)$ describing *anti* loop formation are less than unity and depress the probability of loop formation in a manner analogous to that described above for chain cyclization. In contrast, the angular correlation factors $T_r(-1)$ exceed unity at some chain lengths and hence enhance the formation of *syn* or hairpin loops. The ratio $T_r(-1)/T_r(1)$ exhibits

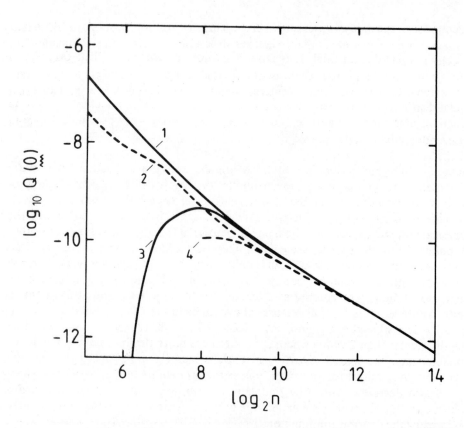

Figure 10. Cyclization probability Q(0) plotted against the number of virtual bonds in a flexible B-DNA helix. Curves 1-4 correspond to different approximations for Q: curve 1, spherical Gaussian with no angular correlations; curve 2, ellipsoidal Gaussian centered at a with no angular correlations; curve 3, spherical Gaussian with angular correlations; curve 4, ellipsoidal Gaussian with angular correlations.

a maximum at $n = 2^7$ and does not attain values of unity until chain lengths 2^{10}.

Summary

The above applications of polymer chain statistics to double helical DNA clearly demonstrate how the severely limited internal rotations of a "rigid" B-DNA backbone are magnified into the enormous flexibility of a long polynucleotide chain. Minor rotational motions within each dimer unit of the chain account satisfactorily for the radii of gyration of DNA spanning a broad range of molecular weights. In addition, the three dimensional spatial distribution functions generated by this set of limited internal rotations follow various macroscopic descriptions previously ascribed to DNA of different chain lengths. If long enough, the "rigid" DNA can condense without the occurrence of severe bends or kinks into compact conformations.[32/47] Indeed, the probability of both cyclization and loop closure based upon this model of limited flexibility parallels the occurrence of such structures in experimental systems.

The detailed motions ascribed to the phosphodiester linkages in the above model, however, are only approximate. The conformational states chosen are realistic in the sense that base stacking is maintained and also that the bases may interact to some extent with the aqueous environment. As a first approximation, all states of the model are equally favored. In a more exact model, flexibility in all the rotatable bonds of the polynucleotide backbone as well as slight variations in valence bond angles and bond lengths must be considered. Furthermore, chain averages should be based upon the relative energies of the various configurations of the helical backbone.

The above computations of polynucleotide average properties are further approximate in the sense that they apply to an ideal unperturbed DNA double helix free of long range excluded volume and electrostatic effects. The average dimensions of very long B-DNA helices subject to these long-range repulsive forces are expected to exceed those of an unperturbed chain.[48] The probabilities of cyclization and loop formation in very long chains will also be lower than the data reported above. Excluded volume effects, however, are negligible in short, more "stiff" DNA helices and are not expected to alter the computed preference for rings and loops in the range of 2^8 - 2^9 residues.

Acknowledgement

The author is grateful to Professor Joachim Seelig at the Biozentrum of the University of Basel, Switzerland for his hospitality during the course of this work and to the J. S. Guggenheim Memorial Foundation for a fellowship. The work was also supported by a grant from the U.S.P.H.S. (GM-20861). Computer time was supplied by the University of Basel Computer Center.

References and Notes

1. Bloomfield, V. A., Crothers, D. M., and Tinoco, I., Jr., *Physical Chemistry of Nucleic Acids*, Harper and Row, New York, Chapter 5 (1974).
2. Schellman, J. A., *Biopolymers 13*, 217 (1974).
3. Kratky, O. and Porod, G., *Rec. Trav. Chim. Pays-Bas 68*, 1106 (1949).
4. Shimada, J. and Yamakawa, H., *J. Chem. Phys. 67*, 344 (1977).
5. Yamakawa, H., Shimada, J., and Fujii, M., *J. Chem. Phys. 68*, 2140 (1978).
6. Erfurth, S. C. and Peticolas, W. L., *Biopolymers 14*, 247 (1975).
7. Johnson, W. C., Jr. and Tinoco, I., Jr., *Biopolymers 7*, 727 (1969).
8. Wells, R. D., Larson, J. E., Grant, R. C., Shortle, B. E., and Cantor, C. R., *J. Mol. Biol. 54*, 465 (1970).
9. Patel, D. J. and Canuel, L., *Proc. Natl. Acad. Sci. USA 73*, 674 (1976).
10. Kearns, D. R., *Ann. Rev. Biophys. Bioeng. 6*, 477 (1977).
11. Griffith, J. D., *Science 201*, 525 (1978).
12. Vollenweider, H. J., James, A., and Szybalski, W., *Proc. Natl. Acad. Sci. USA 75*, 710 (1978).
13. Pilet, J., and Brahms, J. *Biopolymers 12*, 387 (1973).
14. Pilet, J., Blicharski, J., and Brahms, J., *Biochem. 14*, 1869 (1975).
15. Hogan, M., Dattagupta, N., and Crothers, D. M., *Proc. Natl. Acad. Sci. USA 75*, 195 (1978).
16. Bram, S. and Beeman, W. W., *J. Mol. Biol. 55*, 311 (1971).
17. Bram, S., *J. Mol. Biol. 58*, 277 (1971).

18. Sundaralingam, M., in *Structure and Conformation of Nucleic Acids and Protein-Nucleic Acid Interactions*, Sundaralingam, M. and Rao, S.T., Eds., University Park Press, Baltimore, pp. 487-524 (1975).

19. Arnott, S., Chandrasekaran, R., and Selsing, E., in *Structure and Conformation of Nucleic Acids and Protein-Nucleic Acid Interactions*, Sundaralingam, M. and Rao, S. T., Eds., University Park Press, Baltimore, pp. 577-596 (1975).

20. Printz, M. P. and von Hippel, P. H., *Proc. Natl. Acad. Sci. USA 53*, 363 (1965).

21. McConnell, B. and von Hippel, P. H., *J. Mol. Biol. 50*, 297 and 317 (1970).

22. Teitelbaum, H. and Englander, S. W., *J. Mol. Biol. 92*, 55 and 79 (1975).

23. Frank-Kamenetskii, M. D. and Lazurkin, Yu, S., *Ann. Rev. Biophys. Bioeng. 3*, 127 (1974).

24. McGhee, J. D. and von Hippel, P. H., *Biochem. 14*, 1297 and 1568 (1975).

25. In contrast to this interpretation, previous molecular theories of breathing assume the phenomenon to involve conformations significantly perturbed from the classical B-DNA structure. See, for example, Sobell, H. M., Reddy, B. S., Bhandary, K. K., Jain, S. C., Sakore, T. D., and Seshadri, T. P., *Cold Spring Harbor Symp. Quant. Biol. 42*, 87 (1978) and Yathindra, N., in *Proceedings of the International Symposium on Biomolecular Structure, Function, and Evolution*, Srinivasan, R., Ed., Pergamon, London (1979).

26. Kim, S.-H., Berman, H. N., Seeman, N. C., and Newton, M. D., *Acta Crystallogr. B29*, 703 (1973).

27. Lee, C.-H., Ezra, F. S., Kondo, N. S., Sarma, R. H., and Danyluk, S. S., *Biochem. 15*, 3627 (1976) and *16*, 1977 (1977).

28. Cozzone, P. J. and Jardetzky, O., *Biochem. 15*, 4853 and 4860 (1976).

29. Yathindra, N. and Sundaralingam, M., *Proc. Natl. Acad. Sci. USA 71*, 3325 (1974).

30. Olson, W. K., *Biopolymers 14*, 1775 (1975).

31. Olson, W. K., *Macromolecules 8*, 272 (1975).

32. Olson, W. K., *Biopolymers 18*, 1213 (1979).

33. Olson, W. K., *Biopolymers 17*, 1015 (1978).

34. Davis, R. C., and Tinoco, I., Jr., *Biopolymers 6*, 223 (1968).

35. Jordan, R. C., Brant, D. A., and Cesaro, A., *Biopolymers 17*, 2617 (1978).

36. Yevich, R. and Olson, W. K., *Biopolymers 18*, 113 (1979).

37. Yoon, D. Y. and Flory, P. J., *J. Chem. Phys. 61*, 5358(1974).

38. Hardman, N. and Jack, P. L., *Nuc. Acids Res. 5*, 2405 (1978).

39. Deumling, B., *Nuc. Acids Res. 5*, 3589 (1978).

40. Lee, C. S. and Thomas, C. A., Jr., *J. Mol. Biol. 77*, 25 (1973).

41. Pyeritz, R. E. and Thomas, C. A., Jr., *J. Mol. Biol. 77*, 57 (1973).

42. DeLisi, C. and Crothers, D. M., *Biopolymers 10*, 1809 (1971).

43. Rubin, H. and Kallenbach, N. R., *J. Chem. Phys. 62*, 2766 (1975).

44. Jacobson, H. and Stockmayer, W. H., *J. Chem. Phys. 18*, 1600 (1950).

45. Johnson, D. and Morgan, A. R., *Proc. Natl. Acad. Sci., USA 75*, 1637 (1978).

46. Flory, P. J., Suter, U. W., and Mutter, M., *J. Am. Chem. Soc. 98*, 5733 (1976).

47. Additional calculations demonstrate that the random occurrence of a sharp kink in the flexible DNA model is not consistent with the solution properties of B-DNA; $<s^2>$ is reduced substantially, Gaussian behavior is attained at much shorter chain lengths (2^9 - 2^{10}), and small (2^5 - 2^6) loops and rings are highly favored when kinks occur with a probability of 0.01.

48. Flory, P. J., *Principles of Polymer Chemistry*, Cornell University Press, Ithaca, N.Y., Chapter 12 (1953).

Accessible Surface Areas of Nucleic Acids and Their Relation to Folding, Conformational Transition, and Protein Recognition

Charles J. Alden and Sung-Hou Kim

Department of Biochemistry
Duke University Medical Center
Durham, North Carolina 27710

and

Department of Chemistry and Laboratory of Chemical Biodynamics
University of California
Berkeley, California 94720

Introduction

Most biologically functional macromolecules have an intrinsic ability to fold into well defined three dimensional structures in physiological environments. Once they are folded, each molecule presents to its solvent environment a characteristic surface structure, when averaged over a long time period. Although the solvent-accessible surface structure is in dynamic fluctuating states, as demonstrated by the exchange of tritium or deuterium between solvent water and internal protons of a protein,[1] the fluctuation is centered around a defined average structure, with a well defined average "solvent-accessible surface structure."

A three dimensional structure determined by X-ray crystallographic method is a time-averaged (usually for a period of several weeks or longer of data collection) and space-averaged (over million molecules in a crystal) structure. Thus the crystal structure can be considered as a "static" structure representing the average of the population of dynamically fluctuating structures. Based on this "static" structure, one can determine the surface structure that represents the population-averaged surface of the molecule which is accessible to solvent or other interacting ligands.

For proteins, which incorporate a wide spectrum of polar, aliphatic and aromatic side groups and which exhibit a variety of secondary structures, the question of solvent effects as a driving force in chain folding has long been discussed.[2,3] The concept of accessible surface area has been introduced to estimate the relative changes in solvent-accessible surface upon folding polymers from random coils to compact forms and also to assess the surface exposures of different classes of atoms within folded polypeptides.[4-8] For nucleic acids, by contrast, the concept of relative surface exposure has not been explored extensively. Thus we have examined the solvent-

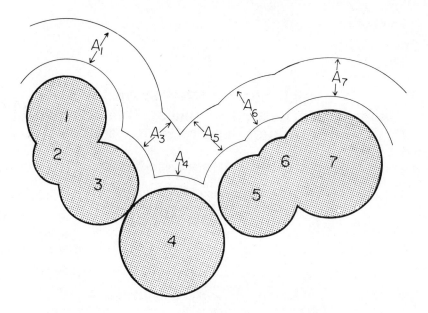

Figure 1. Schematic diagram of accessible surface areas. Around the examined molecule (shaded) is constructed an envelope defining the loci of points for the center of a spherical probe which may just touch the molecule. This construction is sliced by a set of parallel planes, and the perimeter (A) of the surface envelope is calculated for each atom. In the figure two separate envelopes, corresponding to different probe sizes, are shown. Note that as the probe radius increases, the accessibilities to it for atoms within a concavity decrease.

accessible surfaces of a variety of polynucleotides, (a) to evaluate the extent of surface burial on folding, (b) to compare relative exposures between hydrophobic and hydrophilic atoms, (c) to identify the most accessible atoms for intermolecular interactions, and (d) to relate helical conformational transitions to environmental changes. The details of the topics described here have been published.[9]

Calculation of "Static" Accessible Surface Area

The methodology employed in this study is very similar to that described by Lee and Richards[4] in their calculation of polypeptide surface areas. We imagine a spherical "probe" or solvent molecule of radius r_w free just to touch but not penetrate the van der Waals surface of the examined molecule. The closed surface defined by all possible loci for the *center* of the probe is defined as the accessible surface of the molecule, and those portions of the surface for which the probe touches only one atom comprise the accessible surface of that atom (Figure 1). For actual calculation, a molecule is sliced into sections with thickness h, then the accessible surface area for each slice is calculated. The calculation is an approximate integration, for which the *precision* will increase as h decreases. In general, the calculation is repeated several times, with slicings at different orientations, both to test the reproducibility of the calculations and to achieve a credible average value for each surface. The **accuracy** of these calculations depends only upon the values chosen for the van der Waals and probe radii (r_v and r_w), and of course upon the accuracy of the input coordinates. For details, see Lee and Richards[4] and Alden and Kim.[9]

Throughout, we use the terms "accessible surface area," and "exposure," synonymously, but to emphasize different aspects. Note that these surface areas involve contacts with only *one* probe molecule; interaction of the probe with other (bound) solvent molecules is not considered in these calculations

To determine best values for van der Waals radii, we used the intermolecular contact and volume survey of Bondi[10] for carbon, phosphorus, aromatic nitrogen and ester oxygen radii. Values for presumed spherical carbonyl oxygen, amino and methyl groups were derived from the volume decrements for each group, cited in the same survey, by the use of a computer program which calculates volumes in an approximate integral fashion, analogous to the surface accessibility program. The van der Waals radii used are listed in Table I. In some cases, we used modified radii to implicitly include hydrogen atoms for computational simplicity (Table I).

Maximal Accessible Areas of Polynucleotide Components

To gauge the change in surface exposure upon folding a molecule, it was necessary first to determine the maximum *possible* surface exposure of the components of that molecule, corresponding to the unfolded state. For polypeptides, this exposure has usually been determined by calculating the surface exposure of the middle residue of model tripeptides with all conformation angles *trans*.[4] For polynucleotides, with

Table I
Van der Waals Radii Used In This Study

Symbol	Group Type	Radius ($\mathring{A}$) Including hydrogen atoms explicitly	Radius ($\mathring{A}$) Including hydrogen atoms implicitly
ALC	Aliphatic carbon	1.70	2.00
ARC	Aromatic carbon	1.77	1.77
SOX	Sugar oxygen (ring and ester)	1.40	1.40
BOX	Carbonyl oxygen	1.64	1.64
POX	Phosphate oxygen	1.64	1.64
ARN	Aromatic ring nitrogen	1.55	1.55
AMN	Amino nitrogen	1.72	1.75*
PHO	Phosphorus	1.80	1.80
ARH	Aromatic ring hydrogen	1.01	--
ALH	Aliphatic hydrogen	1.17	--
AMH	Amino hydrogen	1.17	--
WAT	Water	1.40	1.40

*1.86 $\mathring{A}$ is suggested as more appropriate.

several variable torsion angles per residue, the all-*trans* conformation may not correspond to the one (of ~50) most exposed conformation; in addition, it is known that two adjacent phosphoester bond torsion angles are reluctant simultaneously to adopt *trans* values.[11] Figure 2 illustrates the accessible surface exposure of the phosphate oxygens for the model case of dimethyl phosphate as a function of the conformation angles about the two ester bonds. Included also are indicators for the corresponding conformation angles about the ester bonds observed in the structure of transfer RNA.[12] We note that the all-*trans* conformation, avoided by polynucleotide phosphates, corresponds to a minimal exposure for the charged phosphate oxygens. In addition, the repulsive lone-pair interactions of the ester ox-

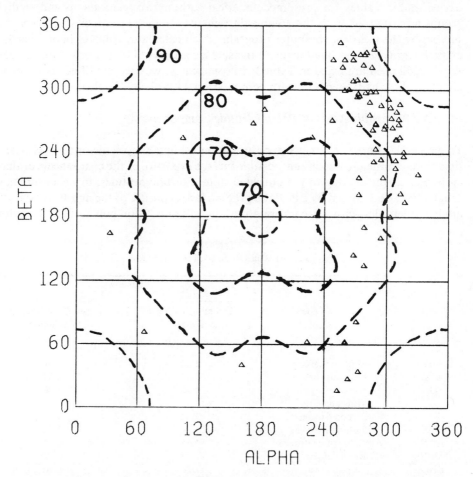

Figure 2. Accessible surface exposure of phosphate oxygens in dimethyl phosphate. The exposure of the phosphate oxygens was calculated for different conformations about the two ester bonds. Dashed lines represent contours of equal exposure; triangles represent the corresponding conformations in the backbone of the observed crystal structure of yeast phenylalanine transfer RNA.[12] The reluctance of nucleotide backbones to adopt an all-*trans* conformation may in part be due to the burial of charged phosphate oxygens that such conformations entail.

ygens are believed to discourage adoption of this conformation in polymers.[13/14] Hence a more elaborate construction was invoked to determine maximal exposures for the individual atoms (see Alden and Kim[9]). Maximal exposure for the deoxyribose atoms were calculated for both the C3'-endo and C3'-exo puckered forms. The total maximum exposures for the carbon and oxygen atoms were seen to be very similar, although the exposure of individual atoms were different for the two forms. Maximal exposures of the nucleoside bases were calculated for the bases assuming the normal *anti* conformation observed in polynucleotides. Table II lists the maximal surface exposures for the DNA components (calculated with explicit hydrogens) and for the RNA components (with implicit hydrogens). While it is quite possible that not all of the atoms in a nucleotide may simultaneously achieve their maximal exposure, this list gives a fair estimate of the relative degree of exposure for the various nucleotide atom types for the unfolded state. Thus we note that the average *maximal* accessible surface area for an unfolded DNA chain is approximately 490 $\AA^2$ per nucleotide, with the bases accounting for nearly half (44 percent) of this total.

Polarity of Exposed Atoms

One difficulty in describing the relative exposures of polar and nonpolar groups is the assignment for the aromatic nitrogen atoms, which are quite polar and frequently form hydrogen bonds within the plane of the base, yet which contribute to the π orbitals of the ring and act as aromatic molecules when approached normal to the plane of the base. Arbitrarily dividing the contribution from such atoms equally among the polar and nonpolar classifications, we note that in the unfolded state polar atoms comprise slightly over half (54 percent) of the exposed surface of DNA, with approximately equal contributions from the bases and the backbone. For the backbone there is an almost equal exposure of hydrocarbon and oxygen atoms, while for the bases 42 percent of the atoms may be classed as non-polar. For the individual nucleoside bases, however, there is a pronounced range of maximal polar group exposure: 69 percent for G, 61 percent for C, 52 percent for T and 45 percent for A.

Table II
Maximum Accessible Surface Areas in $\AA^2$ for
The Components of Double Helical DNA and RNA
for ($r_w = 1.4 \AA$)

DNA backbone with C3'-endo sugars	275.01
DNA backbone with C3'-exo sugars	274.28
RNA	288.93
Adenine	213.66
Cytosine	195.84
Guanine	236.76
Thymine	212.67
Uracil	186.60

Surface of DNA Double Helices

Accessible surface areas for double- helical DNAs were calculated by summing the individual atomic exposures of the middle base-paired residues in various double-stranded, complementary trinucleotides. Helices examined were A- and B-DNA,[15] C-DNA[16] and D-DNA.[17] For each of these helices hydrogen atoms were included in the input coordinates, and areas were calculated with a slicing separation of 0.1 Å. In all cases except D-DNA, several different base sequences were examined, to test for sequence and composition dependence of surface accessibility. Values for (A,T)- and (G,C)- rich polymers were taken as the averages of the four distinct trinucleotide

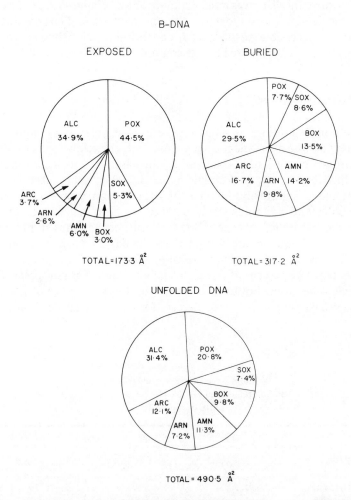

Figure 3. Group type exposures in unfolded and helical B-DNA. Solvent accessible surface areas were calculated using a 1.4Å probe radius for maximally unfolded DNA and B-DNA of average base sequence. The accessible surface buried in folding the extended chain into a double helix, is also shown. In folding of the polymer, the bases become mostly buried while the phosphate oxygens remain nearly fully exposed.

sequences consisting of one type of base pair. The results of these averagings are summarized in Table III.

Overall, the folding of a long DNA molecule into a double helix results in the burial of approximately two-thirds of its maximally exposed surface, a value quite similar to the resulst for globular proteins.[4/7] A comparison of the relative surface exposures of B-DNA with the unfolded molecules is shown in Figure 3. For most atom types, the reduction of accessible surface upon folding is close to their proportion of

Table III
Group Type Exposures in $Å^2$
For Two Residues of DNA Double Helices

(a) A,T Rich Polymers				
	A-DNA	B-DNA	C-DNA	D-DNA
ALC	7.50	11.87	17.39	26.31
ARC	4.77	5.41	4.50	3.96
SOX	23.12	21.34	20.24	22.41
POX	143.92	152.52	152.06	128.12
BOX	17.48	11.90	10.21	8.44
ARN	11.38	8.99	9.12	7.20
AMN	1.40	1.58	1.91	2.34
PHO	0.96	0.36	0.14	0.30
ARH	5.02	2.35	1.40	0.00
ALH	126.47	122.70	125.64	129.17
AMH	7.98	11.28	11.23	11.67
TOTAL	350.00	350.30	353.84	339.92
POS	20.76	21.85	22.26	21.21
NEUT	143.76	142.33	148.93	159.44
NEG	185.48	186.12	182.65	159.27

(b) G,C Rich Polymers			
ALC	6.46	6.36	7.12
ARC	6.32	12.33	11.28
SOX	22.00	14.96	14.10
POX	150.34	155.84	157.06
BOX	12.36	9.18	9.18
ARN	9.28	8.88	9.04
AMN	6.35	3.25	2.68
PHO	0.96	0.36	0.14
ARH	3.16	5.70	4.88
ALH	106.48	100.78	106.98
AMH	30.80	25.33	23.81
TOTAL	354.51	342.97	346.27

the total maximal surface area, with base atoms being buried relatively more than sugar atoms. The sole and striking contrast to this pattern occurs for the phosphate oxygens, whose exposure is reduced only slightly upon transition from the extended to the helical form. Thus the accessible surface in DNA double helices is more polar than for the random coils, with the phosphate oxygens alone accounting for nearly 45 percent of the total accessible surface area. The contribution from the bases now

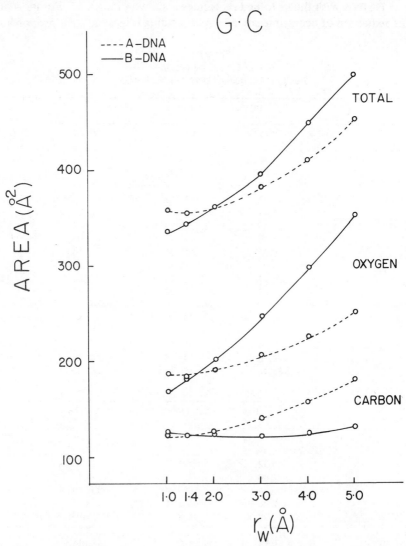

Figure 4. Accessible surface areas for two base-paired residues of poly(dG)•poly(dC) in the A- and B-helical forms, calculated varying the probe radius r_w. For the polymer, the B-DNA form entails relatively greater total exposure, and particularly a greater exposure of phosphate oxygens, for large probe radii. This result is consistent with the experimentally observed result that high water activity encourages adoption of the B-DNA conformation.

accounts for only about a fifth of the total surface in the double-stranded helices, while the sugars account for over a third of the surface.

Considering the observed effect of solvent upon DNA helix conformation, and the pronounced difference in gross shapes of the different forms, the similarities of their total exposures were somewhat surprising. Hence we calculated the surface accessibilities for A- and B-DNA over a wide range of probe radii (1.0 to 5.0Å), to determine if larger probes, such as hydrated metal ions or structured water aggregates, could discriminate between the different DNA shapes. The results of this

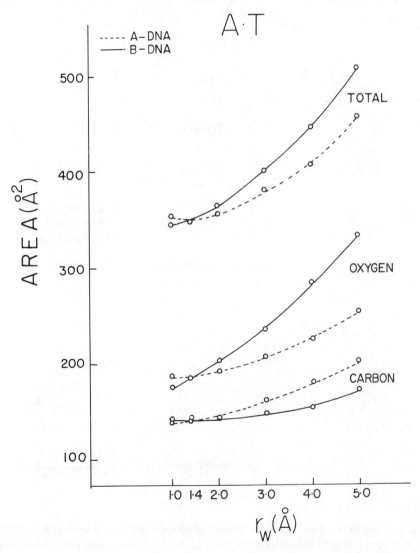

Figure 5. Accessible surface areas for two base-paired residues of poly(dA)•poly(dT) in the A- and B-helical forms, calculated varying the probe radius r_w. Rest of the details same as in Figure 4 legend.

test are shown in Figures 4 and 5. We note that as r_w increases, the *total* surface accessibility of B-DNA increases more rapidly than does that of A-DNA. Moreover, this difference is attributable almost entirely to a greater exposure of the phosphate oxygens in B-DNA, as the aliphatic carbon exposure becomes relatively greater for

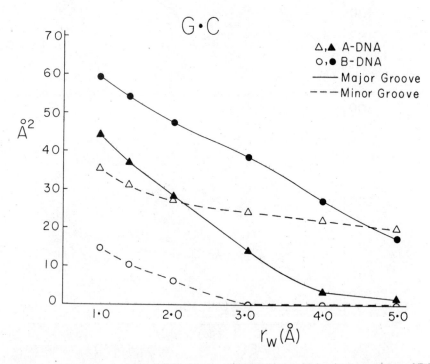

Figure 6. Accessible surface areas for DNA base atoms of the major and minor grooves of A- and B-DNA for G•C base pairs calculated varying the probe radius r_w. As r_w increases to large values, corresponding to extended water complexes or larger amino acid side chains, the only groups with significant exposures are the amino groups of C and G.

A-DNA. Curiously, DNA accessibility curves intersect just at the r_w corresponding to the radius of a single water molecule. On the other hand, when one examines the *base* surface accessibility for larger values of r_w (corresponding to bulky side groups of proteins, structured water, or hydrated metal ions) the accessibility of the major groove of A-DNA decreases abruptly, as does the minor groove exposure of B-DNA (see Figures 6 and 7).

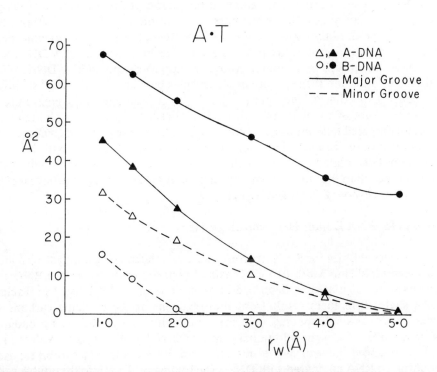

Figure 7. Accessible surface areas for DNA base atoms of the major and minor grooves of A- and B-DNA for A•T base pairs, calculated varying the probe radius r_w. As r_w increases to larger values (see legend for Figure 6), the only group with significant exposures is the methyl group of T.

For probe radii greater than 3 Å, only guanine amino groups in the minor (shallow) groove exhibit significant exposure in A-DNA, while thymine methyl and cytosine amino groups in the major (wide) groove dominate the base exposures of B-DNA. These groups may be expected to be the most significant for recognition by DNA-binding proteins.

For (A,T)-rich polymers, adoption of the A, B, or C helical forms results in virtually identical total group-type exposures to a 1.4Å radius probe molecule, corresponding to a single water molecule. The eight-fold alternating- sequence D-DNA has a slightly smaller total exposure, and a significantly smaller (over 30 Å²) exposure of oxygens, reflecting the virtual collapse of the minor groove in that helix. For (G,C)-rich double helices there is a slightly greater variation (∼12 Å²) in total exposures depending upon conformation. Between different conformations there are pronounced changes in the exposures of some atoms. In A-DNA, C1′ is quite exposed and C3′ is not, while the converse is true for B-DNA, reflecting the difference in sugar puckering modes. The exposures of all other backbone atoms are quite similar for the two helices. There is, however, a marked difference in exposure of the base atoms between the different polymers.

In B-DNA, the exposure of base atoms at the major groove is about 25 Å² (50 percent) greater than in A-DNA, while base atoms in the B-DNA minor groove have only about one-third the exposure of those in A-DNA. In particular, thymine methyl groups in the major groove are twice as exposed as in B-DNA as in A-DNA, while guanine amino groups in the minor groove are far more exposed in A-DNA. Within a given conformation the exposure of a particular base also is influenced slightly by the *sequence* of adjacent base pairs, with the most pronounced difference being a greater exposure of pyrimidines and smaller exposure of purines in alternating sequences compared to homopolymer tracts. However, the sequence-dependent variations in exposure are much less noticeable than are the variations with helix conformation. For isolated water molecules the major groove of A-DNA is only slightly more exposed than the minor groove, whereas in B-DNA, the major groove is five to six tims more exposed than the minor groove.

Surfaces for RNA Double Helix and Single Helix

Surface accessibilities for RNA structures were calculated using nonhydrogen atoms and a coarser slicing width (0.5 Å). Maximal surface exposures for the RNA components were calculated as described in earlier section and are listed in Table II. While slight differences result from the treatment of the aliphatic carbons and amino groups as single spheres, omitting separate hydrogens, the most pronounced difference arises from the additional ribose hydroxyl group. This is evident in the slightly enhanced sugar oxygen exposure, and decreased aliphatic carbon exposure, for unfolded RNA compared with DNA. The folding of RNA into the double helical A-RNA[18] structure results in a burial of approximately 63 percent of the maximal surface. As in the case of helical DNAs, the greatest reduction in exposure comes for the aromatic bases, while the relative contribution of the phosphate oxygens to the

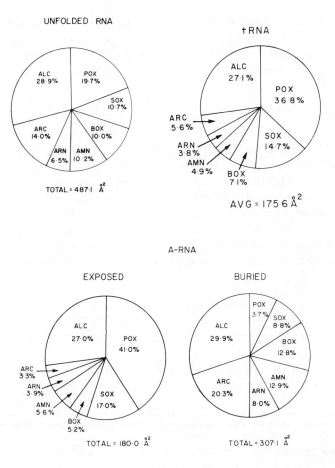

Figure 8. Group type exposures in extended and folded RNA. Solvent accessible surface areas were calculated using a 1.4Å probe radius for maximally exposed RNA (with random base sequence), the crystal structure of yeast phenylalanine transfer RNA,[12] double helical A-RNA (with random base sequence) and the buried surface in A-RNA. Despite the irregular folding and considerable chemical modification of the tRNA, its accessible surface has a distribution very similar to that of A-RNA.

exposed surface is again over double that in the unfolded state. For double-helical A-RNA, polar groups comprise nearly 70 percent of the total accessible surface with sugar oxygens contributing about 10 percent more and aliphatic carbons correspondingly less, than for DNA (Figure 8).

The surface exposure of a single strand of polyribocytidylic acid (rC) in the 11-fold A-RNA conformation was compared with the observed 6-fold poly (rC) conformation obtained from fiber diffraction studies.[19] We note that the total exposure per residue is about 20 Å^2 (10 percent) less in the latter case, with the reduction being due almost entirely to a greater burial of carbon atoms; nitrogen and oxygen atoms have about equal exposures in both cases. This suggests that 6-fold poly (rC) is probably more stable than an A-RNA type conformation for a single strand. Potential energy calculations of dinucleoside monophosphates, extended to helical polymers, without consideration of solvent effects, also indicate a preference for approximately six-fold helices,[20] so there is reason to believe that the fiber structure is representative of a preferred helical form for single stranded polycytidylic acid and is not merely an artifact of lattice interactions in the crystallite.

Transfer RNA

The accessible surface for yeast phenylalanine transfer RNA in the orthorhombic crystal form was calculated using the coordinates of Sussman, *et al.*[12] Only non-hydrogen atoms (with their appropriate radii) were included, and a slicing separation of 0.5 Å was employed. To facilitate computation, the molecule was divided into three overlapping segments; the region of overlap was sufficiently broad that every atom was accompanied by its full set of neighbors in at least one set of calculations. Individual atomic accessible surface areas were averaged for three separate slicing orientations and summed for each residue and atom type and for the whole

Table IV
Backbone Exposures in tRNA Stems and Loops:
Average (and Standard Deviations) per Nucleotide in Å²

	tRNA helical*		tRNA non-helical		tRNA	A-RNA
Ribose C	41.6	(7.5)	32.2	(18.2)	38.2	48.7
Phosphate	72.1	(9.4)	48.6	(17.9)	63.5	75.8
Sugar 0	29.8	(4.6)	17.6	(12.3)	25.2	25.2

*A residue is classified as helical when it adopts a conformation characteristic of a standard double helix *and* when its base is stacked with both neighboring residues. Here residues 2-6, 11-14, 23-31, 35-44, 50-53, and 62-75 are designated as helical, all others (except 1 and 76) as nonhelical.

molecule, as listed in Table IV. The average exposure per residue, 175 Å² is slightly less than the average for an interior residue of normal double helical A-RNA, despite the presence of many substituted groups on the surface and the unstacking of four bases as obseved in the crystal structure of this tRNA (for a review, see Kim[21]); apparently the extensive tertiary interactions more than compensate for any base stacking in the chain folding. Figure 8 shows the relative exposures of various group types in this tRNA; it may be noted that the overall distribution is quite similar to that of an A-RNA: bases are exposed slightly more and phosphate oxygens somewhat less than for the regular polymer. The near identity of alphatic carbon exposures is rather remarkable: it should be noted that the contribution of aliphatic carbons in the modified bases and sugars accounts for nearly 20 percent of the ALC portion, and 5 percent of the total accessible area of this tRNA, just offsetting the greater burial of the ribose carbons in the nonhelical residues compared to A-RNA. Inspection of the list of individual atomic exposures for this tRNA revealed that, without exception, every "modified" aliphatic group is exposed substantially, suggesting that the modification of bases significantly increases (by about 20 percent) the areas of groove surfaces for various proteins to recognize.

The exposure calculations for the phosphate oxygens and ribose carbons of the individual residues of tRNA reveal that the backbone exposures within helical regions are quite uniform and close to those in helical A-RNA, while significant variations are found only at residues forming sharp bends at the ends of stems and in loops. Furthermore, the backbone atoms in loops are found to be *less* exposed than those in the stems.

Two-Thirds of Nucleic Acid Surfaces are Buried on Folding

We have calculated the water-accessible surface areas of double-stranded DNAs, RNA and for one transfer RNA, and compared them to the maximal surface exposures for unfolded polynucleotides. In general, folding of the polymer reduces the surface exposure to about one-third of its maximal possible value, which is similar to the surface reduction calculated for folding of globular proteins.[4,7] Upon folding, there is a noted tendency for nonpolar groups (particularly the aromatic bases) to be buried, as expected, and for the charged phosphate oxygens to maintain near maximal exposure. As is also the case for proteins, those polar atoms which are buried form hydrogen bonds with other polar groups. Thus there is a general tendency for polynucleotides to obey the dictum of "polar out, nonpolar in" when folding from a random chain to a compact structure.

Driving Force in Nucleic Acid Folding is Quite Different from That of Proteins

For an average DNA residue, the surface area that becomes buried on folding maximally extended coils into the double-helical B-DNA form is $317Å^2$, of which $126Å^2$, or about half, is due to nonpolar atoms. Similarly for transfer RNA, neglecting changes in exposure caused by addition of modified groups, the surface area buried upon folding an extended coil into the native structure is about $23,700Å^2$, of which about $12,200Å^2$ corresponds to burial of hydrocarbons. Per nucleotide residue, these values correspond to 311 and $160Å^2$, respectively. The hydrophobic effect of nonpolar surface burial is widely regarded as one of the dominant factors in protein folding. Based on a calibration of amino-acid solubilities *versus* side chain accessibilities, this effect has been estimated to contribute about $-24cal/mole-Å^2$ to the net free energy of protein folding.[7,25] For nucleic acids, the burial of water-accessible area per unit mass ($0.95Å^2$/dalton for DNA, $0.97Å^2$/dalton for tRNA) is quite similar to that calculated for proteins (0.85 to $1.12Å^2$/dalton).[7] In addition, the proportion of polar and nonpolar surface, both exposed and buried, are not greatly different for proteins[8] and for nucleic acids. However, despite these similarities, the thermodynamic properties of the two systems differ markedly: upon chain unfolding, nucleic acids exhibit positive values for entropy and enthalpy,[22-24] whereas those quantities are both negative for protein denaturation. If one assumes a similar energy/area ratio for hydrophobicity in nucleic acids as in proteins, it follows that other forces more than counteract its effect upon the energetic terms. Alternatively, it may be possible that the large dipole moments of the aromatic bases eliminate the hydrophobic effect for their surfaces. In either case, the dominant driving force(s) in nucleic acid folding must differ from that of proteins.

Comparison of the measured free energies, enthalpies and entropis of 19 oligonucleotides[22] with their buried or exposed surface areas yielded only a very approximate correlation. This is a direct consequence of the somewhat surprising result that, within a given helical conformation, the exposed (or buried) water- accessible area for a given base pair shows only minor variation (less than $8Å^2$) with respect to the sequence of the neighboring base pairs, although the thermodynamic parameters for certain sequence isomers may vary by as much as a factor of two. For these oligonucleotides, the free energy/buried area ratio was approximately -1.2 ± 0.6 cal/mole-$Å^2$, which is the same order of magnitude as the corresponding ratio for transfer RNA.[24] Both the small magnitude and the large scatter of these values also suggest that factors other than solvent-base interactions are the dominant determinants of the sequence-specific stability of these nucleotides.

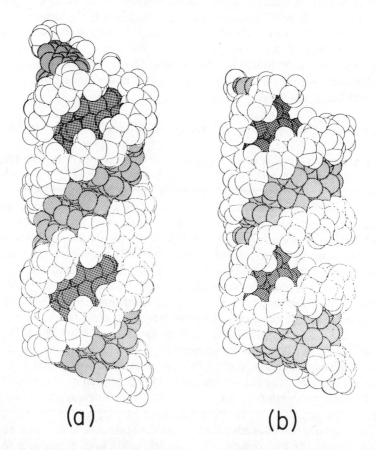

(a)   (b)

Figure 9. Illustrations of A-DNA double helix. View (a) is normal to the helix axis while view (b) represents the helix tilted by 27.0° to reveal maximum groove width. Dark shading indicates bases in the major groove and lighter shading indicates bases in the minor groove. In A-DNA the phosphate oxygens of the two strands are directed toward each other across the deep, narrow major groove while the minor groove is nearly flat. This contrasts with B-DNA (see Figure 10).

Correlation of A → B Helical Transitions to DNA Hydration

A wide variety of experimental evidence has related DNA helical conformation transitions to altered solvent conditions. X-ray diffraction[26] and Raman spectroscopy[27] of DNA fibers, as well as infrared dichroism of DNA films[28/29] and dilute solutions[30] all clearly indicate marked conformational changes at defined levels of relative humidity or water activity. Similarly, gravimetric,[31] infrared,[32] calorimetric,[33] and density gradient centrifugation studies[34/35] clearly demonstrate distinct hydration of DNA. These studies combined indicate that the level of hydration required to maintain DNA in a double-helical structure, and for which all the primary DNA hydration sites are occupied, is about 9 to 10 waters per nucleotide. To convert DNA to

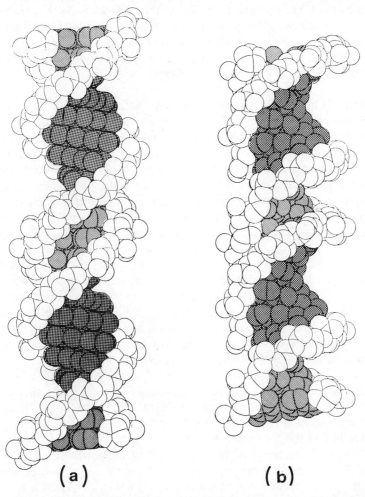

(a) **(b)**

Figure 10. Illustrations of B-DNA double helix. View (a) is normal to the helix axis while view (b) represents the helix tilted by 31.5° to reveal the maximum groove width. The shading is the same as in Figure 9. Note that, unlike A-DNA (Figure 9) the phosphate oxygens are directed outward, delineating two grooves of approximately equal depth but unequal width.

the B helical form requires extra water, for a total of 13 to 18 per nucleotide.[30] These added waters have properties more like bulk water than the primary hydration shell[32] and may be considered as a secondary shell of hydration, not interacting directly with the DNA but forming water-water bridges.[36]

The surface accessibility calculations described above provide support for this mechanism for DNA conformation transition. The virtual identity of the total surface accessibilities for A- and B-DNA at $r_w = 1.4Å$, corresponding to a single water molecule, indicates that the primary hydration layers for both forms are quantitatively similar. However, for larger values of r_w, the increased accessible surface for B-DNA relative to A-DNA, and particularly that portion associated with the phosphate oxygens (Figures 4 and 5), indicates that B-DNA is much more amenable to formation of extended solvent complexes about its charged phosphate groups. Sterically, this interpretation appears justifiable: in B-DNA both phosphate oxygens point radially outward from the helix, whereas in A-DNA one phosphate oxygen is directed inward toward the major groove (Figures 9 and 10).

This difference in hydration may be illustrated another way. For both A- and B-DNA, the volumes enclosed by the van der Waals surface of the molecule are virtually identical: about 450 $Å^3$ per two (paired) nucleotides. Constructing a cylinder about each helix, with radius r equal to the average of the radial phosphate oxygen coordinates plus one water diameter, the volume per double-stranded residue is $V = \pi r^2 h$, where h is the axial rise per residue of the particular helix. Subtracting 450 $Å^3$ from this volume yields the groove volume accessible to water; dividing this volume by the mean volume of one water molecule in liquid (30 $Å^3$) yields the number of waters per residue which may be contained in the grooves. Using r = 11.6 or 12.3Å, and h = 2.56 or 3.38Å, for A- and B-DNA respectively, the number of waters per nucleotide are 10.5 and 19.3 for the two polymers. Again, the calculation is crude but the result is quite compatible with the experimentally observed values of 9-10 and 18 waters per A- and B-DNA residue, respectively.[30] Thus, conversion from the B to the A form upon reduction of water activity (for a DNA capable of existing in either form) may be seen as a response to removal of excess water from the grooves, converting a structure with two concave grooves into a shorter structure with one flatter groove and one narrower groove (Figures 9 and 10).

It is now fairly well established, at least in hydrated films and fibers, that DNA rich in A•T pairs prefer the B conformations, while sequences rich in G•C pairs have a wider range of stability in the A and B forms.[26/29/37] Density gradient centrifugation studies[35] have linked the lower density of A,T-rich polymers to a higher level of hydration; this result is supported by a recent set of theoretical calculations which also indicate that an A•T pair can bind one or two more water molecules than a G•C pair.[37] In light of the multiplicity of coordinated geometries available to hydrated ions complexing with phosphate groups and the base-lined grooves, a unique steric description of a DNA-solvent superstructure appears unlikely. However, the influence of specific base atoms upon the secondary hydration layer (determined by examining atom exposures for $r_w > 1.4Å$) can provide clues about the willingness of

...e environment to allow or enforce a particular DNA conformation. For A•T base ...irs, the greatest exposure is exhibited by thymine methyl groups, and here only in ...e B conformation (over 80 percent of the total base exposure for $r_w \geq 4.0$ Å). Since ...methyl groups are hydrophobic, their exposure in the major groove would tend to encourage water-water aggregation in the major groove and thus pressure the major groove to remain wide. For G•C pairs the most exposed groups are the amino groups of guanine on the minor groove (over 80 percent of the total base exposure in the A form, no exposure in the B form, for $r_w = 4.0$Å) or cytosine on the major groove (over 70 percent of the total base exposure in B-DNA at $r_w \geq 4.0$Å). Both groups are hydrophilic, therefore, extended water-DNA base interactions would result in either the A- or B-helical forms. Thus for (G,C)-rich DNA, the availability or lack of excess water will dictate the DNA conformation: under high water activity the major groove will fill and the phosphates will form more extensive hydration complexes   under low water activity, the major groove will collapse and the phosphate oxygens will become more buried.

Major Groove of B-DNA and Minor (Shallow) Groove of A-DNA Are Primary Recognition Surfaces for Proteins

As can be seen from Figures 6 and 7 as the radius of the probe becomes 3 Å or greater, the *primary* base exposures occur in the major groove of B-DNA and the minor groove of A-DNA. Specifically, the methyl group of thymine and amino group of cytosine provide the most of the accessible area on the major groove in B-DNA, and the amino group of guanine on the minor groove of A-DNA. Since most of the side chains of amino acids are considerably larger than single water molecules the above observation can be considered as suggesting that the *primary* recognition surface of B-DNA is the major groove and that of A-DNA is the minor groove. This is consistent with the observation that protein-DNA contacts occur predominantly in the major groove of DNA in the E. coli RNA polymerase - *lac* promoter complex[39] and in the *lac*-repressor-*lac* operator complex[40] if one interprets protection *as well as* enhancement of methylation as due to close contact of grooves with proteins.

Acknowledgements

This work was supported by grants from the National Institute of Health (CA-15802 and K04-CA-00352) and the National Science Foundation (PCM76-04248). C.J.A. is an N.I.H. research fellow.

References and Footnotes

1. Hvidt, A. and Nielsen, S. O. *Advan. Protein Chem. 21* 287-386 (1966).
2. Kauzmann, W. *Adv. Protein Chem. 14* 1-63 (1959).
3. Tanford, C. *J. Am. Chem. Soc. 84*, 4240-4247 (1962).
4. Lee, B. and Richards, F. M. *J. Mol. Biol. 55*, 379-400 (1971).
5. Shrake, A. and Rupley, J. A. *J. Mol. Biol. 79*, 351-371 (1973).
6. Finney, J. L. *J. Mol. Biol. 96*, 721-732 (1975).

7. Chothia, C. *Nature 254*, 304-308 (1975).
8. Richards, F. M. *Ann. Rev. Biophys. Bioeng. 6*, 151-176 (1977).
9. Alden, C. J. and Kim, S.-H. *J. Mol. Biol.*(in press).
10. Bondi, A. *J. Phys. Chem. 68*, 441-451 (1964).
11. Kim. S.-H., Berman, H. M., Seeman, N. C., and Newton, M. D. *Acta Cryst. B29*703-710 (1973).
12. Sussman, J. L., Holbrook, S. R., Warrant, R. W., Church, G. M., and Kim, S.-H., *J. Mol. Biol. 123*, 607-630 (1978).
13. Sundaralingam, M. *Biopolymers 7*, 821-860 (1969).
14. Newton, M. D. *J. Amer. Chem. Soc. 95*, 256-258 (1973).
15. Arnott, S. and Hukins, D. W. L. *Biochem. Biophys. Res. Comm. 47*, 1504-1509 (1972).
16. Arnott, S. and Selsing, E. *J. Mol. Biol.98*, 265-269 (1975).
17. Arnott, S., Chandrasekaran, R., Hukins, D. W. L., Smith, P. J. C., and Watts, L. *J. Mol. Biol. 88*, 523-533 (1974).
18. Arnott, S., Hukins, D. W. L. and Dover, S. D., *Biochem. Biophys. Res. Comm. 48*, 1392-1399 (1972).
19. Arnott, S., Chandrasekaran and Leslie, A. G. W. *J. Mol. Biol. 106*, 735-748 (1976).
20. Hingerty, B. and Broyde, S. *Nucl. Acids Res. 5*, 127-137 (1978).
21. Kim. S.-H. "Advances in Enzymology" (ed. A. Meister) Vol. 46, pp. 279-315 (1978).
22. Borer, P. N., Dengler, B., Tinoco, Jr., I., and Uhlenbeck, O. C. *J. Mol. Biol. 86*, 843-853 (1974).
23. Edelhoch, H. and Osborne, Jr., J. C. *Adv. Protein Chem. 30*, 183-276 (1976).
24. Privalov, P. L. and Flimonov, V. V. *J. Mol. Biol. 122*, 447-464 (1978).
25. Chothia, C. *Nature 248*, 338-339 (1974).
26. Arnott, S., in "Organization and Expression of Chromosomes" (eds. V. G. Allfrey, E. K. F. Bautz, B. J. McCarthy, R. T. Schimke and S. Tissieres) Dahlem Konforenzen: Berlin (1976).
27. Erfurth, S. C., Bond, P. J. and Peticolas, W. I. *Biopolymers 14*, 1245-1257 (1975).
28. Pilet, J. and Brahms, J. Biopolymers *12*, 387-403 (1973).
29. Brahms, J., Pilet, J., Tran, T. P. L. and Hill, L. R. *Proc. Nat. Acad. Sci. USA 70*, 3352-3355 (1973).
30. Wolf, B. and Hanlon, S. *Biochemistry 14*, 1661-1670 (1975).
31. Falk, M., Hartman, Jr., K. A., and Lord, R. C. *J. Amer. Chem. Soc. 84*, 3843-3846 (1962).
32. Falk, M., Hartman, Jr., K. A. and Lord, R. C. *J. Amer. Chem. Soc. 85*, 387-391 (1963).
33. Privalov, P. L. and Mrevlishvili, G. M. *Biofizika 12*, 22-29 (1967).
34. Hearst, J. E. *Biopolymers 3*, 57-68 (1965).
35. Tunis, M. J. B. and Hearst, J. E. *Biopolymers6*, 1345-1353 (1968).
36. Lewin, S., *J. Theoret Biol. 17*, 181-212 (1967).
37. Wells, R. D., Burd, J. F., Chan, H. W., Dodgson, J. B., Jensen, K. F., Nes, I. F., and Wartell, R. M. "CRC Critical Reviews in Biochem." 305-340 (1977).
38. Goldblum, A., Perahia, D. and Pullman, A. *FEBS Letters 91*, 213-215 (1978).
39. Johnsrud, L. *Proc. Nat. Acad. Sci. USA75*, 5314-5318 (1978).
40. Gilbert, W., Maxam, A. and Mirzabekov, A., in "Control of Ribosome Synthesis" (eds. Kjeldgaard and Maaloe) pp. 139-148 (1976).